Advances in Fossil Energy R&D 2015

Topical Conference at the 2015 AIChE Annual Meeting

Salt Lake City, Utah, USA
8-13 November 2015

ISBN: 978-1-5108-1850-7

Printed from e-media with permission by:

Curran Associates, Inc.
57 Morehouse Lane
Red Hook, NY 12571

CURRAN ASSOCIATES INC. proceedings.com

Some format issues inherent in the e-media version may also appear in this print version.

Printed by Curran Associates, Inc. (2016)

AIChE
120 Wall Street, FL 23
New York, NY 10005-4020

Phone: (800) 242-4363
Fax: (203) 775-5177

www.aiche.org

Additional copies of this publication are available from:

Curran Associates, Inc.
57 Morehouse Lane
Red Hook, NY 12571 USA
Phone: 845-758-0400
Fax: 845-758-2634
Email: curran@proceedings.com
Web: www.proceedings.com

TABLE OF CONTENTS

437552 Optimization Models and Algorithms for Water Supply Chain Network Design and Operations in Shale Gas Production

Abstract:

In this work, we address the optimal design and operations of water supply chain networks for shale gas production [1, 2]. We develop a mixed-integer linear fractional programming (MILFP) model, the objective of which is to maximize the economic efficiency of freshwater in this shale water supply chain [2]. This model simultaneously takes into account the design and operational decisions, including freshwater source selection, multiple transportation modes, and different water management options [3-7]. Water management options include underground disposal, commercial centralized wastewater treatment (CWT), and onsite treatment (filtration, lime softening, thermal distillation) [8]. To globally optimize the resulting MILFP problem efficiently, we present three tailored solution algorithms: a parametric approach [9], a reformulation-linearization method [10], and a novel Branch-and-Bound & Charnes-Cooper transformation method. The proposed models and algorithms are illustrated through two case studies based on Marcellus shale play, in which onsite treatment shows its superiority in improving freshwater conservancy, maintaining a stable water flow, and reducing transportation burden.

References

[1] J. Gao and F. You, "Optimal design and operations of supply chain networks for water management in shale gas production: MILFP model and algorithms for the water-energy nexus," *AIChE Journal,* vol. 61, pp. 1184-1208, 2015.

[2] J. Gao and F. You, "Shale Gas Supply Chain Design and Operations towards Better Economic and Life Cycle Environmental Performance: MINLP Model and Global Optimization Algorithm," *ACS Sustainable Chemistry & Engineering*, 2015, 10.1021/acssuschemeng.5b00122.

[3] J.-P. Nicot and B. R. Scanlon, "Water Use for Shale-Gas Production in Texas, U.S," *Environmental Science & Technology,* vol. 46, pp. 3580-3586, 2012.

[4] C. E. Clark, R. M. Horner, and C. B. Harto, "Life Cycle Water Consumption for Shale Gas and Conventional Natural Gas," *Environmental Science & Technology,* vol. 47, pp. 11829-11836, 2013.

[5] M. Jiang, C. T. Hendrickson, and J. M. VanBriesen, "Life Cycle Water Consumption and Wastewater Generation Impacts of a Marcellus Shale Gas Well," *Environmental Science & Technology,* vol. 48, pp. 1911-1920, 2013.

[6] I. J. Laurenzi and G. R. Jersey, "Life Cycle Greenhouse Gas Emissions and Freshwater Consumption of Marcellus Shale Gas," *Environmental Science & Technology,* vol. 47, pp. 4896-4903, 2013.

[7] A. Vengosh, R. B. Jackson, N. Warner, T. H. Darrah, and A. Kondash, "A Critical Review of the Risks to Water Resources from Unconventional Shale Gas Development and Hydraulic Fracturing in the United States," *Environmental Science & Technology,* vol. 48, pp. 8334-8348, 2014.

[8] J. M. Wilson and J. M. VanBriesen, "Oil and Gas Produced Water Management and Surface Drinking Water Sources in Pennsylvania," *Environmental Practice,* vol. 14, pp. 288-300, 2012.

[9] Z. Zhong and F. You, "Globally convergent exact and inexact parametric algorithms for solving large-scale mixed-integer fractional programs and applications in process systems engineering," *Computers & Chemical Engineering,* vol. 61, pp. 90-101, 2014.

[10] D. J. Yue, G. Guillen-Gosalbez, and F. Q. You, "Global Optimization of Large-Scale Mixed-Integer Linear Fractional Programming Problems: A Reformulation-Linearization Method and Process Scheduling Applications," *AIChE Journal,* vol. 59, pp. 4255-4272, 2013.

436630 Metrics Driven Systems Analysis for Water & Energy Sustainability

Abstract:

This presentation focusses on showcasing how sustainability metrics can be utilized to drive energy and water sustainability outcomes in communities. Communities and local governments play an important role in developing people centric sustainable solutions as they have broad influence over activities that result in energy and water consumption within their boundary and jurisdiction. But most often their decisions are characterized by lack of quantitaitve information regarding the complex and often conflicting interactions between different aspects of the problem including the so called water-energy nexus. The authors will present how carbon footprint and water footprint metrics can be utilized within the framework of scenario modeling to help policy makers evaluate sustainability outcomes and tradeoffs associated with different policy choices. Several interesting results related to the sustainability analysis of San Luis Basin in south-central Colorado will be presented.

428508 An Industrial Ecology Approach for Managing Wastewater from Shale Gas Production

Abstract:

The rapid development of the unconventional shale gas industry offers tremendous economic benefits while simultaneously posing a critical challenge of managing vast quantities of produced and flowback water. Approximately 10-25% of the water injected during hydraulic fracturing returns to the surface as highly contaminated water with total dissolved solids reaching as high as 350,000 mg/Liter with potential detrimental impacts on human health and ecosystems. Sustainable development of the shale gas industry will require economical management of wastewater while minimizing associated environmental impacts. Membrane distillation (MD) is an emerging technology for treatment of high salinity wastewater and offers promise owing to its lower energy requirements and capital investment when compared with existing competing technologies. By utilizing the concept of industrial ecology, the economic and environmental sustainability of the MD technology can be further enhanced by a systems-level integration of MD process with waste heat sources. One such source is the waste heat available at existing natural gas compressor stations. The U.S. Natural Gas (NG) pipeline network is a highly integrated transmission grid consisting of more than 300,000 miles of transmission pipeline and 1,400 Compressor Stations (CS). In most NG CS, a portion of NG is burned to meet the energy demands of the compression process. Concurrently, about two-thirds of the fuel energy is rejected as waste heat mainly in the form of high temperature flue gas.

We present a framework to evaluate the synergies and potential of utilizing waste heat available from existing natural gas compressor stations for a novel MD technology for treatment of saline wastewater in the Marcellus shale gas play. We determine the quantity and quality of waste heat available at existing NG CS based on capacity, thermodynamic constraints, and the inlet feed temperature for MD. We also estimate the amount of wastewater that could be treated at each compressor station in Marcellus shale gas play based on experimentally measured flux rates. A Geographic Information System (GIS) based optimization framework is developed for informing decision regarding optimal wastewater routing from existing shale gas wells to hypothetical MD plants collocated with compressor stations. The results and insights from this work will be helpful in characterizing regional opportunities and constraints for management of wastewater using MD technology and waste heat sources. The implications of the work for informing sustainable management of water-energy nexus for the shale gas industry will be discussed along with a comparison with existing treatment options.

437566 Integrated Optimization of Water Supply Systems Scheduling for Energy Efficient Operations

Abstract:

In this work, we address the simultaneous optimization for the integrated water supply systems scheduling and pump operation to minimize the cost and energy consumption of pumps. The water supply systems are operated to transport water from sources to consumers [1-3]. The operation of the water supply systems is expensive because of large amount of water demands in the industry process and the daily life. Optimal scheduling of the water supply systems helps obtain remarkable savings by coordinating the pumping with time varying electricity tariffs [4]. The characteristics of the pumps can be described by a set of differential equations [5]. A great amount of energy can be saved if the operation of the pumps can be optimized. We integrate and optimize the water supply system scheduling and the pump operation simultaneously. We first formulate the integrated problem into a mixed-integer dynamic optimization (MIDO) model, which is then reformulated into a mixed-integer nonlinear programming (MINLP) full space model by discretizing the differential equations [6-8]. We propose an efficient flexible recipe method which can reduce the computational complexity significantly to solve the integrated problem [9, 10]. The flexible recipe method uses a set of discrete points to approximate the detailed dynamic optimization problems. In case study, we use a real water supply network system to demonstrate the applicability of the proposed full space modeling framework and the efficiency of the flexible recipe solution method.

References

[1] M. M. Eusuff and K. E. Lansey, "Optimization of water distribution network design using the shuffled frog leaping algorithm," *Journal of Water Resources Planning and Management,* vol. 129, pp. 210-225, 2003.

[2] A. Vasan and S. P. Simonovic, "Optimization of water distribution network design using differential evolution," *Journal of Water Resources Planning and Management,* vol. 136, pp. 279-287, 2010.

[3] D. L. Boccelli, M. E. Tryby, J. G. Uber, L. A. Rossman, M. L. Zierolf, and M. M. Polycarpou, "Optimal scheduling of booster disinfection in water distribution systems," *Journal of Water Resources Planning and Management,* vol. 124, pp. 99-111, 1998.

[4] B. Ulanicki, J. Kahler, and H. See, "Dynamic optimization approach for solving an optimal scheduling problem in water distribution systems," *Journal of Water Resources Planning and Management,* vol. 133, pp. 23-32, 2007.

[5] K. E. Lansey and K. Awumah, "Optimal pump operations considering pump switches," *Journal of Water Resources Planning and Management,* vol. 120, pp. 17-35, 1994.

[6] Y. Chu and F. You, "Integration of scheduling and control with online closed-loop implementation: Fast computational strategy and large-scale global optimization algorithm," *Computers & Chemical Engineering,* vol. 47, pp. 248-268, Dec 20 2012.

[7] H. Shi, Y. Chu, and F. You, "Novel Optimization Model and Efficient Solution Method for Integrating Dynamic Optimization with Process Operations of Continuous Manufacturing Processes," *Industrial & Engineering Chemistry Research,* vol. 54, pp. 2167-2187, 2015.

[8] H. Shi and F. You, "A Novel Adaptive Surrogate Modeling based Algorithm for Simultaneous Optimization of Sequential Batch Process Scheduling and Dynamic Operations," *Submitted,* 2015.

[9] Y. Chu and F. You, "Integrated Scheduling and Dynamic Optimization of Sequential Batch Processes with Online Implementation," *AIChE Journal,* vol. 59, pp. 2379-2406, Jul 2013.

[10] Y. Chu and F. You, "Integrated Scheduling and Dynamic Optimization of Complex Batch Processes with General Network Structure Using a Generalized Benders Decomposition Approach," *Industrial & Engineering Chemistry Research,* vol. 52, pp. 7867-7885, Jun 12 2013.

Optimization of Integrated Water and Multi-Regenerator Membrane Systems

Musah Abass and Thokozani Majozi[*]

School of Chemical and Metallurgical Engineering, University of Witwatersrand, 1 Jan Smuts Avenue, Braamfontein, Johannesburg, 2000, South Africa

Abstract (ID # 408816)

Water and energy are key resources in the process industry. The depletion of freshwater and energy resources coupled with stringent environmental regulations on effluent discharge limits have called for innovative designs for sustainable use of water and energy. This can be achieved through process integration techniques that are environmentally benign and economically feasible. This work proposes a robust water network superstructure optimisation approach for the synthesis of a multi-regenerator network for simultaneous water and energy minimisation. Two types of membrane regenerators are considered for this work, namely, electrodialysis and reverse osmosis. In each of the membrane regenerators, a detailed design model is developed and incorporated into the water network model. This presents a rigorous design and accurate cost representation as compared to the "blackbox" approach. The presence of continuous and integer variables, as well as nonlinear constraints renders the problem a MINLP. The developed model is applied to a pulp and paper case study to demonstrate its applicability, assuming a single contaminant scenario. The model was solved in GAMS using a solver BARON and the results indicate a 43.7% freshwater reduction, 50.9% decrease in wastewater generation and 47% savings in total annualised cost.

Keywords: Sustainable, Synthesis, Optimization, Reverse-osmosis, Electrodialysis.

1. Introduction

Water and energy are important resources for the development and wellbeing of humanity. They are key components in the process industry and great amounts of each resource are consumed to produce the other. Process integration is often employed in order to achieve a holistic water network for water and energy minimisation. This is done through an integrated water network that is open for direct reuse, recycle and regeneration reuse/recycle for sustainable cost-effective water and energy usage.

Water pinch techniques and mathematically based optimization techniques are the two main approaches used for optimal synthesis of water networks in process industry. Wang and Smith (1994a, b) presented the seminal work in water pinch analysis which sets to target the reuse and regeneration reuse/recycle of wastewater in order to reduce freshwater consumption. The work of Wang and Smith (1994a, b) has since developed into advanced graphical, tabular and heuristic methods for water network synthesis.

Takama *et al.* (1980) presented a superstructure optimisation approach for water minimisation in a petrochemical refinery based on a fixed mass load. The work of Takama *et al.* (1980) was later extended by several other researchers (Doyle and Smith 1997, Ahmetovic and Grossmann 2010) in order to obtain optimal freshwater usage in their respective water network superstructure. Notable among them are the work of Quesada and Grossmann (1995), Karrupiah and Grossmann (2006) and Faria and Bagajewicz (2011) who present algorithms to obtain feasible solutions in cases of the more complex nonlinear programming (NLP) and mixed integer

[*] Corresponding author: Tel: +27 11 717 7384; Fax: +27 82 456 1500
Email address: Thokozani.Majozi@wits.ac.za;

nonlinear programming (MINLP) problems that are in many cases a challenge to solve. Mathematical optimization approaches are however computationally intensive.

Tan *et al.* (2009) presented a water network superstructure with a single membrane partitioning regenerator which allows for possible reuse/recycle. The work of Tan *et al.* (2009) considered the "blackbox" approach which uses linear cost functions for the membrane regenerators. This does not give an accurate cost representation of the water network. Khor *et al.* (2011) addressed the gap on the work of Tan *et al.* (2009) by developing a detailed model representation for water network regeneration synthesis using a MINLP optimization framework. The work of Khor *et al.* (2011) was, however, limited to a single regenerator with a fixed design. In a more recent development, Yang *et al.* (2014) proposed a unifying approach by combining multiple water treatment technologies capable of treating all major contaminants. The work focused on unit specific short cut cost functions in order to gain detailed understanding of trade-offs between removal efficiency of treatment units and the cost of the units, as well as the impact on the unit design.

The membrane technologies adopted in this work involve electrodialysis (ED) and reverses osmosis (RO). Reverse osmosis (RO) is a pressure driven membrane separation process that selectively allows the passage of one or more species through the membrane unit. The application of membrane systems as regenerators generally requires enormous amount of energy. Most published work, however, uses linear cost functions or "blackbox" representation for membrane partitioning regenerators (Tan *et al.,* 2009, Khor *et al.,* 2012). This does not give an accurate cost representation of the membrane systems. There is, therefore, a need for a technique that would cater for energy minimization through detailed synthesis of membrane regeneration systems in order to obtain optimal variables that affect the operation and economics of the regenerator unit.

The main objective of this work is to develop a systematic technique that would simultaneously optimize energy and water consumption within a water network superstructure. The technique is based on an integrated water and membrane regeneration network superstructure. The membrane regenerators in this representation are ED and RO. The choice of regenerators is motivated by the increasing demand and use of membrane technology for water treatment in the process industry and also the varying potential of different membrane systems to treat specific waste ranges. Detailed synthesis of the membrane regeneration systems is conducted to determine optimal operating conditions for efficient energy usage in terms of costs. The detailed model of the regenerators is incorporated in the overall water network objective function in order to minimize freshwater and energy consumption and also give a true cost representation as compared to the "blackbox" method. It is also demonstrated that treating the removal ratio as an additional degree of freedom allowed for better performance of the integrated system. This is contrary to what is presented in literature where the removal ratio is generally treated as a parameter.

2. Problem statement

The problem statement in this work can be stated as follows:

Given:

(i) A set of water sources, *J*, with known flowrates and contaminant concentrations.
(ii) A set of water sinks, *I*, with known flowrates and known maximum allowable contaminant concentrations.
(iii) Membrane regeneration units with the potential for parallel/series connection for partial treatment of wastewater from sources for reuse/recycle.
(iv) A freshwater source, *FW*, with known concentration and variable and unlimited flowrate.

(v) A wastewater sink, *WW*, with maximum allowable contaminant concentration and variable and unlimited flowrate.

Determine:

(i) The minimum freshwater intake, wastewater generation, the energy consumed in the ED and RO units, and the total annualised costs for ED (TAC_e) and RO (TAC_r).
(ii) Optimal water network configuration.
(iii) Optimum design variables of the regenerators.

3. Model formulation

Based on the problem statement, water network superstructure in Figure 1 is developed. The superstructure representation is an extension of the work by Khor *et al.* (2011). The superstructure in this work incorporates multiple regenerators which are open for parallel and series connection as well as recycle and reuse of both permeate and reject streams from regenerators.

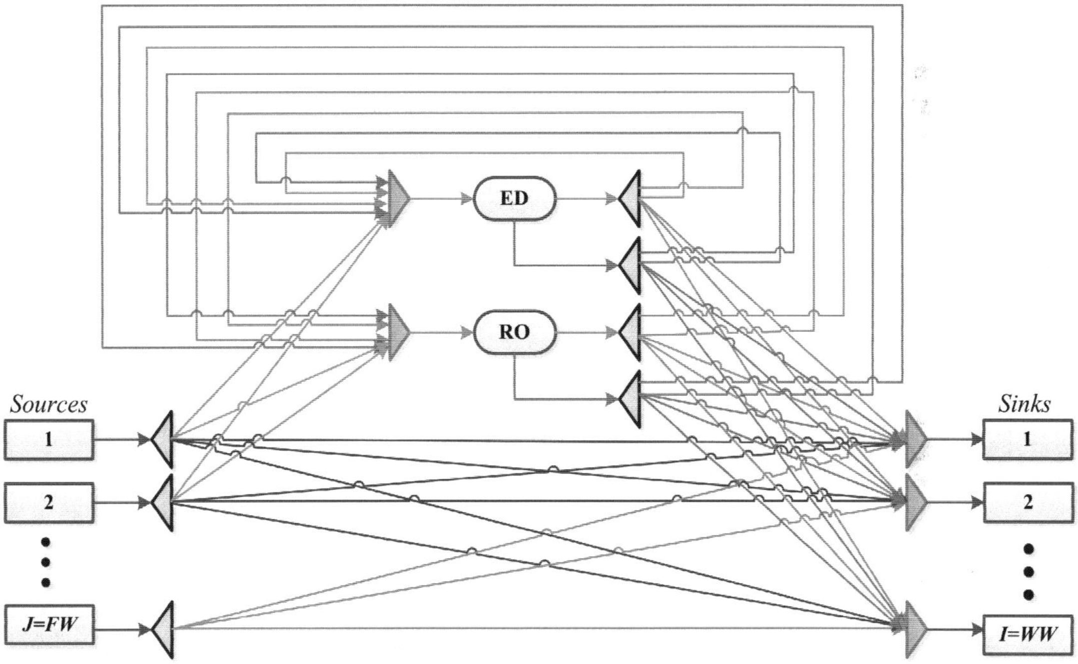

Figure 1: General water network superstructure with multiple membrane regenerators

3.1 *The objective function*

The objective function minimises the overall annualised cost of the water network. This includes total annualised costs of ED and RO units, freshwater cost, wastewater treatment cost, as well as capital and operating costs of piping interconnections. Constraint (1) represents the objective function of the overall water network superstructure. A 1-norm Manhattan distance is considered for all piping interconnections. All pipes are assumed to be of the same material properties and as a result the carbon steel pipes parameters of p and q are adopted for the piping costs.

$$Obj = min \left(\begin{array}{l} (C_{water} FW + C_{waste} WW) AOT + TAC_e + TAC_r + \\ AF \left[\begin{array}{l} \sum_{j \in J} \sum_{i \in I} D^a_{(j,i)} \left(\frac{pQ^a_{(i,j)}}{3600 v} + qY^a_{(j,i)} \right) + \sum_{j \in J} D^b_j \left(\frac{pQ^{ro}_j}{3600 v} + qY^{ro}_j \right) \\ \sum_{j \in J} D^c_j \left(\frac{pQ^{ed}_j}{3600 v} + qY^{ed}_j \right) + \sum_{i \in I} D^{ds}_i \left(\frac{pQ^{ds}_i}{3600 v} + qY^{ds}_i \right) + D^{dr} \left(\frac{pQ^{dr}}{3600 v} + qY^{dr} \right) \\ + D^{de} \left(\frac{pQ^{de}}{3600 v} + qY^{de} \right) + \sum_{i \in I} D^{es}_i \left(\frac{pQ^{es}_i}{3600 v} + qY^{es}_i \right) + D^{er} \left(\frac{pQ^{er}}{3600 v} + qY^{er} \right) \\ + D^{ee} \left(\frac{pQ^{ee}}{3600 v} + qY^{ee} \right) + \sum_{i \in I} D^{fs}_i \left(\frac{pQ^{fs}_i}{3600 v} + qY^{fs}_i \right) + D^{fr} \left(\frac{pQ^{fr}}{3600 v} + qY^{fr} \right) \\ + D^{fe} \left(\frac{pQ^{fe}}{3600 v} + qY^{fe} \right) + \sum_{i \in I} D^{gs}_i \left(\frac{pQ^{gs}_i}{3600 v} + qY^{gs}_i \right) + D^{gr} \left(\frac{pQ^{gr}}{3600 v} + qY^{gr} \right) \\ + D^{ge} \left(\frac{pQ^{ge}}{3600 v} + qY^{ge} \right) \end{array} \right] \end{array} \right) \quad (1)$$

The resulting mathematical model is an MINLP. The nonlinear terms are due to the presence of bilinear terms in mass balance equations and power terms in the cost functions of regeneration units.

4. Illustrative Example

The developed mathematical model is verified and applied to a pulp and paper case study adopted from Chew *et al.* (2008). The MINLP model was implemented in GAMS 24.2 using the general purpose global optimisation solver BARON which obtains a solution by using a branch-and-reduce algorithm. Table 1 shows the basic data for the plant water network which comprises five water sources including the freshwater source, and five water sinks including the wastewater sink.

Table 1: Basic data for water sources and sink

	Sources, *j*			**Sinks, *i***	
j	Flowrate (t/h)	Concentration (mg/L)	*i*	Flowrate (t/h)	Max. concentration (mg/L)
1	2.07	0.0002	1	3.26	0.00057
2	0.34	0.0051	2	0.34	0
3	0.024	0	3	1.34	6.16×10^{-6}
4	7.22	0.0083	4	7.22	0
FW	—	0	*WW*	—	0.01

The case study was applied to three different scenarios in order to ascertain the benefits of incorporating detailed models of the membrane regeneration units in the constraints of the water network.

(i) Case 1 considered a model of the water network without regeneration.
(ii) Case 2 considered a model of water network with regeneration units based on the "blackbox" approach.
(iii) Case 3 considered detailed synthesis of the regeneration units incorporated into the overall water network objective function.

The optimal results for all cases are presented in Table 2. The results showed significant reduction in freshwater consumption, wastewater generation as well as the total cost of the water network for case 2 as compared to the based case.

Table 2: Optimal results of water network based on the case study

	Scenario1	**Scenario 2**		**Scenario3**	
		Fixed RR	Variable RR	Fixed RR	Variable RR
Freshwater use (t/h)	18.30	11.42	9.83	11.64	10.3
% of freshwater savings		37.5	46.3	36.4	43.7
Wastewater generated (t/h)	15.79	8.89	7.3	9.11	7.75
% of wastewater saved		43.7	53.7	42.3	50.9
Total cost of water network($)	$1.170x10^6$	$5.97x10^5$	$5.08x^5$	$6.34x10^5$	$6.20x10^5$
CPU time	0.06	687.50	2764.20	865	16709.78

The results of case 3 are displayed in Figure 2. Case 3 showed significant reduction in water network cost as well as freshwater consumption and wastewater reduction as compared to the direct water network model without regeneration. However, there was an increase in the total water network cost compared to the case 2. This is as a result of case 3 being a true representation of the total water network as it incorporates a detailed design of the membrane regeneration units and gives an accurate expression of the regeneration cost. From the results in Table 2 it is evident that variable removal ratio in case 3 presents the optimal water network configuration as presented in Figure 2, since the model is allowed to choose the performance parameters of the membrane regenerators. This resulted in 43.7 % reduction in freshwater consumption, 50.9% reduction in wastewater generation and 45 % savings in the total annualised water network cost as compared to scenario 1. The model also showed 34% and 41% savings in energy for both ED and RO units respectively.

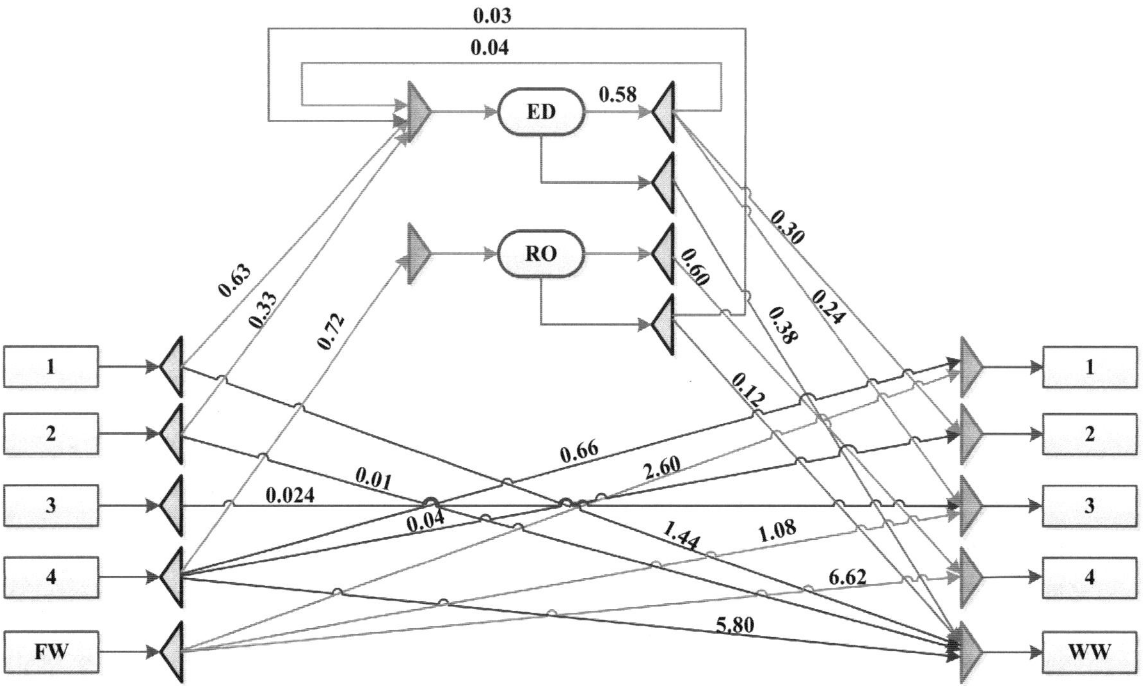

Figure 2: Optimum water network configurations for scenario 3 (Variable RR)

5. Conclusion

This work has addressed the synthesis of a multi-membrane regeneration water network by proposing a MINLP optimisation model that incorporates detailed mathematical models of ED and RO units within a water network. The developed water network allows for direct reuse/recycle and regeneration reuse/recycle. The developed MINLP model is applied to a pulp and paper industry case study and solved using GAMS/BARON. The results showed that setting the removal ratio as variable yields optimal configuration as compared to the fixed removal ratio. The model also showed that incorporating a detailed model of the regeneration unit into the overall model yields accurate expression of the regeneration cost as compared to the ''blackbox'' model. It also gives optimal operating variables of the regeneration units for minimal energy usage.

6. Nomenclature

6.1 Sets

$J = \{j|j = \text{water source}\}$

$I = \{i|i = \text{water sink}\}$

6.2 Parameters

AOT	Annual operating time
C_{waste}	Unit cost for waste
	treatment
C_{water}	Unit cost for freshwater
p, q	Parameter for carbon

	steel piping
v	Velocity of fluid in pipes
$D^{a}_{j,i}$	Manhattan distance from source j to sink i
D^{b}_{j}	Manhattan distance from source j to RO unit
D^{c}_{j}	Manhattan distance from source j to ED unit
D^{ds}_{i}	Manhattan distance from permeate of ED unit to sink i
D^{dr}	Manhattan distance from permeate stream of ED unit into ED unit
D^{es}_{i}	manhattan distance from reject stream of ED unit into sink i
D^{fs}_{i}	Manhattan distance from permeate stream of RO unit into sink i
D^{gs}_{i}	Manhattan distance from reject stream of RO unit into sink i

6.3 Continues variable

FW	Freshwater flowrate
WW	Wastewater flowrate
$Q^{a}_{j,i}$	Flowrate from source j to sink i
Q^{b}_{j}	Flowrate from source j to RO unit
Q^{c}_{j}	Flowrate from source j to ED unit
Q^{ds}_{i}	Flowrate from permeate stream of ED into sink i
Q^{es}_{i}	Flowrate from reject stream of ED unit into sink i
Q^{fs}_{i}	Flowrate from permeate stream of RO unit into sink i
Q^{gs}_{i}	Flowrate from reject stream of RO unit into sink i
TAC_e	Total annual cost of RO unit
TAC_r	Total annual cost of ED unit

6.4 Integer Variables

N_m	Number of reverse osmosis modules
N	Number Electrodialysis cell pairs

6.5 Binary Variables

$$y^{a}_{j,i} = \begin{cases} 1 \leftarrow \text{Existence of piping interconnection between source and sink,} \\ 0 \leftarrow \text{Otherwise} \end{cases}$$

$$y^{b}_{j} = \begin{cases} 1 \leftarrow \text{Existence of piping interconnection between source and RO unit,} \\ 0 \leftarrow \text{Otherwise} \end{cases}$$

$$y^{c}_{j} = \begin{cases} 1 \leftarrow \text{Existence of piping interconnection between source and ED unit,} \\ 0 \leftarrow \text{Otherwise} \end{cases}$$

$$y^{ds}_{i} = \begin{cases} 1 \leftarrow \text{Existence of piping interconnection between permeate of ED unit and sink,} \\ 0 \leftarrow \text{Otherwise} \end{cases}$$

$$y^{ds}_{i} = \begin{cases} 1 \leftarrow \text{Existence of piping interconnection between permeate of ED unit and ED unit ,} \\ 0 \leftarrow \text{Otherwise} \end{cases}$$

$$y^{es}_{i} = \begin{cases} 1 \leftarrow \text{Existence of piping interconnection between reject of ED unit and sink ,} \\ 0 \leftarrow \text{Otherwise} \end{cases}$$

$$y_i^{fs} = \begin{cases} 1 \leftarrow \text{Existence of piping} \\ \quad \text{interconnection between} \\ \quad \text{permeate of RO unit} \\ \quad \text{and sink ,} \\ 0 \leftarrow \text{Otherwise} \end{cases} \qquad y_i^{gs} = \begin{cases} 1 \leftarrow \text{Existence of piping} \\ \quad \text{interconnection between} \\ \quad \text{reject of RO unit} \\ \quad \text{and sink ,} \\ 0 \leftarrow \text{Otherwise} \end{cases}$$

7. Acknowledgement

The authors would like to thank the National Research Foundation (NRF) for funding this work under the NRF/DST Chair in Sustainable Process Engineering at the University of the Witwatersrand, South Africa.

8. References

Ahmetovic, E., & Grossmann, I. E., (2010). Strategies for Global Optimization of Integrated Process Water Networks. *ESCAPE* 20.

Chew, I. M. L., Tan, R., Ng, D. K. S., Foo, D. C. Y., Majozi, T., Gouws, J. (2008). Synthesis of Direct and Indirect and Interplant Water and Network. *Industrial Engineering Chemical Research*. 47(2008) 9485 – 9495.

Doyle, S. J & Smith, R. (1997). Targeting Water Reuse with Multiple Contaminants. *Process Safety and Environmental Protection,* 75(3) 181 -189.

Faria, D. C., & Bagajewicz, M. J., (2010). On the Appropriate Modelling of Process Plant Water Systems. American Institute of Chemical Engineers Journal, 53(3) 666 - 689.

Karrupiah, R., & Grossmann, I. E., (2006). Global Optimization for the Synthesis of Integrated Water and Systems in Chemical Processes. Computers & Chemical Engineering, 30: 650 – 673.

Khor, C. S., Foo, D. Y., El-Halwagi, M. M., Tan, R. R., & Shah, N. (2011). A Superstructure Optimization Approach for Membrane Separation –Based Water Regeneration Network Synthesis with Detailed Nonlinear Mechanistic Reverse Osmosis Model. Industrial & Engineering Chemistry Research, 50(11), 13444 – 13456.

Quesada, I. & Grossmann, I. E., (1995). Global Optimization of Bilinear Process Networks with Multicomponent Flows. Computers and Chemical Engineering, 19(12), 1219-12421.

Takama, N., Kuriyama, T., Shiroko, K. & Umeda, T. (1980). Optimal Planning of Water Allocation in Industry. Computers and Chemical Engineering, 4(80), 251 – 258.

Tan, R. R., Ng, D. K. S., Foo, D. C. Y., Aviso, K. B. (2009). A Superstructure Model for the Synthesis of Single-Contaminant Water Networks with Partitioning Regenerators. *Process Safety and Environmental Protection*, 87(9), 197 – 205.

Yang, L., Salcedo-Diaz, R., Grossmann. E. I. (2014). Water Network Optimization with Wastewater Regeneration Models. *Industrial &Engineering Chemistry Research*. 53(2014)17680 – 17695.

Wang, Y. P. & Smith, R. (1994a) Wastewater Minimization. *Chemical Engineering Science*, 49(94), 981 – 1006.

Wang, Y. P. & Smith, R. (1994b) Design of Distributed Effluent Treatment Systems. *Chemical Engineering Science,* 49(94), 3127– 3145.

424452 An MINLP Model for the Water Management in Shale Gas Operations

Abstract:

This work presents a mathematical programing approach for the optimal management of flowback water in shale gas wells in order to meet water demands in hydraulic fracturing operations considering a fixed scheduling for the completion phases of the wells. The objective function is aimed to determine the minimum total costs for the water network constituted by capital and operational costs. Capital costs are associated to the equipment acquisition of storage units, treatment technologies and disposals; whereas the operating costs considers the fresh water cost, operating cost for water treatment and the transportation costs for all transport trajectories required. Additionally, the total amount of fresh water used (environmental concern) is quantified and minimized; in this context, the methodology accounts for the flowback water reuse once it is treated and stored with the purpose to reduce the fresh water consumption, the wastewater streams and the costs. The model involves the seasonal limitations for the fresh water availability, which is an important challenge for some of the existing shale plays.

The proposed design considers that all the flowback water is treated and the outlet streams satisfy the restrictions for the pollutant concentrations in order to be reused and/or the environmental regulations to be disposed. To carry out this task, only treatment units with tested effectiveness in flowback water coming from shale gas wells have been considered. Also, the given scheduling for the fracking operations of each well taken into account can include the simultaneous operation for several fracking crews. The mathematical formulation corresponds to an MINLP multi-period model. Finally, the example problem uses updated information taken of technical reports from some of the most important existing shale regions (Marcellus and Barnette).

422129 Ionic Liquid Composite Membranes for CO2/Light Gas Separations

Abstract:

A novel approach for performing these separations is to use room temperature ionic liquids (RTILs) in various morphologies as membranes. In comparison to conventional polymers, they perform gas separations due to solubility differences. They can be converted to polymerizable molecules. This allows promising RTILs to be prepared as membranes. Thus, they can be converted to various morphologies as membranes while maintaining the inherent selectivity of the material. They can be prepared as polymer films, composite structures with ionic liquid within the structure and gels. Polymer films have been tested for gas separations up to 40 bar feed pressure. The conditioning or morphology change due to the incorporation of gases such as CO_2 at these high pressures is reversible. This is an important advantage in comparison to conventional polymers. The wide range of morphologies and chemical structures allow for finding materials that not only have the desired physicochemical properties but also the mechanical properties needed to produce viable membranes.

Mixed matrix membranes (MMMs) look to be a very promising candidate for commercialization and large scale implementation. These MMMs consist of three individual components: polymerizable room temperature ionic liquids (poly(RTILs), normal RTILs, and zeolite particles. These components synergize to produce enhanced gas transport. The polymer matrix allows for facile and economical fabrication, while the zeolite provides excellent separation performance. The RTIL prevents defect formation within the membrane by acting as a wetting agent to help the polymer matrix interface aptly with the zeolite. We have investigated how varying the three components in our MMMs affects membrane permeability and selectivity. This initial study identified the optimal type of zeolite particles. More in depth analysis of the ionic liquid revealed the significance of molar volume in the RTIL and free volume in the poly(RTIL) on the membrane's gas separation ability. Previous work established the correlation between increasing RTIL molar volume and decreasing CO_2 solubility selectivity. Current work shows how the polymerized analogues (poly(RTILs) display a similar trend with free volume. This resulted in the creation of MMMs made with smaller molar and free volume RTILs and poly(RTILs) that performed above Robeson's 2008 upper bound for CO_2/CH_4. With this knowledge we aim to devise a systematic design approach for creating state of the art MMMs.

427983 Scale-up of Amine-Containing Membranes for CO2 Capture and Separation

Abstract:

Scale-up of Amine-Containing Membranes for CO_2 Capture and Separation

Witopo Salim[1], Varun Vakharia[1], Dongzhu Wu[1], Yuanxin Chen[1], Lin Zhao[1], Yang Han[1], and W.S. Winston Ho[1,2]

[1]William G. Lowrie Department of Chemical and Biomolecular Engineering, The Ohio State University, 151 West Woodruff Avenue, Columbus, OH 43210-1180, USA

[2]Department of Materials Science and Engineering, The Ohio State University, 2041 College Road, Columbus, OH 43210-1178, USA

Abstract

Amine-containing membranes for CO_2 capture and separation were scaled up to fabricate 14-inch wide membranes by using a continuous roll-to-roll machine equipped with a thin-film-coating assembly. Aided by a material balance equation, three variables, including the coating knife gap setting, substrate rolling speed, and coating solution concentration, were identified as the critical factors to control the membrane selective layer thickness. For the CO_2 separation from a synthesis gas stream containing CO_2 and H_2S, i.e., pre-combustion CO_2 capture in IGCC (integrated gasification combined cycle) technology or hydrogen purification for fuel cells, the thicker amine-containing membranes with around 15 ± 2 microns of selective layer thicknesses were fabricated by using a gap setting of 17 mils, a rolling speed of 0.5 ft/min (set by the drying capacity of the drier, i.e., its length), and an amine coating solution of 18 wt%. For the post-combustion CO_2 capture from flue gas, the same thin-film-coating assembly machine was used to fabricate thinner amine containing membranes with < 250 ± 20 nm of selective layer thicknesses by using various rolling speeds and amine-containing solution concentrations at a fixed gap setting of 0.45 mils. The fabrication of thinner membranes was improved further by using an insert in the coating solution trough to control the coating solution delivery rate and a smooth coating knife base to reduce the thickness variation. From the gas transport measurements, the flat-sheet samples of the scale-up membranes for both applications exhibited similar performances compared to the membranes synthesized in lab scale. CO_2 permeance of approximately 200 GPU and CO_2/H_2 selectivity > 180 for the thicker membranes for the purification of hydrogen and synthesis gas along with CO_2 capture (e.g., pre-combustion capture), and CO_2 permeance of about 800 GPU and CO_2/N_2 selectivity > 140 for the thinner membranes for CO_2 capture from flue gas. Techno-economic analysis of the post-combustion CO_2 capture membrane process using the thin membrane resulted in about \$41/tonne CO_2 captured (based on 2007 dollar). The scale-up membranes can be used to prepare pilot-scale membrane modules, either as spiral-wound or plate-and-frame configuration, for testing with the acid gas mixtures or flue gas.

433693 Microencapsulation of Advanced and Novel CO2 Sorbents

Abstract:

We present an application of microencapsulation for advanced solvents and materials as carbon dioxide sorbent systems. Current research into advanced technical solvent systems has focused on the key elements of kinetics of absorption, material carrying capacity of carbon dioxide, loading and unloading times, and solvent viscosities, among other properties. While some of these properties may be optimized against current carbon capture schemes, there still exist hurdles towards their implementation.
To overcome these shortcomings, we present a motif of encapsulation of these advanced solvents, such as CO2BOLS and NOHMS, in a CO2 permeable shell material, in the form of a microcapsule. Issues of slow absorption, interaction with infrastructure, and high viscosities of materials can be minimized using this technique, and optimization of specific surface area and viscosity changes are engineered into the absorption system.

We present current work and progress on encapsulation of these systems are their performance against conventional absorption systems and techniques.

419917 Novel Membrane Absorption-Based Processes for CO2 Capture

Abstract:

This presentation will consider novel membrane absorption and stripping based processes and devices for CO_2 absorption and stripping for carbon capture and sequestration. Two feed streams are considered: (1) lower temperature (L-T) post-shift reactor synthesis gas at high pressures and temperature up to 100°C where He was used as a surrogate for H_2; (2) simulated flue gas at 40-50°C.

An advanced pressure swing membrane absorption (PSMAB) technology utilizing a hollow fiber membrane contactor and a cyclic absorption-desorption process is being developed to separate H_2 from the low temperature (L-T) post-shift reactor synthesis gas and simultaneously obtain a purified CO_2 stream. Studies have been carried out with 40% CO_2-rest He feed gas mixture coming in at temperatures up to 100^0C and pressures varying from 100 to 250 psig (689-1723kPag). The nonvolatile absorbent consists of a solvent and a suitable amine; the solvent is either polyethylene glycol (PEG) 400 or an ionic liquid (IL), 1-butyl-3-methylimidazolium dicyanamide ([bmim][DCA]). Studies were carried out with microporous hydrophobized polyether ether ketone (PEEK) hollow fiber membrane modules of small sizes having different characteristics. Reducing the dead volume on the module tube side and decreasing the fiber packing density greatly enhanced gas absorption and improved product qualities while increasing temperature had a negative impact on product qualities with absorbents without the amine. Adding a nonvolatile amine such as polyamidoamine (PAMAM) dendrimer Gen 0 to the ionic liquid or PEG 400 improved CO_2 absorption capability especially at higher temperatures. The measured CO_2 concentration in the CO_2-rich product stream from the less-than-optimum membrane modules using such mixed absorbents went up to 93-94%. A model has been developed and numerically solved to predict the cyclic process performance of pure IL as the absorbent. Solubility and diffusion coefficient data for the gases needed in this model were determined for the pure IL absorbent using a pressure-decay-dual transducer apparatus at temperatures of 323, 353, 363, and 373 K and at pressures up to 1.38 MPa (~200 psig). The CO_2/He solubility selectivity was determined for various compositions.

Studies for CO_2 separation from flue gas employed advanced hollow fiber membrane contactors and a nonvolatile absorbent circulating between an absorber and a stripper. The loaded absorbent is heated before it enters the stripper. Preliminary studies were also implemented with a novel integrated absorber-stripper. The observed separation behavior will be illustrated. The deficiencies of the design will be pointed out. The issues of energy consumption and capital costs are of great importance here.

References:

1. T. Mulukutla, G. Obuskovic, K. K. Sirkar, "Novel Scrubbing System for Post-combustion CO_2 Capture and Recovery: Experimental Studies", J. Membrane Sci., 471, 16 (2014).
2. J. Chau, G. Obuskovic, X. Jie , T. Mulukutla, K. K. Sirkar, Solubilities of CO_2 and helium in an ionic liquid containing poly (amidoamine) dendrimer Gen 0, *I&EC Res.*, **52**, 10484 (2013).
3. Chau, G. Obuskovic, X. Jie, K. K. Sirkar, Pressure swing membrane absorption process for shifted syngas separation: Modeling vs. experiments for pure ionic liquid, *J. Membrane Sci.*, **453**, 61 (2014).
4. X. Jie, J. Chau, G. Obuskovic and K. K. Sirkar, Enhanced pressure swing membrane absorption process for CO_2 removal from precombustion shifted syngas with dendrimer-ionic liquid mixtures as absorbents, *Ind. Eng. Chem. Res.*, **53**(8), 3305 (2014).

419948 The Influence of Water on Polymer/Silica Hollow Fiber Sorbents for CO2/N2 Sorption

Abstract:

The Influence of Water on Polymer/Silica Hollow Fiber Sorbents for CO_2/N_2 Sorption

Ying Labreche, Steven K Burgess, Ryan P. Lively, Christopher W. Jones*, William J. Koros*

Georgia Institute of Technology, 311 Ferst Drive NW, Atlanta, GA 30332-0100, USA Abstract Rapid Temperature Swing Adsorption (RTSA) processes offer potential techniques for the capture of CO_2 from flue gas mixtures. Understanding the interactions between the numerous gas phase components and the sorbent material is necessary to develop high performance sorbents for this application. Moisture is commonly present in flue gas applications and can interfere with or alter the mechanisms involved in the adsorption of CO_2 and other gases, so understanding effects of moisture is of particular importance. In this study, we consider the influence on equilibrium water vapor sorption and CO_2 capacity of different amine loaded silica/polymer hollow fiber composites. The advantage of different polymer binders (cellulose acetate vs. Torlon, a poly(amide-imide)) for the silica particles, as the well the impact of an amine dispersant, glycerol, are discussed in detail.

430522 CO2 Capture from Flue Gas By PSA: Bench Scale Demonstration of a Novel Structured Adsorbent

Abstract:

A cost effective pressure swing adsorption (PSA) process is under development for post-combustion CO_2 capture from coal fired power plants. Several research groups from around the world are showing that PSA is still a viable option, as no separation process has emerged as a clear winner for CO_2 capture. PSA is also attractive because it requires only electricity; it does not require any steam from the power plant. This project is being carried out with funding from DOE NETL and four collaborative partners:

The key challenge to the success of this project is two-fold: 1) an attrition resistant and low pressure drop structured adsorbent must be developed based on commercial zeolite that is compatible with the high velocities associated with shorter PSA cycle times, and 2) a PSA cycle configuration must be developed in concert with the structured adsorbent so that the resulting PSA process delivers an exceptional performance at reduced capital and operating costs. Since the rapid PSA concept is conceptually simple but difficult to implement, a select team has been assembled to carry out this ambitious project and demonstrate the feasibility of the PSA concept for CO_2 capture at the bench scale. This team includes researchers from USC, W. R. Grace, Catacel, and Battelle.

To date, a key development includes the fabrication of a structured adsorbent that boosts a bed density of ~ 220 kg of adsorbent (including binder) per m^3 of column. This coated, metal foil, parallel channel, corrugated structure was designed to fit snugly inside 1.5" ID by 18" long column using three, 6" sections. Prior to these cylindrical sections being inserted inside the SS column, the metal foil was coated with zeolite crystals to a thickness of ~ 50 microns and rolled to form the parallel channel, corrugated structure. The SS column was designed to fit directly into a novel 1-bed PSA system. This 1-bed system is capable of mimicking all the steps of a very sophisticated multi-bed PSA cycle schedule. This 1-bed system is being used to demonstrate the concentration of a 15 vol% CO_2 in nitrogen mixture to over 95 vol% CO_2 with a recovery exceeding 90%. Results obtained from this 1-bed system using a column filled with commercial zeolite beads will be compared directly with those obtained from the structured adsorbent column. This presentation will present the latest experimental results obtained from this effort.

435950 Staged Membrane Configurations for CO_2 Capture

Abstract:

Anthropogenic emissions of carbon dioxide (CO_2) from fossil-fuel electric power plants potentially can lead to undesirable global climate change. Consequently, significant research effort has focused on reducing the carbon intensity of fossil fuel use through energy conservation, energy efficiency improvements and Carbon Capture and Storage (CCS).

CCS involves CO_2 separation from flue gas, transportation to a storage location and permanent storage in an appropriate geological reservoir. The US Department of Energy (DOE) has challenged researchers to develop separation processes capable of 90% CO_2 capture at 95% purity, and subsequent compression to 140 bar for transport and storage, without increasing the cost of electricity (COE) by more than 35%. Membrane processes are one of many options under development.

Because of the relatively low CO_2 concentration in post-combustion flue gas, most of the reported process configurations for membrane systems have been sought to generate affordable CO_2 partial pressure driving forces for permeation. This can be achieved by feed compression, vacuum permeation, internal gas recycle, and feed air sweep configurations. Additionally, most process designs involve CO_2 concentration in a two-stage configuration followed by cryogenic separation and liquefaction. These schemes utilize highly selective membranes. While such membranes may reduce energy consumption, the required membrane area will be higher than if lower selectivity, higher intrinsic permeability materials are used.

Process designs that utilize membrane materials with higher permeabilities and lower selectivites are considered. Such materials offer the potential to reduce membrane capital cost at the expense of greater energy consumption. Moreover, the low pressure ratios being considered for membrane capture processes may not be able to exploit economically the potential of higher selectivity materials to concentrate CO_2.

A super-structure for staged membrane capture processes with final cryogenic concentration is proposed which accommodates the previously proposed configurations and the new configurations for lower selectivity materials. The superstructure allows for two types of CO_2 recycle loops: one through the boiler and the other after the boiler; the post-boiler recycle loop avoids undesirable reductions in oxygen concentration. The tradeoff between capital and operating costs is evaluated by determining the levelized cost of electricity for various embodiments of the superstructure to identify the optimal design.

436706 Is Social and Academic Interest in Carbon Capture and Storage Declining?

Abstract:

This study investigates the evolutionary trends of the international collaborations among the research community of carbon capture and storage (CCS) by looking at the collaboration network of countries publishing on CCS. The study also examines the evolution of major research topics in this field by exploring the co-occurrence network of keywords associated with CCS publications.

The analysis is carried out based on the CCS related publications during 1980 - 2013. The study elaborates how both international collaboration network and knowledge structure of the field have notably developed and interlinked over the years especially after 2005 during which almost 94% of the publications appeared. The study shows a clear connection between CCS research and the global negotiations on climate change mitigation. It seems that as we have reached the end of Kyoto Protocol binding period (2012), and following the failures in new strong agreements (specially the 2009 Copenhagen meeting), the social and academic interests in CCS have declined.

417929 Reductive Calcination: Process Integration in Mineral Processing

Abstract:

Reductive Calcination: Process Integration in Mineral Processing

Georg Baldauf-Sommerbauer, Susanne Lux, Matthäus Siebenhofer, Institute of Chemical Engineering and Environmental Technology, Graz University of Technology, NAWI Graz, Graz, Austria

Inorganic carbonates are mainly found in minerals and brines. They have been mined, purified and used by mankind for hundreds of years. Many of the carbonates are used as raw materials for the production of metal oxides which are used as bulk products (e.g. cement, refractories) or fine chemicals (e.g. catalysts). Processing of carbonate based minerals preferably starts with their decomposition in oxidizing atmosphere. From the thermodynamic point of view the decomposition of carbonates is driven by temperature (and pressure). The by-product of thermal decomposition (=calcination) is carbon dioxide. This carbon dioxide is emitted into the atmosphere, leading to commonly acknowledged greenhouse gas problems. Depending on the type of mineral relatively high temperatures have to be applied. The temperature is maintained by combustion of fossil fuels, leading to even more carbon dioxide emissions.

Our approach is to combine the CO_2-emitting step with methanation in a reactor concept, enabling single step conversion of hydrogen with process carbon dioxide to methane. This would lead to less carbon dioxide emissions by lowering the operation temperature of the calcining step, and by coupling the endothermic calcining with the exothermic methanation reaction. A thermodynamic analysis of the data for the decomposition of several carbonates in hydrogen atmosphere to produce methane was performed in the temperature range of 20-500 °C. This analysis revealed that from the first and second group elements only magnesium carbonate provides negative standard free energy of calcination in the range of -55 kJ/mol. Several transition metal carbonates (Mn, Fe, Co, Ni, Zn, Cd, Pb, Cu, Ag) show a negative standard free energy of calcination in the range of -50 to -120 kJ/mol. Analysis of thermodynamics revealed that for all carbonates the standard free energy of calcination is less negative with increasing temperature, thus facilitating the reaction at lower temperatures.

Experimental validation of calcination in hydrogen atmosphere confirmed the technology of single step calcination plus conversion of carbon dioxide into methane and carbon monoxide. The product gas composition and conversion can easily be adjusted by altering the temperature and pressure set point. The calcination step draws significant advantages from reduction of the carbon dioxide partial pressure with the heat carrier hydrogen, while conversion of carbon dioxide to methane and carbon monoxide can be performed in an optimum temperature window.

The BioTechnology Institute

Enzyme Based Strategy for Energy Efficient Carbon Capture and Storage

Presented at AIChE Annual Meeting in Salt Lake City, USA
November 9th, 2015

Shunxiang Xia and Ping Wang*
Collaborator: Jungbae Kim

Department of Bioproduct and Biosystem Engineering
UNIVERSITY OF MINNESOTA

Outline

Introduction

Enzyme based strategy establishment

- pH as driven force
- Enzyme activity and stability improvement
- Carbon capture capacity improvement

Conclusion

UNIVERSITY OF MINNESOTA

The Carbon Capture and Storage

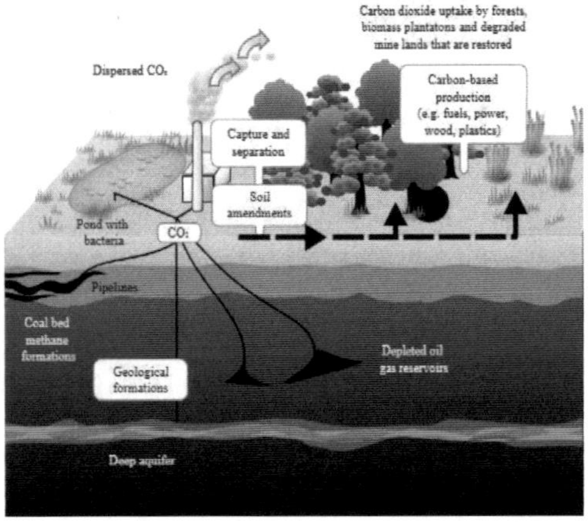

Natural CCS, interrupted dramatically by human activities

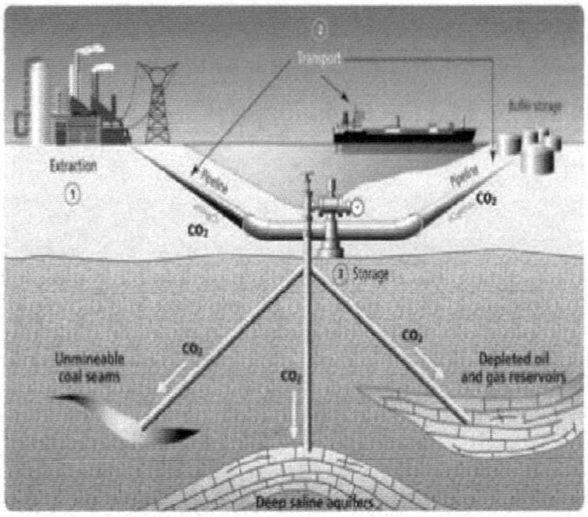

Artificial CCS, increasing carbon fixation and balancing the carbon flux

CO_2 fixation **CO_2 Transport** **CO_2 Storage (Sequestration)**

UNIVERSITY OF MINNESOTA

Chemicals Based Carbon Capture

$$C_2H_4OHNH_2 + CO_2 + H_2O \underset{\text{hot}}{\overset{\text{cold}}{\rightleftharpoons}} C_2H_4OHNH_3^+ + HCO_3^-$$

Monoethanolamine (MEA), Diethanolamine (DEA) and Methyldiethanolamine (MDEA)

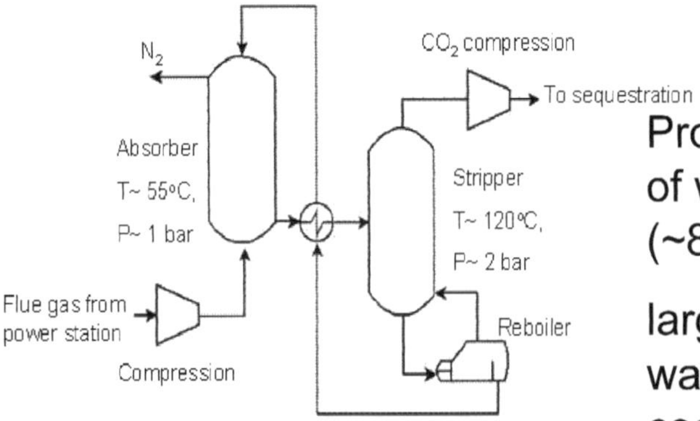

Process involves large amount of water (~80% w/w in the solution)

large portions of the energy are wasted during the heating–cooling cycles

UNIVERSITY OF MINNESOTA

Native Carbon Fixation Pathway

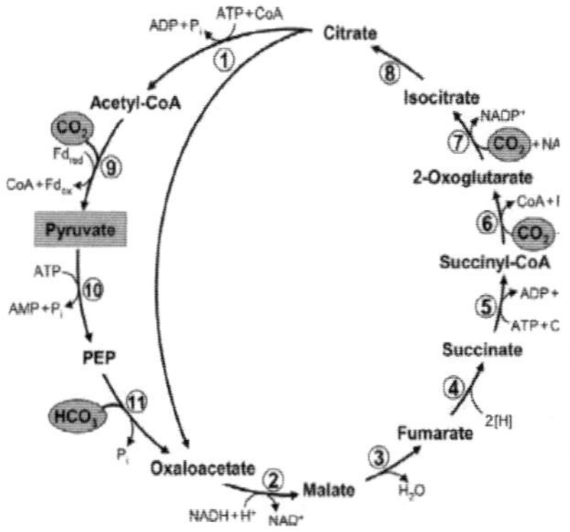

The Reductive Tricarboxylic Acid Cycle
(C2-C3,C3-C4, C5-C6)
Chlorobium limicola, Desulfobacter hydrogenophilus

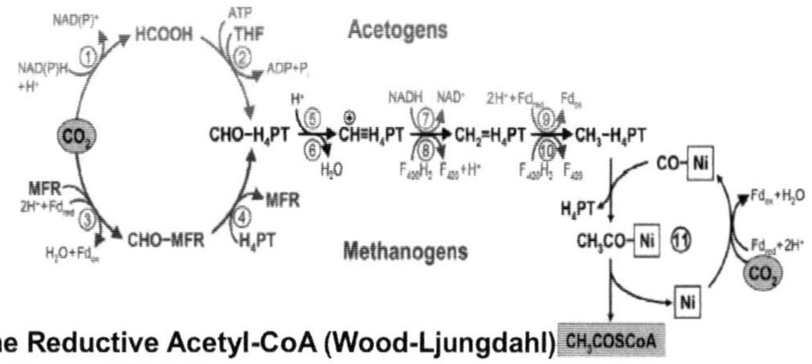

The Reductive Acetyl-CoA (Wood-Ljungdahl) Pathway (C0-C1)
Clostridium

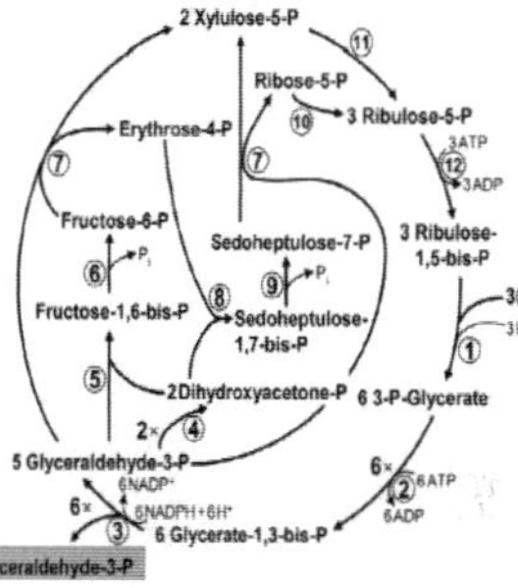

The Reductive Pentose Phosphate (Calvin-Benson) Cycle
Most common in the nature

Ivan A. Berg*, APPLIED AND ENVIRONMENTAL MICROBIOLOGY, Mar. 2011, p. 1925–1936

UNIVERSITY OF MINNESOTA

RTCA Cycle in The Nature

$$\alpha\text{-Ketoglutarate} + CO_2 + NADP_{red} \leftrightarrow \text{D-Isocitrate} + NADP_{ox} + H_2O$$

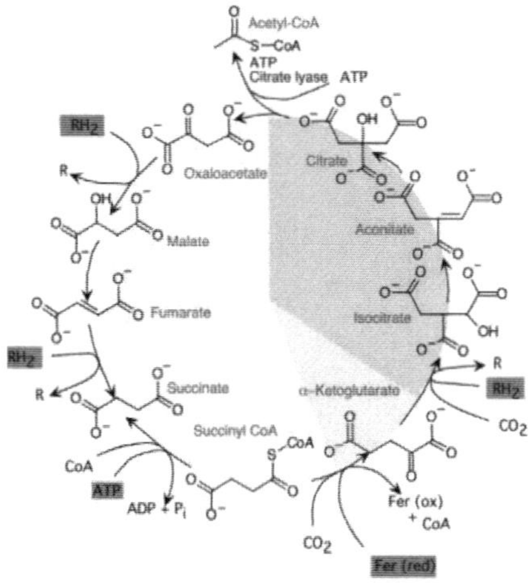

Scheme of reductive TCA cycle

http://news.mit.edu/

Produce carbon compounds from carbon dioxide and water
In reverse of citric acid cycle
Isocitrate dehydrogenase catalyze an important step

UNIVERSITY OF MINNESOTA

Enzyme Based Carbon Capture

Reversible enzyme reaction promise high efficient carbon capture

Manipulation of thermodynamic driving forces for carbon conversion vis pH control

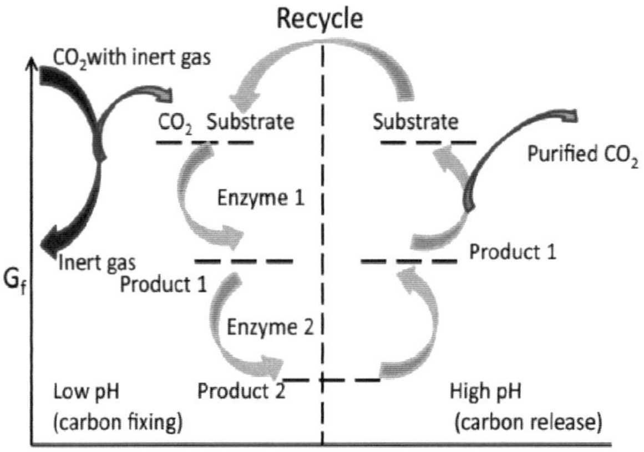

UNIVERSITY OF MINNESOTA

Theoretical Consideration of Reversibility

$$\Delta G_{i,f}(pH,I) = \Delta G^{O}{}_{i,f}(I=0) + RT\left[N_H(i)\ln 10^{-pH} - \alpha\frac{(z_i{}^2 - N_H(i))I^{1/2}}{1+BI^{1/2}}\right]$$

Debye–Hückel equation

	ΔG_f (kJ/mol)				
	pH 5	pH 6	pH 7	pH 8	pH 9
Ketoglutarate	-679.25	-656.42	-633.58	-610.75	-587.92
Isocitrate	-1019.33	-989.91	-960.5	-931.08	-901.67
NADPH	-59.05	89.36	237.77	386.18	534.59
NADP+	-108.72	33.98	176.68	319.38	462.08
CO_2(tot)	-564.61	-554.49	-547.1	-541.18	-535.8
H_2O	-178.49	-167.07	-155.66	-144.24	-132.83
$\Delta_r G^0$ (kJ/mol)	-3.63	-1.45	3.43	9.81	16.71

UNIVERSITY OF MINNESOTA

Reaction Equilibrium Detection

$$Ke = \frac{[Cit][NADP_{ox}]}{[Keto][NADP_{red}][CO_2(tot)]}$$

Reaction equilibrium constant shifted by a factor of ~500 fold when pH changed between 5 and 9

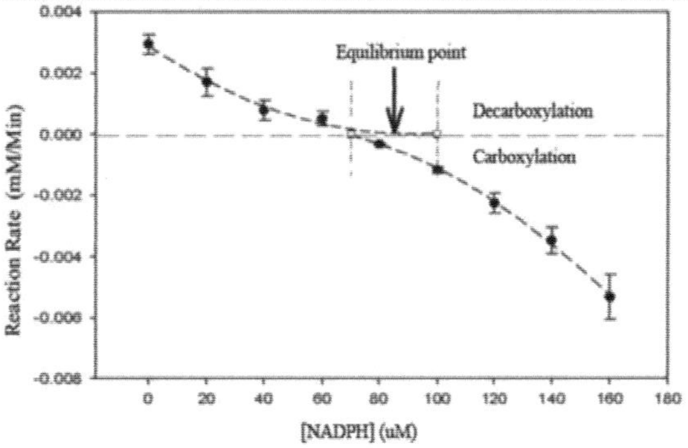

	pH 5	pH 6	pH 7	pH 8	pH 9
[NADP+] (uM)	100	10	10	1	1
[NADPH] (uM)	85±15	60±20	260±20	100±40	400±100
Ke	42.02±6.30	5.95±1.49	1.37± 0.098	0.36±0.10	0.089±0.018

UNIVERSITY OF MINNESOTA

Comparison Between Theoretical Prediction and Experimental Determination

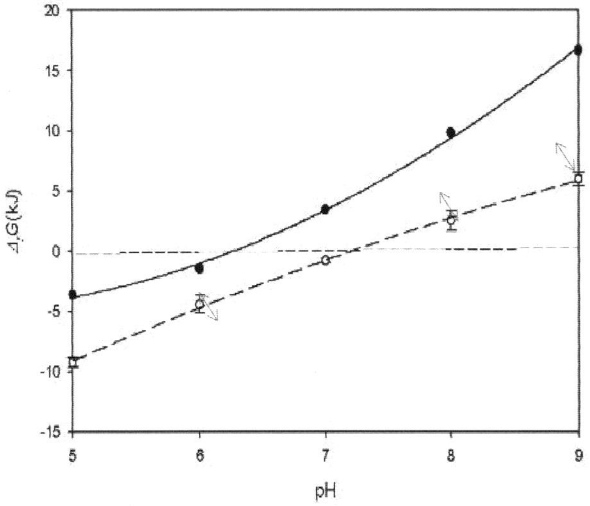

The effect of pH on Gibbs free energy of reaction. Filled circles are $\Delta_r G^0$ (theoretical prediction) and open circle are $\Delta_r G'$ (experimental determination)

Both $\Delta_r G^0$ *and* $\Delta_r G'$ increase with higher pH

Gap between the $\Delta_r G^0$ *and* $\Delta_r G'$

Effect of activity coefficient (Γ)

$$\Delta_r G^0 = \Delta_r G' - RT\ln(\Gamma)$$

Where

$$\Gamma = \frac{\gamma_{Cit}\gamma_{NADP_{OX}}}{\gamma_{keto}\gamma_{NADP_{Red}}\gamma_{CO_2(tot)}}$$

UNIVERSITY OF MINNESOTA

Improvement of Enzyme Activity

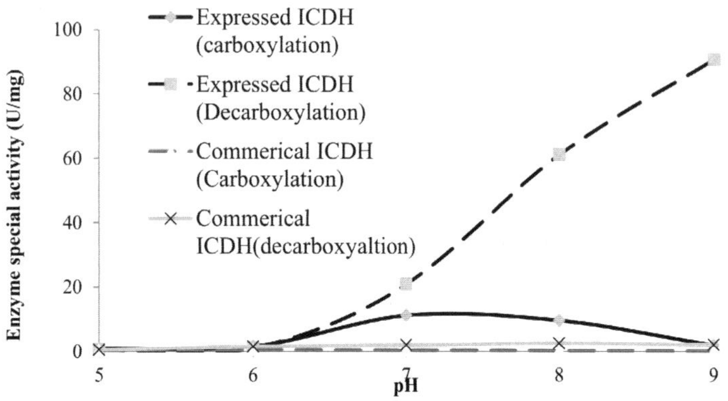

The comparison of activities of commercial (from **porcine heart)** and expressed ICDH (from ***Chlorobium limicola)***

Different source of ICDHs show different activity and optimum pH

The structure of monomer of ICDH

UNIVERSITY OF MINNESOTA

ICDH immobilization through Adsorption

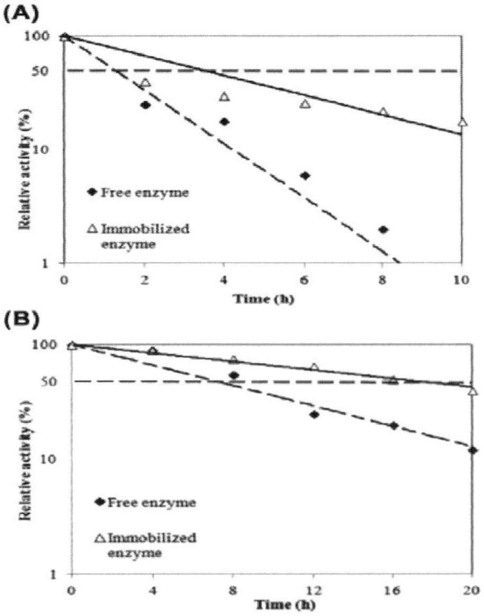

Mesocellular silica foam (MSF) was selected for enzyme immobilization

High area : volume ratio

Small mass transfer limitation

Mild interaction between enzyme and scaffold

Free (square) and immobilized (triangle) enzyme relative activity under pH 6 (B) and pH 9 (A)

UNIVERSITY OF MINNESOTA

Carbon Capture Cycle and System Reuse

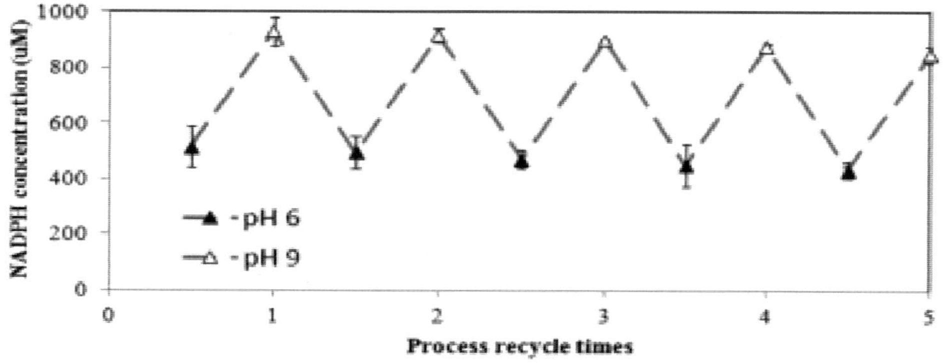

Cycle repeated between pH 6 and 9
Similar equilibrium constant got each cycle
Slight lose of cofactor stability

UNIVERSITY OF MINNESOTA

Cascade Enzymatic Reactions

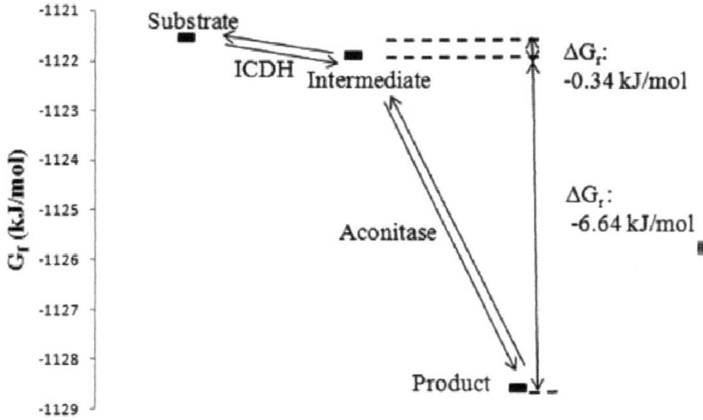

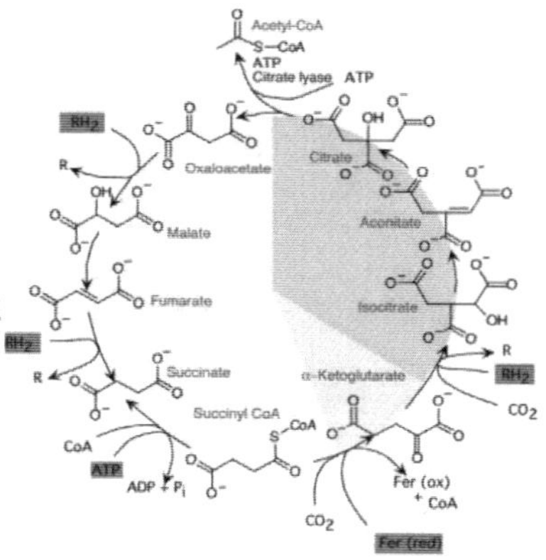

Both ICDH and Aconitase reactions are reversible and controlled by pH. By further converting isocitrate to citrate, carbon capture capacity was increased.

UNIVERSITY OF MINNESOTA

Reversibility of Cascade Reactions

α-Ketoglutarate + CO_2+ $NADP_{red}$ ↔ D-Isocitrate + $NADP_{ox}$ + H_2O

Citrate

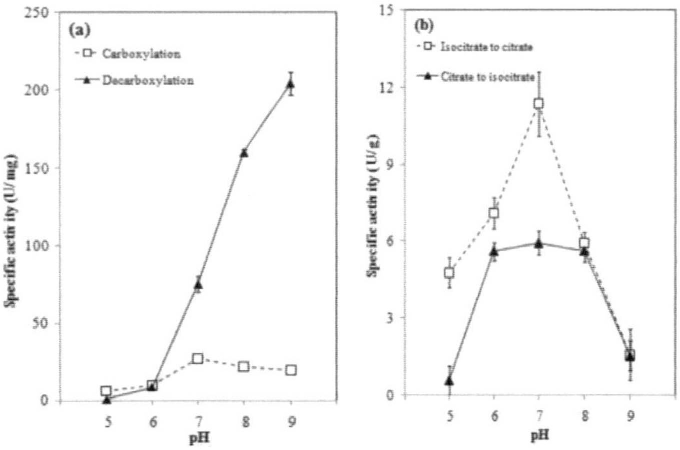

Compared with ICDH, the reversibility of Aconitase reaction was not pH sensitive

UNIVERSITY OF MINNESOTA

The Enhancement of Carbon Capture Capacity with Aconitase

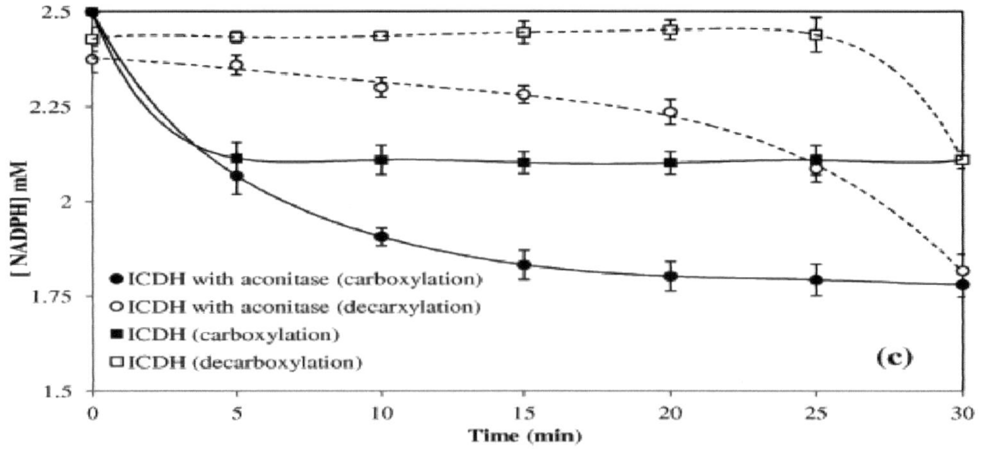

UNIVERSITY OF MINNESOTA

Conclusion

- Manipulation of thermodynamic driving forces for carbon conversion via pH control.
- Enzyme stability and activity were improved by enzyme screening and immobilization.
- Coupled cascade enzymic reactions greatly enhanced carbon capture capacity.

UNIVERSITY OF MINNESOTA

Publications

Xia, Shunxiang, et al. "Biocatalytic carbon capture via reversible reaction cycle catalyzed by isocitrate dehydrogenase." *Biochemical and biophysical research communications* 452.1 (2014): 147-150.

Xia, Shunxiang, et al. "Cascade enzymatic reactions for efficient carbon sequestration." *Bioresource technology* 182 (2015): 368-372.

Xia, Shunxiang, Ping Wang, and Jungbae Kim. "pH sensitivity of carbon dioxide conversion catalyzed by isocitrate dehydrogenase." *Greenhouse Gases: Science and Technology* 5.1 (2015): 102-107.

UNIVERSITY OF MINNESOTA

Acknowledges

The authors thank supports from the Biocatalysis Initiative grant from UMN, and the 'International Collaborative R&D Program' of the Korea Institute of Energy Technology Evaluation and Planning (KETEP, #20118510020020)

UNIVERSITY OF MINNESOTA

428027 Geological Labs on Chip (GLoCs): New Tools for Investigating Key Aspects of CO2 Geological Storage

Abstract:

Underground CO_2 geological storage in deep saline aquifers represents a mediation solution for reducing the anthropogenic CO_2 emissions. Consequently, this kind of storage required adequate scientific knowledge to evaluate injection scenarios, estimate reservoir capacity and asses leakage risks. In this context, we have developed and used high pressure / high temperature microfluidic tools to investigate the different mechanisms associated with CO_2 geological storage in deep saline aquifers. The silicon-Pyrex 2D porous networks can replicate the reservoir p,T conditions ($25 < T < 50°C$, $50 < p < 10$ MPa), geological and topological properties.

This talk will first highlight the strategy to access to global characteristics of our porous media such as porosity and permeability, which are later compared to numerical modelling results. The presentation will then focus on the use of GLoCs to investigate each trapping mechanisms at pore scale. The direct optical visualization and image treatments allow us to follow the evolution of the CO_2/brine phase distribution of the injected CO_2 within the reservoir, including displacement mechanisms and pore saturation levels. Finally, we will present some ongoing work aiming at coupling GLoCs to spectroscopy and optical characterization to get information about dissolution and mineralization trapping such as Raman spectroscopy or the integration of optic fibers to our HP/HT microsystems. Beyond CO_2 geological storage investigations, the GLoCs could also find wider applications in geological-related studies such as Enhanced Oil Recovery (EOR), shale gas recovery or geothermal energy.

434250 Thermal Decomposition of Methane Hydrates Combined with Carbon Dioxide Sequestration

Abstract:

Thermal Decomposition of Methane Hydrates combined with CO_2 sequestration

Formation of methane hydrates followed by recovery of methane gas from the hydrate phase was investigated using thermal stimulation and combined thermal stimulation with CO_2 injection in a laboratory reactor. A large-scale laboratory reactor of internal volume 59.3 liters was used to duplicate the permafrost conditions in which hydrates form naturally. The range of initial hydrate saturation values studied was 10 % to 50 % by pore volume and the range of heating rates studied was 0 W to 100 W. In the case of thermal stimulation combined with CO_2 exchange, the range of CO_2 injection rates studied was 0 ml/min to 580 ml/min. The test matrix was designed to investigate the effect of initial hydrate saturation and heating rate on the gas recovery and the thermal efficiency of the process. The CO_2 injection was combined with thermal stimulation to study the enhancement in the gas recovery efficiency and to study the amount of CO_2 sequestered. The concept of CO_2 injection is similar to the fluid injection techniques used in enhanced oil recovery for expelling the desired product from the pores of the matrix through the production wells.

Currently, the field tests make use of a hot fluid injection (such as steam) for raising the temperature of the hydrate bearing sediment, however these methods suffer from the loss of heat during transit from ground to the hydrate sediment. This work made use of a concept of down-hole combustion in which, part of the methane gas released from dissociation of methane hydrates itself will be used as a fuel for in situ combustion. An electrical point heat source was used for simulating combustion in this work. The major purpose of the work with CO_2 injection is to match the flow rates of CO_2 gas released during the in situ combustion process. For example, generation of heat at the rate of 100 W by burning methane will produce the CO_2 at the rate of 155 ml/min. A testing matrix was developed and executed to study the effect of methane recovery and thermal efficiency over a range of initial hydrate saturations (30 % & 50 %), heating rates during dissociation (20, 50 & 100 W) and CO_2 injection rates (110 ml/min & 580 ml/min). Results from those tests demonstrate that, at a *constant initial hydrate saturation*; total recovery of methane increases with an increase in the heating rate and, the thermal efficiency decreases with an increase in heating rate. At a *constant heating rate*, total recovery of methane as well as thermal efficiency increased with an increase in initial hydrate saturation. For tests that combined CO_2 injection and thermal stimulation during dissociation, at a constant value of initial saturation and constant heating rate; an increase in CO_2 injection rate increased the total recovery of the methane. CO_2 injection tests also gave higher recovery values in less time than their corresponding counterparts in the pure thermal stimulation tests. As a representative example, at 50 % saturation and 100 W heating, injection of 580 ml/min CO_2 gave nearly 69 moles of methane in 18 hours as compared to 62 moles in 45 hours for a zero CO_2 injection test.

The motivation for this work comes from the need to utilize unconventional natural gas resources. Natural gas extraction from the hydrates allows sequestration of carbon dioxide helping reduce the carbon footprints and contribute to reducing overall greenhouse gas emissions. Currently, the conventional gas extraction and recovery methods are cheaper than the gas hydrate recovery methods and it is expected that this will change as, more efficient methods are being developed in gas hydrate field. It is required to focus the research on a lab scale experiments to provide economically feasible guidelines for large-scale production tests.

437859 Water Desalination By Shock Electrodialysis

Abstract:

The development of cost and energy efficient water purification systems, specifically desalination systems, is one of the most critical engineering challenges today. Recently, desalination has been used to solve the drought crisis in California, but it is costly and energy intensive [1]. Shock electrodialysis is a newly developed method that uses ion concentration polarization (ICP) zones and deionization shocks in porous media near an ion selective membrane to purify water. In this project, we demonstrate experimentally the feasibility of shock electrodialysis in water desalination. An overlimiting current, a current that is more than the diffusion limited current, is applied through a weakly charged porous medium (a silica glass frit with pore sizes ranging from 0.75 to 1 micron) sandwiched between two cation-selective membranes (Nafion), producing a sharp concentration gradient or "shock"[2]. The cations move through the cation-selective membrane toward the cathode while the anions get trapped on the cation-selective membrane on the anode side. This movement of ions produces a sharp concentration gradient with high ion concentration near the anode side and low ion concentration near the cathode side. Flowing saltwater through the porous medium and adding a splitter that splits the flow into two streams at the end can separate the high ion concentration (brine) stream and the low ion concentration (fresh) stream. This separation allows for continuous desalination. Using this device, we were able to remove up to 99+% of ions from NaCl feed solutions of various concentrations (1 mM, 10mM, and 100mM) at a flowrate of 75 uL/min. Keeping the flow rate constant at 75 uL/min and varying the concentrations revealed that plots of the percentage of ion removal versus the applied current divided by the limiting current for all three concentrations overlapped, agreeing with the theory. By varying the flow rate and keeping the concentration of NaCl constant at 10mM, it was seen that at lower flowrates, there was more ion removal at the same applied current, agreeing with the theory because a lower flowrate allows the solution to spend more time in the device which allows for more ion removal. The 37.5 uL/min and 75 uL/min flowrates achieved 99+% ion removal at the highest applied current (4080 uA) while the 150 uL/min flowrate achieved 90% ion removal. Overall, preliminary results show good desalination factors, reasonable efficiency, and unique separation capabilities. These unique separation capabilities stem from the feed water flowing through the glass silica and the large electric field applied. The small pore size in the glass silica frit is able to separate out large particles and the large electric field has the potential for disinfection; both of these qualities may find applications in compact water treatment and brine concentration.

[1] Roth, S.; California's last resort: Drink the Pacific; Desert Sun; 2015. http://www.desertsun.com/story/news/environment/2015/04/20/californias-last-resort-drink-pacific/26081355/

[2] Mani, A.; Zangle, T.; Santiago, J.G. On the Propagation of Concentration Polarization from Microchannel-Nanochannel Interfaces Part I: Analytical Model and Characteristic Analysis. Langmuir 2009, 25 (6), 3898-3908

437635 Exergy Analysis of a Power Plant in Abu Dhabi (UAE)

Abstract:

The first objective of this research was to conduct an exergy analysis for a power-generation plant in Abu Dhabi (UAE) in order to determine the main source of its irreversibilities. Our results indicated that the combustion chamber was the main factor (70.2%) of the total exergy destruction of the plant. On the other hand, the compressor had the lowest contribution towards exergy destruction (12.4%). In the second part of this investigation, Aspen Hysys with the Soave-Redlich-Kwong (SRK) equation of state were utilized in order to simulate the effects of high temperatures and absolute humidity of ambient air on the exergy efficiency of the different equipment in the power plant. Our data showed that the temperature of ambient air had more negative effects on the performance of the plant than its absolute humidity. The exergy efficiency of the combustor and compressor decreased with temperature while the exergy efficiency of the turbine increased with temperature. Finally, a comparative study on performance of the power plant between design conditions (T=288K, RH=60%) and summer conditions (T=316K, RH=50%) in Abu Dhabi indicated that the power plant lost 4.66 % of its net power output and 4.61 % of its exergy efficiency.

438525 Insights into the Hydrothermal Stability of ZSM-5 Under Relevant Biomass Conversion Reaction Conditions

Abstract:

Zeolite catalysts used for the conversion of carbohydrates to renewable platform chemicals (e.g., glucose to HMF) in the condensed phase are shown to be sensitive to the presence of inorganic salts which alter the zeolite surface chemistry. Although zeolites such as ZSM-5 have been shown to be stable in de-ionized water, the presence of NaCl in the reaction mixture at dilute and saturated levels enhances the hydrolysis of the Si-O-Al bridges and the release of Al3+ species. These dissolved Al3+ species catalyze the conversion of glucose through homogeneous catalytic processes and obscure the apparent reactivity of the zeolite catalyst. Comparative XRD, solution-state NMR, and ICP-MS show that the reactive aluminum species in solution are leeched in a time-limited process from Si-O-Al bonds involved in extra-framework species and/or at defect sites without affecting the zeolite lattice

EFFECTS OF SEASON AND HEATING MODE ON IGNITION AND BURNING BEHAVIOR OF TEN SPECIES OF LIVE FUEL MEASURED IN A FLAT-FLAME BURNER SYSTEM

Samantha A. Smith, Jonathan R. Gallacher, and Thomas H. Fletcher

Abstract

The effects of season and heating mode on ignition and burning behavior of living vegetation were studied in a flat-flame burner system with a radiant panel. The goal was to identify what plant characteristics have the greatest influence on burning behavior and to understand the effects of heating mode on ignition and burning. Experiments were performed for ten species over a two-year period, with 25 runs completed each month for each species. A flat flame burner (1000°C, 10 mol% O_2) and radiant panel (50 kW/m^2) provided the convection and radiation sources, respectively. Time-dependent mass, surface temperature and flame characteristics were measured. The effect of moisture content on ignition time varied with species. Ignition time showed a strong dependence on heating mode, with broadleaf species showing a much stronger response to added radiation than non-broadleaf species. No samples exposed to the radiant panel alone ignited. Heating mode and moisture content affected flame characteristics and the relationship depended upon season and species. The observed relationship between ignition time and moisture content did not agree with the relationship observed in wet wood. Moisture content alone did not adequately describe seasonal changes in ignition behavior. However, by including other factors in the prediction model (e.g. surface area and mass), it is possible to predict live-fuel ignition. This implies that plant function plays a role in the burning behavior of live fuels.

1. Introduction

Ignition of wood and other cellulosic fuels has been studied for over 100 years. Research has largely focused on requirements for ignition, conditions during burning, and predictive modeling techniques (including rate of spread calculations). The ultimate goal in wildland fire research is two-fold: (1) to understand the physical phenomena that occur within wildland fires, and (2) to develop models that can predict wildland fire behavior. This study presents some of the results from a two-year project to study ignition and burning phenomena relative to seasonal changes in plant growth and the mode of heat transfer used for fuel heating.

Ignition can occur when a fuel sample is heated to the point where pyrolysis rates are high enough to support a gaseous flame *and* a gas phase flammable mixture exists. Researchers often measure ignition temperature and time. These values are then used as empirical estimates of the surface temperature at which pyrolysis rates can support a continuous flame and the time required to reach that temperature, respectively. The bulk of the work has focused on ignition temperature. Experimental conclusions to date are mixed. Babrauskas [1, 2] compiled the results of ignition temperature experiments on wood fuels and foliage, respectively. After eliminating the experiments in which the fuel sample was pressed against a hot surface, the reported ignition temperatures ranged from 200-530°C for wood and 201-450°C for foliage. Babrauskas noted the large amount of scatter in the data and suggested that, in addition to variations in experimental setup and measurement techniques, sample condition (e.g. moisture content and size) and species could affect ignition temperature.

Wildland fire observations that species burn differently support Babrauskas's postulate that plant species could be one source of variation in measured ignition temperatures [3]. However, results by Susott [4] showed that material ground from various plant species has the same heat of combustion and similar TGA (thermogravimetric analysis) pyrolysis mass release curves, and should therefore burn similarly. Thus, one possible explanation for the observed differences in ignition properties is the shape and structure of the plant and the effect shape has on heat and mass transfer. However, this explanation has not been tested experimentally. Most empirical correlations used to predict ignition behavior, particularly for live fuels, are species specific [5-8]. Work must be done to understand the differences in ignition behavior between various species.

Investigation of the effect of moisture content on ignition has been studied extensively and supports Babrauskas's postulate that sample condition affects ignition. Most of the work has focused on dead or woody fuels, with the result that moisture increases both ignition time and temperature [5, 6, 9-11]. The results for live fuels are fewer and less consistent [3, 12-15]. Many fire spread models use moisture content as a predictor for ignition. These models have varied success with dead fuels but little success with live fuels [13]. One possible reason for the difference in ignition between live and dead fuels and the performance of fire spread models in predicting fire behavior in live fuels is the difference in water storage between live and dead plants and the resulting treatment of water storage in fire spread models. Water evaporation in dead fuels has been assumed complete in fine fuels once the sample temperature passes 100°C [16, 17], but Fletcher et al. [3] showed there is still a significant amount of moisture in live fuels when ignition occurs. Pickett [18] showed water release still occurring at surface temperatures in excess of 200°C and Prince [14] showed significant differences in the temperature profiles of live and dead foliage during ignition and burning even with the same moisture content. Work by McAllister et al. [13] showed significant differences in the ignition behavior between live and dead pine needles. These differences have led researchers to postulate that there is significant interaction between the free water and the cells in live plants that does not occur in dead plants [13, 19]. Finney et al. [20] postulated that water release in live fuels is not complete until the breakdown of the cellular structure occurs. Still other work has been done indicating root structure [21], plant dry mass [22], and chemical composition [13, 23] could have a larger effect on ignition of live fuels than moisture content, though results are mixed in work to quantify the effect of chemical composition (e.g. [15]). Thus, a fundamental understanding of the physical processes that drive live fuel combustion is both absent and necessary if predictive models are to be developed.

Another of Babrauskas's postulates was that differences in ignition temperature arise due to differences in experimental setup. This is particularly true when looking at the different heat transfer mechanisms. Much of the work looking at heat transfer mode has focused on dead and woody fuels [10, 17, 24-31]. Experiments performed by Rothermel [17] showed fuel pre-heating in no-wind and backing-fire situations, leading to the conclusion that radiation drives fire spread. This conclusion and his resulting fire spread model established a precedent that radiation is the dominant form of heat transfer in wildland fires. However, other experiments have shown that significant amounts of pyrolyzates are not formed at the fuel temperatures associated with radiant pre-heating [32]. Anderson [33] concluded that radiant heat flux can provide no more than 40% of the energy required for sustained fire spread. Still other work has shown flame propagation to depend strongly on direct flame contact with un-burned fuel [34, 35]. Many researchers have concluded that radiation heat transfer dominates in large fires [24, 31, 36, 37] and fires with little to no wind in homogeneous fuel beds [28, 38-40], but the relative effect of radiation and convection for fires outside these conditions is still unknown [28, 40]. Only a limited amount of work has been performed for live fuels and foliage [13, 41]. Work to quantify the contributions of radiation and convection in live-shrub combustion is necessary to understand the basic theory of fire spread and to develop a model that accounts for both mechanisms of heat transfer.

The aim of this project is to explore the effect of season and heating mode on ignition and burning behavior to better understand what physical processes drive fire spread in live shrubs.

2. Methods

Combustion experiments were performed on ten species every month for two years. Each month 25 replicates were performed for each species. The ten species consisted of three species from California, three from Florida, two from Utah, and two from Montana. The plants from California consisted of chamise (*Adenostoma fasciculatum*), Manzanita (*Arctostaphylos glandulosa*), and Ceanothus (*Ceanothus crassifolius*). Gallberry (*Ilex glabra*), fetterbush (*Lyonia lucida*), and sand pine (*Pinus clausa*) originated from Florida, while big sagebrush (*Artemisia tridentata*), and Gambel oak (*Quercus gambelii*) are from Utah. Finally, lodgepole pine (*Pinus contorta*) and Douglas-fir (*Pseudotsuga menziesii* var. *glauca*) are from Montana. Since most of the samples had to be collected from other locations, they were placed in a sealed plastic bag and shipped overnight, where they were placed away from light with the bag still closed until that species could be tested.

Experiments were performed using a water cooled, flat-flame burner (FFB) and radiant panel (see Figure 1). The FFB produced post-combustion gases at 1000 °C and 10 mol% O_2 while the radiant panel provided 50 kW m^{-2} at the sample location. A glass cage prevented ambient air entrainment. In order to perform the experiments, the sample was suspended within the cage using an alligator clip and holding rod. This rod was placed on a Mettler Toledo XS204 Cantilever mass balance, which recorded any differences in mass as the specimen burned or experienced pyrolysis. Each month, experiments were performed using three heating cases, namely convection, radiation, and combined convection plus radiation. In all cases there was a visual camera placed in front of the sample and an infrared camera placed above the sample (see Figure 1).

Transient flame, mass, and surface temperature data were collected during each run. Flame characteristics were extracted from these data, including maximum flame height (cm), time to maximum flame height (s), ignition temperature (°C), maximum ignition temperature (°C), ignition time (s), and burn out time (s). There were also many measurements of the specimen itself, such as the density, surface area, moisture content, length, width, thickness, etc.

Both single and multiple parameter prediction models were developed to describe each of the flame characteristics listed above, namely time to ignition (t_{ig}), surface temperature at ignition (T_{ig}), maximum surface temperature at ignition ($T_{ig,max}$), maximum flame height (MFH), time to maximum flame height (t_{MFH}) and burnout time (t_{BO}). Ignition has been shown to be a local phenomenon [19]. Surface temperature has also been shown to have large spatial variations during the burning of live foliage [19]. However, many ignition temperature measurements do not measure the spatial variation in surface temperature. Correlations for both average surface temperature at ignition and maximum surface temperature at ignition (assuming the location of maximum temperature corresponds to the localized ignition point) were developed. One-parameter models were developed using simple linear regression. The F-statistic, p-value, and confidence interval on the slope were used to determine if the slope term and regression model were significant.

Forward and backward stepwise regression was used to develop correlations with multiple parameters. To do this, 500 correlations were developed by randomly selecting four parameters for use in each model and by randomly selecting a transformation of the data for those parameters. To develop the models, the parameters were first assigned to groups based on type of measurement to reduce the chance of multicollinearity. The groups were: (1) moisture content, relative moisture content and water mass; (2) fresh mass and dry mass; (3) length, width and surface area; (4) needle length, stem diameter, thickness and density; and (5) lipid content, volatile fraction, fixed carbon and ash content. For each correlation, four of the five groups were randomly chosen and then a parameter from each of those four groups was randomly selected. Once the parameters were identified, a data transformation was applied to the data for each parameter. The transformations were: (1) no change, (2) square root of the data, (3) square of the data, (4) natural log of the data, and (5) inverse of the data. Once the data and data transformations were assigned, stepwise regression was performed to develop a correlation to fit the selected data to each of burning characteristics listed above. Once the 500 models were created, the model with the highest adjusted R^2 for each burning characteristic for each species was identified.

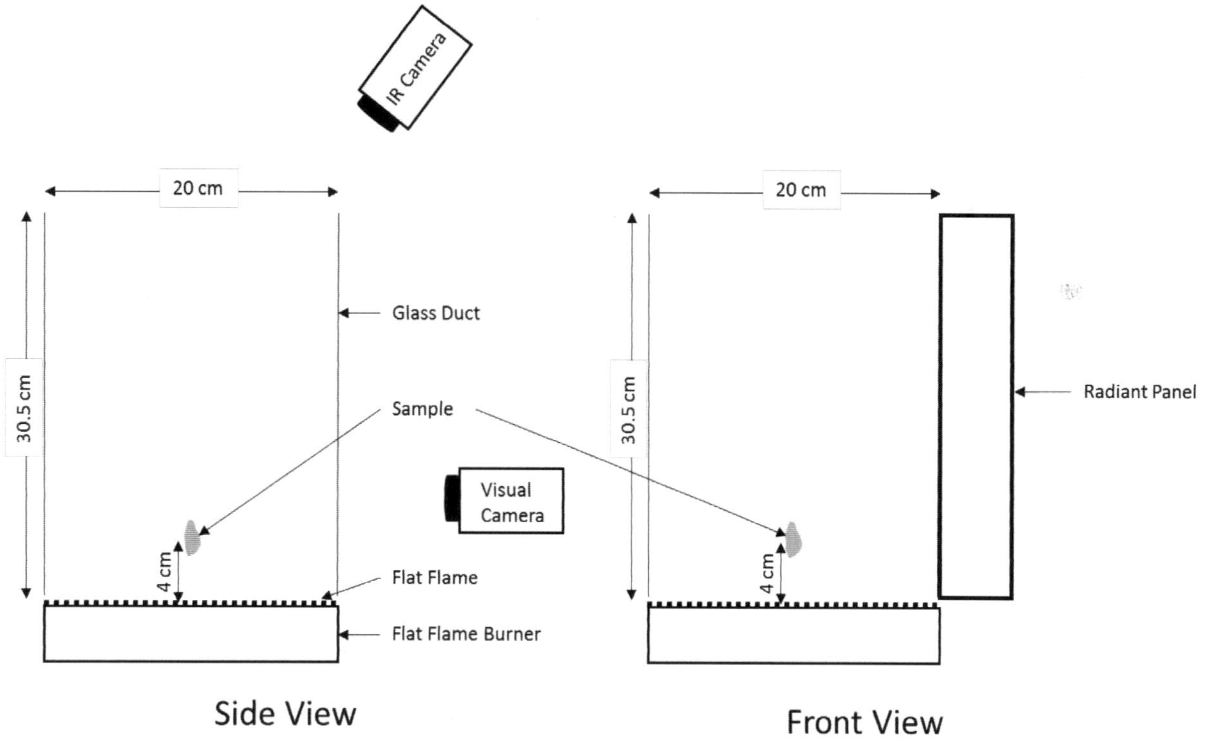

Figure 1. Experimental apparatus: flat flame burner, radiant panel, IR camera, video camera, glass cage to prevent ambient air entrainment, sample location, mass balance and sample holding rod. Post-flame gases are 1000 °C and 10 mol% O_2.

3. Results and Discussion

It was predicted that an increase in the moisture content of a species would result in a longer ignition time and would closely resemble the burning of wet wood. This was not the case. Table 1 illustrates how live fuels responded to different burning modes during the year for different flame characteristics. If there was a statistically significant positive or negative slope at the 95% confidence level, then a P or N was indicated, respectively. Zeros indicate that there was no statistically significant relationship. Overall, the results from the live fuels were different from the expected behavior with wet wood illustrating that one-parameter models do not work for live fuels. Ceanothus, Gambel Oak, Douglas-fir, and Chamise, however, did exhibit a positive correlation to the wet wood behavior at the 95% confidence level.

Table 1. Significance of yearly trends by species

	Ignition Time (s) vs MC		**MFH (cm) vs MC**		**Ignition Devol (%/s) vs MC**		**Ignition Devol (%/s) vs Ignition Time (s)**	
Species	Conv*	Comb**	Conv	Comb	Conv	Comb	Conv	Comb
Manzanita	0	0	0	0	0	0	N	0
Ceanothus	P	P	0	0	0	0	0	0
Douglas-fir	P	0	N	0	0	0	0	0
Gambel Oak	P	P	N	N	0	0	0	0
Fetterbush	0	0	0	0	0	0	N	0
Gallberry	0	0	0	0	0	0	N	0
Sand Pine	0	P	0	0	0	0	0	0
Chamise	P	P	0	0	0	0	0	0
Sagebrush	0	0	0	0	0	0	0	0
Lodgepole Pine	0	0	0	0	0	0	0	0
Wet Wood (expected)	P	P	N	N	P	P	P	P

*Conv means convection-only experiments; **Comb means combined convection and radiation

The models created using stepwise regression were much better than the one-parameter models. Table 2 shows the adjusted R^2 values for the flame characteristics measured with these new models. The average adjusted R^2 value for all the models is 0.406. However, the models for temperature account for much less of the data variation than the models for the other flame characteristics; the average adjusted R^2 value for the non-temperature models in 0.569. This is partly due to the fact that samples bend and move during burning and thus the entire sample surface was not visible to the IR camera during the run.

Table 2: Adjusted R^2 values when regressing flame characteristics using stepwise models.

Species	t_{ig}	T_{ig}	$T_{ig,max}$	MFH	t_{MFH}	t_{BO}
Manzanita	0.80	0.527	0.568	0.740	0.919	0.747
Ceanothus	0.794	0.481	0.372	0.743	0.905	0.676
Douglas-fir	0.303	0.571	0.692	0.342	0.254	0.50
Gambel Oak	0.664	0.472	0.186	0.432	0.86	0.81
Fetterbush	0.537	0.379	0.058	0.431	0.417	0.459
Gallberry	0.733	0.152	0.223	0.683	0.743	0.511
Sand Pine	0.434	0.294	0.247	0.58	0.624	0.639
Chamise	0.554	0.286	0.305	0.617	0.502	0.449
Sagebrush	0.449	0.285	0.247	0.403	0.589	0.403
Lodgepole Pine	0.489	0.315	0.309	0.384	0.36	0.30

Figure 2 illustrates how ignition time and moisture content vary throughout the year for Douglas-fir and Manzanita with different heating modes (convection-only, and combined convection and radiation). It is observed that there is no statistically significant relationship between moisture content and ignition time. Manzanita had approximately the same moisture content in July and November, but the ignition times in July were about half as long as the ignition times in November. Douglas-fir moisture content doubled from May to September, yet there was no significant change in the ignition time. These examples point toward the initial observation that the single parameter model of just moisture content cannot provide an accurate prediction for live fuels (see Table 1). Since the moisture content isn't the only thing that is impacting the burning of the live fuels, it is logical to consider the plant's physiological structure and how that can change the way it burns. The structure of the plant does change throughout the seasons and that can change how the species stores the water and how the leaves are shaped. These changes can help account for the changes that are not accounted for by just the water in the plant. It is important to note that, while a study of how physiological factors impact the burning of the live fuels is needed to develop a deeper theoretical understanding of live fuel combustion, it is still possible to predict burning behavior with statistical models that use the same inputs as current fire spread models.

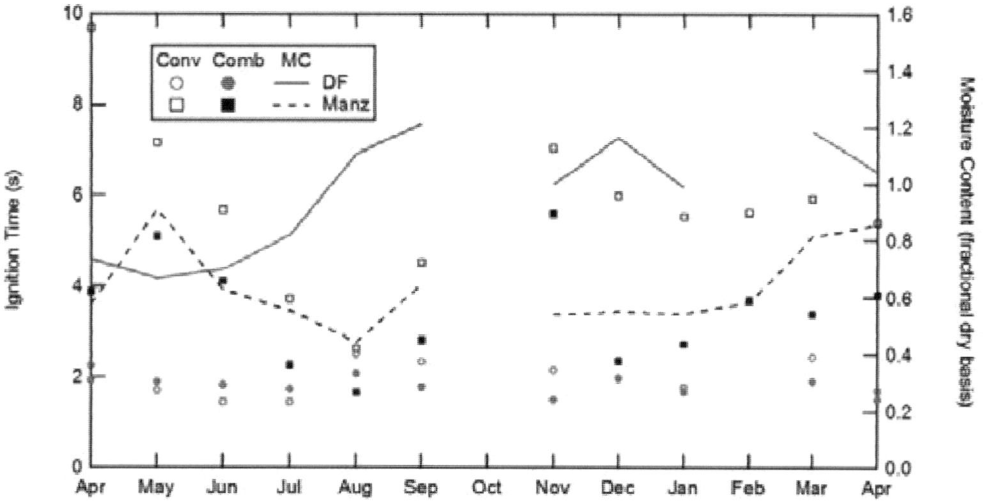

Figure 2. Ignition time and moisture content versus month for manzanita and Douglas-fir. Closed symbols indicate radiation and convection ignition while open symbols indicate convection-only ignition.

In addition to the models that were produced, there was data collected on the different heating modes and how they impact the burning of the live fuels. Table 3 shows the percentage of months there was a statistically significant difference between the convection-only and convection and radiation runs for each species. There is a noticeable difference between the broadleaf species and the needle species for the ignition time and time to maximum flame height due to the added radiation. Radiation alone was never enough to ignite fuel samples without a pilot ignition source. The heat did, however, cause the samples to undergo pyrolysis and then char oxidation.

Table 3. Effect of heating mode on ignition variables. Table entries indicate the percentage of months that radiation and convection ignition differed significantly convection-only ignition at a 95% confidence level

Species*	Ignition Time	Time to Max Flame Height	Maximum Flame Height	Fraction Left at Ignition	Fraction Loss at Ignition
Manzanita	92	83	83	25	42
Ceanothus	100	91	100	45	18
Gambel oak	83	100	83	50	17
Fetterbush	100	100	8	17	25
Gallberry	92	100	0	17	58
Dougals-fir	9	27	27	0	9
Sand pine	25	25	8	0	17
Chamise	33	25	17	17	0
Sagebrush	22	22	22	11	11
Lodgepole pine	50	25	13	38	13

*The first five species listed are broadleaf species, the other five are non-broadleaf.

4. Summary and Conclusions

Ignition and burning behavior for ten live fuels were studied in a flat-flame burner apparatus. Experiments were performed over a two-year period to test the effect of season (specifically moisture content) on ignition and burning behavior. The effect of heat transfer mode was also tested by performing some of the seasonal experiments with a convection-only heat source, some with a radiation-only heat source, and some with both heat sources.

Results comparing ignition and burning characteristics with moisture content and season were mixed. Ceanothus, Gambel oak, Douglas-fir and chamise all exhibited a positive correlation between ignition time and moisture content while the other six species show no correlation at a 95% confidence level, indicating a simple relationship between moisture content and ignition is not adequate to describe ignition in live fuels. However, linear stepwise models capture much of the variability in ignition behavior. The ignition behavior of live fuels in different seasons but at the same moisture content was different. These results suggest a significant difference between live fuels and dead fuels that isn't explained by moisture content. An in-depth study involving careful combustion experiments and detailed physiological measurements is necessary to develop a theoretical understanding of fire spread in live fuels. In the absence of a theoretical understanding, simple statistical models can still be developed that describe fire behavior accurately and that use as inputs the same information currently used in most fire models.

Heating mode results were clearer. Ignition occurred in none of the unpiloted radiation-only experiments. These results do not indicate radiation was unimportant, but rather that additional modes of heat transfer were needed. Inclusion of a radiant flux in the convective environment of a flat-flame burner significantly decreased observed ignition times for broadleaf species but not for non-broadleaf species. Maximum flame height time showed the same dependence as ignition time for all species while maximum flame height showed the opposite dependence, except for fetterbush and gallberry which showed almost no difference between convection-only and radiation with convection experimental runs. Mass loss rate and fraction remaining at ignition were affected much less by the differences in heating than other ignition characteristics.

5. References

[1] Babrauskas, V., J. of Fire Protection Eng., 12 (2002) 163-189.
[2] Babrauskas, V., Ignition handbook, Fire Science Publishers, 2003.
[3] Fletcher, T. H., Pickett, B. M., Smith, S. G., Spittle, G. S., Woodhouse, M. M., Haake, E.Weise, D. R., Comb. Sci. and Tech., 179 (2007) 1183-1203.
[4] Susott, R. A., For. Sci., 28 (1982) 404-420.
[5] Xanthopoulos, G.Wakimoto, R. H., Canadian J. of Forest Research, 23 (1993).
[6] Dimitrakopoulos, A. P.Papaioannou, K. K., Fire Tech., 37 (2001) 143-152.
[7] Smith, S. G., Effects of moisture on combustion characteristics of live California chaparral and Utah foliage, Brigham Young University (2005).
[8] Shen, C., Application of fuel element combustion properties to a semi-empirical flame propagation model for live wildland Utah shrubs, Brigham Young University (2013).
[9] Gill, A. M.Moore, P. H. R., Ignitability of leaves of Australian plants, A Contract Report to the Australian Flora Foundation, Centre for Plant Biodiversity Research, 1996.
[10] Moghtaderi, B., Novozhilov, V., Fletcher, D. F.Kent, J. H., Fire Safety J., 29 (1997) 41-59.
[11] Catchpole, W. R., Catchpole, E. A., Tate, A. G., Butler, B. W.Rothermel, R. C., A model for the steady spread of fire through a homogeneous fuel bed, 2002.
[12] Engstrom, J. D., Butler, J. K., Smith, S. G., Baxter, L. L., Fletcher, T. H.Weise, D. R., Comb. Sci. and Tech., 176 (2004) 1577-1591.
[13] McAllister, S., Grenfell, I., Hadlow, A., Jolly, W. M., Finney, M.Cohen, J., Fire Safety J., 51 (2012) 133-142.
[14] Prince, D., Measurement and modeling of fire behavior in leaves and sparse shrubs, Brigham Young University (2014).
[15] Alessio, G. A., Penuelas, J., De Lillis, M.Llusia, J., Plant Biology, 10 (2008) 123-128.
[16] Albini, F. A., Proc. Comb. Inst., 11 (1967) 553-560.
[17] Rothermel, R. C., A mathematical model for predicting fire spread in wildland fuels, Research Paper INT-115, USDA Forest Service, 1972.
[18] Pickett, B. M., Effects of moisture on combustion of live wildland forest fuels, Brigham Young University (2008).

[19] Prince, D.Fletcher, T. H., A combined experimental and theoretical study of the combustion of live vs. dead leaves, Combustion Institute - Western States Section (WSS/CI), 2013.
[20] Finney, M. A., Cohen, J. D., McAllister, S. S.Jolly, W. M., Int. J. of Wildland Fire, 22 (2013) 25.
[21] Pellizzaro, G., Duce, P., Ventura, A.Zara, P., Int. J. of Wildland Fire, 16 (2007) 633-641.
[22] Jolly, W. M., Hadlow, A. M.Huguet, K., Int. J. of Wildland Fire, 23 (2014) 480-489.
[23] Pyne, S. J., Andrews, P. L.Laven, R. D., Introduction to Wildland Fire, 2nd ed, Wiley, 1996.
[24] Simms, L., Comb. Flame, 4 (1960) 293-300.
[25] McCarter, R. J.Broido, A., Pyro., 2 (1965) 65-85.
[26] Simms, L.Law, M., Comb. Flame, 11 (1967) 377-388.
[27] Pagni, J., J. of Heat Trans., 97 (1975) 153-155.
[28] Morandini, F., Santoni, P. A.Balbi, J. H., Fire Safety J., 36 (2001) 519-543.
[29] Dupuy, J. L., Maréchal, J.Morvan, D., Comb. Flame, 135 (2003) 65-76.
[30] Reszka, P.Torero, J. L., Exp. Therm. and Fluid Sci., 32 (2008) 1405-1411.
[31] Silvani, X.Morandini, F., Fire Safety J., 44 (2009) 279-285.
[32] Cohen, J. D.Finney, M. A., VI International Conference on Forest Fire Research, (2010).
[33] Anderson, H. E., Heat transfer and fire spread, in Research Paper INT-69. 1969, USDA Forest Service.
[34] Vogel, M.Williams, F. R., Comb. Sci. and Tech., 1 (1970) 429-436.
[35] Carrier, G. F., Fendell, F. E.Wolff, M. F., Comb. Sci. and Tech., 75 (1991) 31-51.
[36] Balbi, J.-H., Rossi, J.-L., Marcelli, T.Santoni, P.-A., Comb. Sci. and Tech., 179 (2007) 2511-2537.
[37] Paudel, A., Application of one dimensional turbulence (odt) to model fire spread through biomass fuel bed, Brigham Young University (2013).
[38] Morvan, D.Dupuy, J. L., Comb. Flame, 127 (2001) 1981-1994.
[39] Morvan, D.Dupuy, J. L., Comb. Flame, 138 (2004) 199-210.
[40] Sullivan, A. L., Ellis, P. F.Knight, I. K., Int. J. of Wildland Fire, 12 (2003) 101-110.
[41] Stocks, B. J., Alexander, M. E., Wotton, B. M., Stefner, C. N., Flannigan, M. D., Taylor, S. W., Lavoie, N., Mason, J. A., Hartley, G. R., Maffey, M. E., Dalrymple, G. N., Blake, T. W., Cruz, M. G.Lanoville, R. A., Canadian J. of Forest Research, 34 (2004) 1548-1560.
[42] ASTMStandardD4442-07, Test methods for direct moisture content of wood and wood-based materials, in ASTM International. 2007: West Conshohocken, PA.
[43] Weber, R. O., Prog. Energy Combust. Sci., 17 (1991) 67-82.
[44] Sullivan, A. L., Int. J. of Wildland Fire, 18 (2009) 349-368.

438326 Characterization and Particle Sizing of the Composition of E-Cigarette Aerosol

Abstract:

In recent years, electronic cigarettes, or e-cigarettes, have rapidly gained immense popularity throughout the world and specifically the United States. Despite the increase in use, little is known about the health effects of these vaporizers, compared to traditional cigarettes. The e-cigarette uses a wick and heated coil to atomize e-liquid, creating the aerosol inhaled by the user. E-liquids are composed of propylene glycol, glycerin, usually nicotine, and various flavoring agents. E-cigarettes were customized in order to simulate a puff on the device by pushing a known volume of air through. An Automated Puffer Machine (APM) was designed to control power delivered to the coil, puff volume, puff frequency, and puff duration. An Electrostatic Classifier (EC) and Condensation Particle Counter (CPC) then sampled a diluted puff, analyzing particle size and concentration. A Poly Urethane Foam (PUF) sampler was used to separate and trap the chemical components of the e-cigarette emission. The PUF was then placed in a volatile solvent to dissolve the particulate and vapor components, and the solvent was analyzed using gas chromatography with mass spectroscopy. The PUF sample was used because is closely emulates what a user would normally inhale.

Work has also been done to quantify and describe the aerosol released by electronic cigarettes in order to create profiles of the particles released into the atmosphere. Examples include work to quantify particle size distribution and chemical content of the aerosol in large gas chambers and looking at particle size and distribution with rapid sampling. Particle size of the aerosol was a bimodal distribution, with main peak averages of 30 nanometers and 80 nanometers. Research isolated the components of the e-liquid attempting to sort which vaporized component contributed to the collected particle size profile. Further research attempted to determine the amount of nicotine available within both the vapor and particle components of the aerosol and identify the exact composition of the aerosol created by an e-cigarette when the e-liquid is vaporized.

This study has found that a variety of e-liquids ranging in nicotine contents and flavors consistently produce a bimodal distribution of aerosol particle sizes. Furthermore, this study concludes that the proportion of components effects the concentration of particles of the aerosol.

438306 Ammonia Removal from Aquaculture Stocking Water

Abstract:

A continuous membrane system capable of maintaining low levels of total ammonia nitrogen (TAN, 3, and 4 +) in aquaculture holding and transport vessels is being developed. Ammonia is the major metabolite of fish, therefore, adequate TAN removal is necessary to maintain the health and productivity of aquatic organisms. High levels of ammonia result in the damage of gas exchange tissues, leading to increased mortality and decreased average harvest weight. Membrane systems are desirable for their scalability, lack of compaction and channeling, as in packed beds, and their ability to stabilize additives used during functionalization. Zeolite 13X demonstrates efficient 4 + exchange with + in aqueous solutions. Zeolite 13X is inexpensive and functions well as an ion exchange material with high ammonium exchange capacities. In order to improve the stability of zeolite 13X in water and develop a continuous TAN removal scheme, mixed-matrix membranes have been fabricated in the presence and absence of porogens. Porogens have been used to investigate the effect of pore size and pore density on the membrane structure and the membrane performance. The membranes with porogens exhibited higher flux but lower ammonia removal due to decreased residence time. Porogen-free membranes had a higher ammonia removal capacity due to longer residence time. The ammonia removal capacity was found to correlate well with the amount of zeolite 13X incorporated. An optimal formulation will maximize flux and available zeolite surface area, minimize residence time, and retain large ammonia loadings after a large volume of liquid has been treated.

438394 Impact of Chemical Dopants and Passivation Schemes on Carbon Nanotube Sheet Conductivity

Abstract:

Carbon nanotubes (CNTs) are being investigated as an alternative to conventional metallic conductors. Improving the electrical performance of bulk CNT materials through the use of chemical dopants is an active area of research. The present work focuses on doping efficacy and dopant stability under various environmental conditions. CNT sheet material was doped with a series of compounds of varying structure, and a 3 to 8-fold increase in electrical conductivity was observed after doping with potassium tetrabromoaurate, iodine, or iodine monobromide. The general trend for the doped samples in ambient conditions was a steady decrease in electrical conductivity over several months. These experiments served as an initial assessment of doped CNT sheet stability for conventional conductor applications and demonstrate the necessity towards passivation schemes for this emerging technology. Different encapsulation strategies are currently being explored to retain the conductivity of the films when doped. The conductivity of these samples is being systematically measured while storing them under ambient conditions. Consideration for the encapsulant type, deposition technique, and solvent effects (which may dissolve the dopants from the CNT network) will be characterized. A summary detailing the progress to date will provide an overview of both the best methods to enhance electrical conductivity in CNT sheets as well as the suitable passivation schemes for practical operation.

438530 Îœfilms: Dynamics of Thin-Films Under Physiological Fluids and Shear Flow

Abstract:

Thin-film drug-delivery is an effective alternative to conventional tablets, capsules and liquids as it improves medicinal bioavailability, efficacy, dosage accuracy and onset of effectiveness. Typically, it consists of a thin, dissolving film that orally administers drugs. The significantly large surface area of the thin-film causes rapid disintegration, releasing drugs into systemic circulation via absorption through the oral mucosa. Thin-films are currently studied under static conditions that do not account for the influence of dynamic variables like shear stress due to blood flow, changes in electric potential and interactions with biological fluids on drug delivery. Here we seek to study films under dynamic physiological conditions that closely mimic those within the body, using a microfluidic chip called μFILMS. In the talk, we show our initial results, chiefly construction of the μFILMS chip using additive manufacturing, analysis of the output signals and biochemical/mechanical adjustment of the chip parameters to optimize the visualization and/or characterization of the films. The construction of the μFILMS chip consists of a Transwell® static membrane placed between two glass slides, forming two distinct

438444 Photocatalytic Methanol Reforming on TiO2

Abstract:

Hydrogen gas has great potential for use in transportation fuels and industrial applications, but as it is currently produced commercially from steam reforming of unrenewable natural gas. An alternative source of hydrogen could be accessed from the reforming of sustainably-grown biomass. Through the use of a photocatalyst, biomass can be decomposed, producing clean-burning fuel. Titanium dioxide (TiO_2) is a metal oxide semiconductor of particular interest in photocatalysis due to its activity under ultraviolet illumination and its stability under catalytic reaction conditions. Ordinary TiO_2 (anatase, rutile) only absorbs UV wavelengths. Since a much larger fraction of solar irradiation is in the visible range, we aim to modify TiO_2-based photocatalyst to interact with visible light. Supporting metal nanoparticles on the surface of TiO_2 is known to cause a red shift in the absorbance of TiO_2. We have supported copper, nickel, silver, platinum, and palladium on TiO_2 through insipient wetness impregnation. The metal-modified TiO_2 was tested in two types of in situ reactors; one inside a simple kinetic reactor, and the other in a UV-Visible Spectrophotometer. The samples were then reacted under UV and visible light. Gaseous mixtures of methanol, oxygen and steam were passed over the catalysts while the temperature of the reaction was externally controlled. The products of the reaction were identified by gas chromatography. The addition of each of the metals shifted the absorbance of the TiO_2 to longer wavelengths. Future work includes further characterization of the materials, as well as varying the particle size of the metals and the support, and doping the support with hydrogen, carbon, and nitrogen.

438531 Room Temperature Shape Memory Polymers

Abstract:

Temperature controlled shape memory polymers can be synthesized via a facile method using long chain fatty acid and cross-linked natural rubber. For this experiment, rubber was swollen in liquid stearic acid (75 °C). Different weight percent loadings of acid were achieved by differing swelling times in one hour increments from 1 hour to 5 hours. The normal shape memory cycle consist of increasing the temperature of the swollen polymer above the melting point of the acid, stressing the polymer, then immediately quenching the heat in room temperature water. The rubber will hold stress without relaxing until heated above the melting point of stearic acid again, thus returning the rubber to its original state. Obtaining shape memory behavior at temperatures lower than 75 °C is desirable to mitigate safety concerns and reduce heating required, improving process economics. Investigation of the room temperature shape memory behavior was conducted through modifying the shape memory cycle. It was found that stearic acid swelled natural rubber exhibits a 30% fixity over a 24 hour period. The stress can be relaxed by heating the swelled rubber above the melting point of stearic acid.

417807 Lessons Learned from the Secarb Integrated CO2 Capture and Storage Project

Abstract:

The United States Department of Energy (DOE) established seven Regional Carbon Sequestration Partnerships (RCSPs) in 2003. The Southeast Regional Carbon Sequestration Partnership (SECARB), whose lead organization is the Southern States Energy Board (SSEB), represents 13 States within the south eastern United States.The SECARB Anthropogenic Test has entered into an efficiency enhancement phase for the CO_2 capture unit and a post-injection monitoring phase for the CO_2 geologic storage reservoir.

The CO_2 capture unit is located at Alabama Power Company's James M. Barry Electric Generating Plant (Plant Barry) in Bucks, Alabama.Alabama Power is the second largest subsidiary of Southern Company, serving homes, businesses and industries in the southern two-thirds of Alabama. The capture facility, equivalent to 25 megawatts (MW), utilizes Mitsubishi Heavy Industries America's (MHIA) KM-CDR™post-combustion amine capture technology. The capture unit utilizes MHIA's proprietary KS-1™ solvent to achieve high levels of CO_2retention with significant reductions in energy penalty.

Southern Company is working with Mitsubishi Heavy Industries on continued testing at the 25 MW CO_2 capture and compression plant at Plant Barry. Most recently, a high efficiency system (HES) was installed to support a DOE/National Energy Technology Laboratory-funded project. The HES is designed to recover heat from the inlet flue gas and CO_2 regenerator. This heat is then used directly in the Rankine cycle by preheating boiler feedwater. The HES will demonstrate an increased efficiency of the CO_2 capture unit and will have the following ancillary benefits:

- Reduced water usage in the flue gas desulfurization system;
- Better SO_3 capture by condensation on fly ash;
- Better ESP performance through reduced gas volume and lower ash resistivity; and
- Better Hazardous Air Pollutants capture performance due to condensation enhancement.

The SECARB project is the largest integrated carbon capture, transportation and storage project at a pulverized coal-fired power plant in the USA. It has resulted in the deployment of the world's largest CO_2 capture and compression plant for a coal fired boiler in Texas. This new commercial facility will have a CO_2capture capacity of 4,776 metric tons per day.

From August 2012 until September 2014, a total of 114,104 metric tons of CO_2captured at Plant Barry were transported by pipeline and stored underground in a deep, saline geologic formation within the Citronelle Dome located in Citronelle, Alabama. The injection target is the lower Cretaceous Paluxy Formation which occurs at a depth of 2,865 meters (9,400 feet). The injection wells and surface facilities are located on property which is owned fee simple by the field's operator, Denbury Onshore, LLC.

Field monitoring data indicates that the CO_2 has remained within the permitted Area of Review and no negative seismic anomalies are present above or in the caprock, thus indicating the CO_2 is contained within the injection zone. Post injection monitoring will continue for a period of three years.

423777 New Advanced Aqueous Amine Solvents for CO2 Capture

Abstract:

In the efforts to improve upon the aqueous MEA benchmark for post-combustion CO_2 capture, a number of new and advanced aqueous amine solvents have been proposed. While several of these solvents have advanced to slipstream testing, there are still great opportunities and needs for innovation in the design and synthesis of amine compounds that have reduced energy penalty, high CO_2 capacities, low volatility, low viscosity, enhanced stability and are economically viable for scale-up. In this presentation, we detail our recent work in developing libraries of new amines that possess these desirable features. Results indicate viability for further scale-up of these solvents in continuous absorber-stripper processes.

432498 Aminosilicone Carbon Capture Process Development

Abstract:

Liquid aminosilicone CO_2 capture solvents have been developed that have the potential to significantly reduce the cost of carbon capture from pulverized coal power plants. Compared to aqueous amines, aminosilicones have the potential to reduce CO_2 capture cost through higher CO_2 absorption capacity, greater thermal stability, and reduced solvent losses due to lower volatility. A process utilizing aminosilicone dissolved in a glycol co-solvent will be demonstrated at the National Carbon Capture Center in 2015. This demonstration will generate engineering data needed to progress the technology toward commercialization. In parallel, a bench-scale project is focused on developing a process utilizing a phase-changing aminosilicone, which has the unique behavior of solidifying upon reaction with CO_2. Economic analysis of these processes integrated with a coal-fired power plant suggests that aminosilicone-based CO_2 capture is substantially more efficient and less costly than aqueous amine CO_2 capture. Discussion of current process development activities and economic analyses for the pilot-scale solvent process and the bench-scale phase-change process will be presented.

428617 Anti-Foaming Study for Physical Solvents for Pre-Combustion CO2 Capture

Abstract:

The pre-combustion capture process involves converting fossil fuel into syngas, primarily a mixture of carbon monoxide and hydrogen, usually by gasification. During pre-combustion, coal is first transformed into hydrogen and CO_2. In a CO_2 separation unit, the carbon dioxide is separated from the hydrogen using a physical solvent, such as commercial Selexol solvent. This process has the potential of capturing and removing 90 percent of CO_2emissions from power plants - major contributors to global warming.

We previously explore the use of hydrophobic physical solvent, hybrid-PDMS, for pre-combustion carbon capture. Solvent showed compatible CO_2 solubility to Selexol at different operating conditions. However, foaming issue prevent its use for pre-combustion carbon capture. To solvent this problem, we investigated adding a none-volatile ionic liquid, allyl pyridinum Tf_2N, in hybrid-PDMS. A 1-liter Continuous Stirred Tank Reactor (CSTR) equipped with a 2-liter gas buffer tank has been used to measure CO_2, H_2, and CH_4isotherms at different temperatures. For gas solubility measurement, a dynamic pressure-step method was applied.

This study has established an understanding of the influence of second non-volatile solvent on hybrid-PDMS solvent physiochemical properties, including foaming, gas solubility, CO_2 selectivity, viscosity, and so on.

421974 Molecular Modeling of Non-Aqueous CO2 Capture Solvents

Abstract:

Guanidium-based carbon dioxide binding organic liquids (CO_2BOLs) are strong non-nucleophilic bases well suited for binding CO_2 to their aliphatic alcohol group. They become ionic upon reaction with CO_2 and revert to their non-ionic form by thermal removal of CO_2. Post-combustion capture of carbon dioxide can be performed with solvents that exhibit high CO_2 binding capacities, are non-volatile at operating conditions, and have low viscosity. CO_2BOLs meet the first two requirements, but their high viscosities in their CO_2-bound state render them costly for industrial applications. CO_2BOLs are still very attractive candidates for CO_2 capture because they require no additional solvents and are liquid at post-combustion temperature and pressure conditions, reducing thermal regeneration cost. In this work, candidate CO_2BOL molecules are studied, in collaboration with experiment, toward the design of low-viscosity CO_2BOL compounds. *Ab initio* electronic structure calculations were performed for accurate molecular properties including charges, as well as to obtain CO_2 adsorption energies. To simulate CO_2BOLs mixtures in their liquid state, classical molecular dynamics were used at varying carbon loadings. Viscosities were then calculated using Green-Kubo relations. We have found that the hydrogen-bonding network in the liquid state, both intra-molecular and between different molecules plays a key role in determining and tuning viscosities. Based on this, a reduced model was developed that predicts the CO_2BOL viscosity given a few molecular structural parameters and CO_2 loading. With the reduced model, a high number (100+) of candidate CO_2BOLs is screened. A list of promising candidate molecules is in the process of being synthesized and their viscosities experimentally measured.

428966 Regulatory Process Control of an Advanced Post-Combustion Amine Scrubbing System

Abstract:

Regulatory Process Control of an Advanced Post-Combustion Amine Scrubbing System

Matthew S. Walters, Thomas F. Edgar, Gary T. Rochelle

McKetta Department of Chemical Engineering, The University of Texas at Austin, 200 E. Dean Keeton St. Stop C0400, Austin, TX 78705

Keywords: amine scrubbing, dynamic modeling, pilot plant validation, process control

---

Amine scrubbing is a leading technology for capturing CO_2 from fossil fuel power plant emissions. The in-and-out intercooled absorber[1] and advanced flash stripper[2] process configuration (Figure 1) using concentrated piperazine solvent has been proposed to reduce the capital and operating costs associated with separating CO_2 from flue gas and compressing it for geological sequestration. A low-fidelity dynamic model of this advanced process has been developed and implemented in MATLAB® for the purpose of screening regulatory process control strategies. The low-fidelity model has been validated with both a high-fidelity model and pilot plant data, and it is valid for simulating the system near the design conditions.

The primary process control objective of an amine scrubbing plant will likely depend on environmental regulations, electricity markets, and enhanced oil recovery (EOR) economics. Two possible objectives are proposed here: delivering a constant rate of CO_2 to an EOR facility and maintaining a 90% removal rate in the absorber. We consider a capture unit integrated with a base-loaded coal-fired power plant that experiences disturbances in boiler load of ≤ 5%. Steam is extracted from the power plant turbine intermediate pressure/low pressure (IP/LP) crossover. The steam extraction valve is assumed to be 100% open at design conditions, which limits steam availability during a disturbance. The off-design turbine operation is predicted using Stodola's Ellipse Law.

In the constant CO_2 delivery case, the compressor is flow controlled to meet the specified CO_2 rate and the CO_2 inventory in the solvent is allowed to vary. In the case where removal rate from the flue gas is controlled, two choices for manipulated variables are examined: the solvent circulation rate (set by the lean solvent pump) and the CO_2 delivery flowrate (set by the compressor speed). After assigning manipulated variables for level and water balance control, the remaining variables are paired with self-optimizing process variables.

With constant CO_2 delivery, the removal rate in the absorber floats for a period of several hours when a disturbance is introduced. Because the process variables spend long periods away from steady state, the steam extraction rate is also floating. The process is not optimal when it is away from steady state, and there is an energy penalty associated with the transient period. There is also a clear tradeoff between electricity produced and the amount of CO_2 available for EOR. This scenario may require an oversized compressor to avoid reaching the capacity limit as the pressure of the stripper drops. In the event of a sustained disturbance, the CO_2 inventory may become depleted and cause the solvent to approach the solid solubility limit. No oscillations in process variables were observed in this scenario.

As a result of the heavy material and energy recycle in the process, maintaining a constant removal rate in the absorber is challenging with only feedback control[3]. Manipulating solvent rate to control removal with a small process inventory leads to significant oscillations in process variables. The oscillations could possibly be reduced by less aggressive tuning or a larger lean storage tank; however this will make the response much slower and a large inventory is likely not acceptable. These observations are consistent with previous work that investigated using solvent rate to control removal on a simple absorber and reboiled stripper with monoethanolamine[4,5]. Without any knowledge of the absorber, a feedback controller that manipulates compressor speed must be tuned non-aggressively to assure stability and is subject to slow oscillations as the system comes to steady state. A model-based supervisory controller is proposed to improve upon this strategy by making control moves that take into account the heavy material recycle.

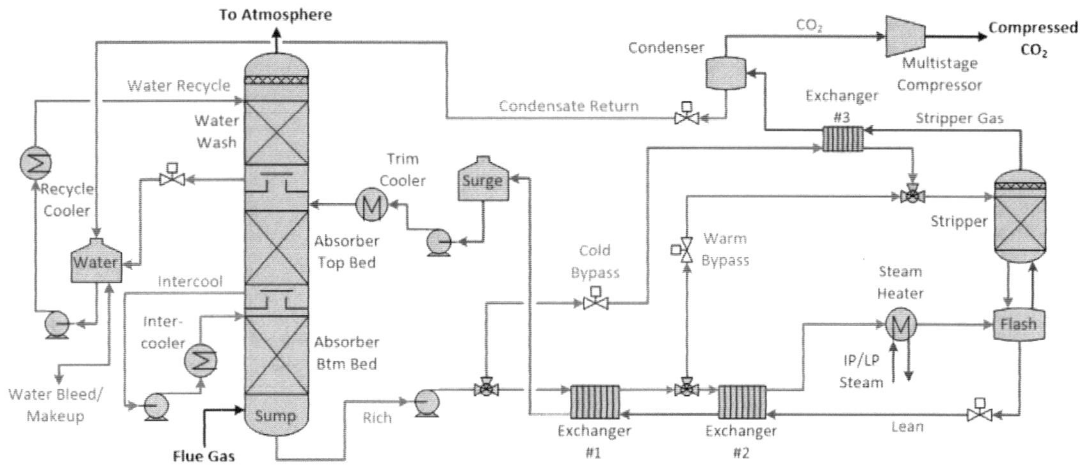

Figure 1. In-and-out intercooled absorber with advanced flash stripper configuration.

References

[1] Sachde D, Rochelle GT. Absorber Intercooling Configurations using Aqueous Piperazine from Sources with 4 to 27% CO_2. *Energy Procedia* 2014;63:1637-56.

[2] Lin Y-J, Madan T, Rochelle GT. Regeneration with Rich Bypass of Aqueous Piperazine and Monoethanolamine for CO_2 Capture. *Ind Eng Chem Res* 2014;53:4067-74.

[3] Walters MS, Edgar TF, Rochelle GT. Dynamic Modeling, Validation, and Time Scale Decomposition of an Advanced Post-Combustion Amine Scrubbing Process. *Energy Procedia* 2014;63:1296-1307.

[4] Ziaii SF. *Dynamic Modeling, Optimization, and Control of Monoethanolamine Scrubbing for CO_2 Capture*. The University of Texas at Austin. Ph.D. Dissertation. 2012.

[5] Ceccarelli N, van Leeuwen M, Wolf T, van Leeuwen P, van der Vaart R, Maas W, Ramos A. Flexibility of Low-CO_2 Gas Power Plants: Integration of CO2 Capture Unit with CCGT Operation. *Energy Procedia* 2014;63:1703-1726.

424541 Dynamics of Absorption of CO2 into Sprayed Solvents

Abstract:

Low gas phase pressure drop, less susceptibility to corrosion due to the absence of internals, and greater tolerance to the use of precipitating solvents are some of the favorable attributes associated with spray columns, particularly with regard to CO_2 absorption. Specifically, there is currently a need for robust spray absorption data for the CO_2-MEA system, especially regarding droplet size data for this system. Droplet size measurements are particularly critical to ascertain the surface area available for absorption. This presentation discusses absorption rates in terms of overall mass transfer coefficients, drop size data, and planar surface area for the CO_2-MEA system quantified in a 0.2 m ID, lab-scale column. Phase Doppler Interferometry (PDI) is used to characterize the drop size distributions. The effects of L/G ratio, inlet loading, MEA concentration, and gas-liquid contact height on the performance of a full-cone nozzle are elucidated. The dropsize data and surface area quantification indicate that the free MEA content dominates spray absorption. This work provides insight about spray absorption, especially with regard to CO_2 and amine systems.

425527 Integrated Market and Engineering Modelling and Optimization of Low-Carbon Electricity Generation

Abstract:

Fossil fuels will undoubtedly remain a core part of the world's energy system for the foreseeable future, and possibly beyond the end of the century. However, in this timeframe, the global energy landscape is expected to significantly change as we transition to a low carbon economy in an effort to avoid the effects of dangerous climate change.

A key aspect of this change will be the increasing penetration of intermittent renewable electricity into the worlds energy system – albeit to different extents in different locations. Therefore, it will be vital that the decarbonized thermal power plants which are operating in this period are designed to operate in an integrated fashion with intermittent renewable generators, as opposed to in an antagonistic manner.

In this contribution, we present an integrated assessment of the UK's electricity market in the period to 2050. To achieve this, we combine dynamic investment and dispatch models which provide insight to the likely installed generation capacity in the UK with detailed plant-scale models of the decarbonized power plants likely to be in operation in this period, focusing on super-critical coal-fired power plants (SCCPP) and open cycle gas turbines (OCGT) integrated with post-combustion CO_2 capture[1].

We consider how changing market landscapes and dynamics necessitate different operating patterns which in turn drive different approaches to the optimal design and operation of these plants.

[1] Mac Dowell, N. et al., Int. J. GHG Con., 2013, 12, 247-258, Mac Dowell, N. et al., Int. J GHG Con., 2013, 13, 44 – 58, Mac Dowell and Shah, CACE, 2015, 74, 2015, 169 - 183

431482 Upscaling a Devolatilization Model for Oxy-Combustion Simulations Using Validation and Uncertainty Quantification

Abstract:

Large-eddy simulations of utility-scale coal boilers are intended to accelerate the deployment of next-generation oxy-combustion and AUSC designs. While the chemical percolation devolatilization (CPD) model is known to be quite accurate for predicting devolatilization rates of multiple coal types, it is far too computationally expensive for the scale of simulation intended. In this study, we present a methodology for up-scaling a first-order-with-yield model from CPD using the newest techniques in validation and uncertainty quantification. The approach incorporates any uncertainty in the parameters input into CPD and any additional uncertainty in the quantities of interest output from the CPD model. While the resulting models provide extremely good fits for the four coal types we have explored, the methodology must be repeated for each coal of interest.

433517 Thermodynamic Modeling of MEA-Based CO2 Capture Process with Uncertainty Quantification and Validation with Steady-State Data from a Pilot Plant

Abstract:

The US DOE's Carbon Capture Simulation Initiative (CCSI) is focused on the development of state of the art process models to accelerate the development and commercialization of post-combustion carbon capture system technologies. For developing a "gold standard" process model of the MEA-based CO_2capture systems, a rigorous thermodynamic framework is developed and integrated with the rigorous properties models developed by the authors. The model is validated with the steady-state data from a pilot plant.

Although deterministic thermodynamic models exist for the MEA-CO_2-H_2O system, a stochastic model considering uncertainty quantification is also developed in this work. The accuracy of the results from design and optimization studies is contingent upon accurate representation of uncertainty bounds on simulation outputs. For the deterministic model, vapor-liquid equilibrium (VLE), heat capacity, and heat of absorption data from the open literature are used to calibrate the parameters in the e-NRTL thermodynamic framework. A Bayesian methodology is used to develop the stochastic model by considering uncertainty in the model parameters. A surrogate response surface model is developed using Multivariate Adaptive Regression Splines (MARS) to reduce the computational expense of the methodology. Using the experimental data, the Bayesian method provides informed posteriors. The posterior parameter distributions from the thermodynamic model are propagated through the overall process model to determine uncertainty in prediction of outputs of interest (e.g. energy input required for solvent regenerator, CO_2capture efficiency).

This model is validated with high quality steady-state data collected over a wide operating range from the National Carbon Capture Center in Wilsonville, Alabama. The key manipulated variables such as solvent flowrate and reboiler steam flowrate and disturbance variables such as the flowrate and CO_2 concentration of the inlet flue gas were varied widely in these test runs. Validation studies with the NCCC data help to narrow down the model uncertainty. The study shows that a systematic uncertainty quantification approach using rich experimental data can be very valuable to improve the accuracy of model results.

433707 A Reduced-Order Building Approach to Simulation-Based Optimization of Complex Energy Systems

Abstract:

Data obtained through complex simulations are routinely used to build models and inform process decisions. For many applications, use of the full fidelity system simulation is simply too expensive. A number of techniques can be used to generate reduced-order models of the original system. However, techniques commonly used, such as artificial neural networks and support vector machines, result in surrogate models that are too complex and overfit the data. A simple algebraic surrogate model makes the relationship between inputs and outputs available in closed form, and is much more amenable to the tasks of optimization, control, and prediction.

We describe an approach to complex energy system optimization based on the ALAMO software for building simple yet accurate surrogate models of simulations. ALAMO is a computational methodology that has been developed to learn algebraic surrogate models from exogenous data [1]. In order to model nonlinear behavior present in process data, ALAMO performs a number of nonlinear transformations of input variables to populate a regression basis set. An optimization-based subset selection method then identifies the optimal subset of regression variables according to a model fitness metric such as Mallows' Cp, Akaike's information criterion, or Bayesian information criterion. The resulting models consist of a linear combination of nonlinear transformations of input variables, and their simple algebraic form can help provide insight on the system at hand. ALAMO has been used to successfully generate surrogate models for a bubbling fluidized bed carbon capture reactor. These algebraic surrogates were then used to find the optimal configuration of reactors and regenerators in a chemical process superstructure for a CO_2 capture system.

References

[1] Cozad, A., N. V. Sahinidis, and D. C. Miller, Automatic learning of algebraic models for optimization, AIChE Journal, **60**, 2211-2227, 2014.

434598 Optimization Under Uncertainty with Rigorous Process Models

Abstract:

Optimization under uncertainty (OUU) allows for uncertain parameters in the process model to be systematically accounted for during optimization, which reduces the risk of poor performance far from the expected values (e.g., average) of the uncertain parameters. This is especially important for design and analysis of carbon capture systems, which must be designed using models with uncertain parameters (as many systems are novel and industrial scale validation data are not available yet) and to effectively operate in a variety of scenarios due to load following. A popular approach for process design under uncertainty involves formulation of a two-stage stochastic (e.g., bi-level optimization) problem, where design variables (e.g., flowsheet topology, size and number of units, etc.) are determined in the first step and operating variables (e.g., flowrates, stream pressures, etc.) are determined in the second step, after the uncertain parameters are realized. This allows for recourse, or reaction in response to additional information regarding uncertain parameters. Traditionally, stochastic programs for process design (or planning/scheduling) are typically solved in algebraic modeling environments such as GAMS or AMPL, which automatically calculate exact derivations and interface with powerful, large-scale optimization algorithms. However, for process development, commercial flowsheet simulation software packages (e.g., AspenPlus, gPROMs, etc.) are preferred. This paper focuses on computationally tractable formulations for process optimization under uncertainty with rigorous models and process simulators. The methods are demonstrated in a solid sorbent based post combustion CO_2capture system design case study.

A novel approach for OUU is developed, where the OUU problem is treated as a bi-level optimization problem, as described above. In an outer loop, the design variables are determined using a derivative free optimization algorithm (e.g., BOBYQA), and in the inner loop, the optimal operating variables are determined for each scenario using the Sequential Quadratic Programming (SQP) algorithms built into Aspen Customer Modeler (ACM) or gPROMs. Scenarios represent the discrete joint probability distribution for the unknown parameters. For example, assume there are two unknown parameters for a CO_2 capture system design problem: flue gas flowrate f (related to net power output) and a CO_2 adsorption kinetic parameter k. Each scenario would consist of a value for f and $k$along with the probability of this scenario occurring. The bi-level structure allows for the inner problems to be decoupled and solved in parallel. Optionally, the approximation of the joint probability distribution for continuous uncertain parameters may be enhanced using response surface models, which reduces the number of required scenarios and decreases computational demands. The methods described in this paper are part of the Framework for Optimization and Quantification of Uncertainty (FOQUS) software package developed as part of the Carbon Capture Simulation Initiative (CCSI), which supports solution of up to 1,000 inner optimization problems in parallel using Turbine (another software package in the CCSI toolset). Interfaces with standard chemical engineering process simulation software (e.g., AspenPlus, ACM, gPROMs and Excel) are established with the SimSinter software package (also in the CCSI toolset).

The methods presented in this paper are demonstrated in a retrofit design case study of a solid sorbent-based post combustion CO_2 capture system for a 650 MW pulverized coal supercritical power plant. Several kinetic parameters, power output/flue gas flowrate, and capture level are considered as uncertain parameters. The capture system model and the inner optimization problem are solved in ACM. The ACM model consists of approximately 150,000 equations which result from discretization of partial differential algebraic equation adsorber and regenerator bed models. The inner optimization problem has six operating decision variables (cooling water flowrates, sorbent recirculation rate, steam injection rate, and CO_2 recycle fraction), and requires approximately 3 CPU-minutes to solve on average. The outer problem considers of 17 design variables, including the dimensions/size of equipment, number of parallel units, and heat exchange areas. The problem formulation seeks to minimize the cost of electricity while satisfying CO_2 capture/recovery specifications. These results are compared against deterministic optimization of the nominal (e.g., expected value) case, and the benefits of optimization under uncertainty are discussed. Finally, the graphical interface to these methods through FOQUS will be presented.

434879 Dynamic Modeling of Steam Thermal Power Plants for Real-Time Optimization

Tuesday, November 10, 2015: 8:30 AM
250E (Salt Palace Convention Center)

Chen Chen[1], Kyle D. Such[1], Zhiquan Zhou[2], Xinsheng Lou[2], Shizhong Yang[2], Olutoye Akinjiola[2], Carl Neuschaefer[2] and George M. Bollas[1], (1)Department of Chemical and Biomolecular Engineering, University of Connecticut, Storrs, CT, (2)Alstom Power Inc., Windsor, CT

The variable market demand, and the increasing safety and environmental regulations imposed on power plants, force operators to develop dynamic simulation tools, to optimize system operation, scheduling and maintenance, and test control strategies , using test-beds of dynamic models of power plants.[1–3] In this work, we explore such simulation environments and in particular the Modelica language and the Thermal Power library of the Dymola software for dynamic simulation, real time optimization and control of a 600 MW subcritical power plant with reheat cycle[4] (Fig. 1). The detailed modeling steps, steady-state validation and transient analysis (ramping of load) are discussed. Conventional control designs are also successfully incorporated in the system model. It is shown that the Dymola environment provides a robust test-bed for controller tuning and stability analysis.

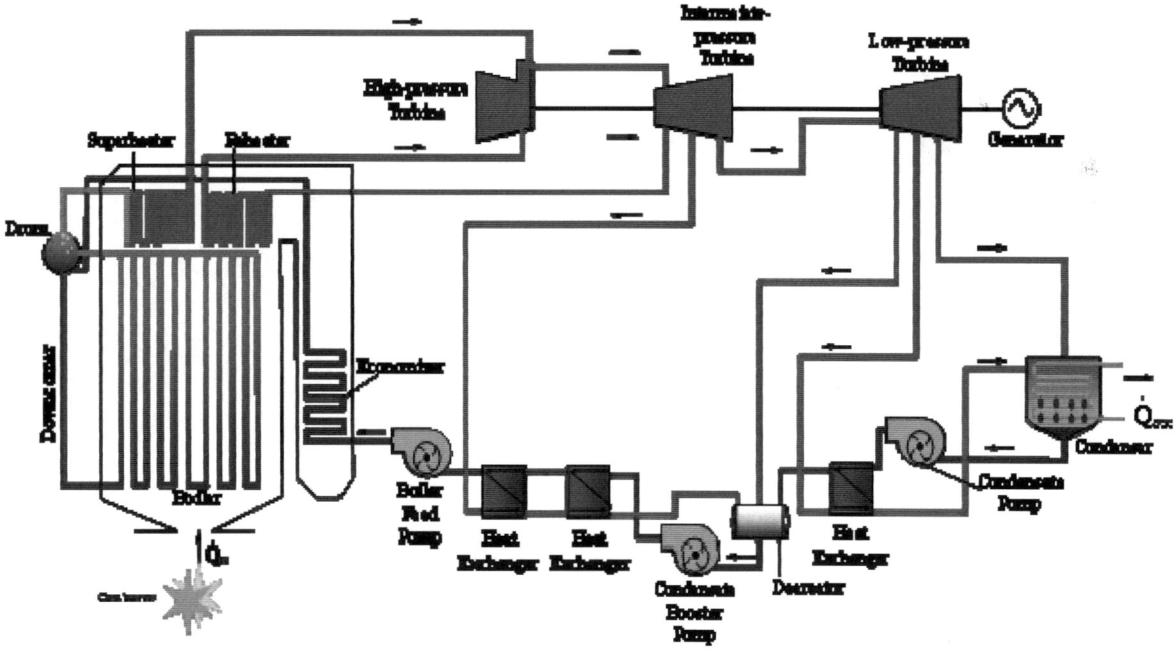

Figure 1: Reheat regenerative cycle, 600-MW subcritical-pressure fossil power plant.

The Functional Mockup Interface (FMI) is utilized to bridge the models developed in Dymola with the optimization functionality of Matlab, for the real-time optimization application.[5] The target of real-time power plant optimization is to seek for maximum profit/efficiency of the plant by pushing the (steady state) system operating point towards optimum time-varying operating trajectories, while satisfying safety and operability constraints during the transient between steady states. Here, the work flow for dynamic optimization using the FMI Toolbox is presented. The optimization capabilities of the complementary tool chain, which includes Dymola, FMI and Matlab, were successfully demonstrated in case studies of nominal efficiency optimization of the whole plant with respect to admissible plant inputs.

References

1. Lu, S. Dynamic modelling and simulation of power plant systems. *Proc. Inst. Mech. Eng. Part A J. Power Energy* **213,** 7–22 (2005).

2. Lawal, A., Wang, M., Stephenson, P. & Obi, O. Demonstrating full-scale post-combustion CO_2 capture for coal-fired power plants through dynamic modelling and simulation. *Fuel* **101,** 115–128 (2012).

3. Lawal, A., Wang, M., Stephenson, P. & Yeung, H. Dynamic modelling of CO_2 absorption for post combustion capture in coal-fired power plants. *Fuel* **88,** 2455–2462 (2009).

4. Singer, J. G. *Combustion Fossil Power: A Reference Book on Fuel Burning and Steam Generation.* (Combustion Engineering, 1991).

5. Åkesson, J., Årzén, K.-E., Gäfvert, M., Bergdahl, T. & Tummescheit, H. Modeling and optimization with Optimica and JModelica.org—Languages and tools for solving large-scale dynamic optimization problems. *Comput. Chem. Eng.* **34,** 1737–1749 (2010).

434346 Potential for Solid-Oxide Fuel Cell Technologies to Minimize Water Use in the Electric Power Sector

Abstract:

The U.S. Department of Energy (DOE) provides a worldwide leadership role in the development of advanced fossil fuel-based energy conversion technologies, with a focus on electric power generation with carbon capture and storage (CCS). As part of DOE's Office of Fossil Energy (FE), the National Energy Technology Laboratory (NETL) implements research, development and demonstration (RD&D) programs that address the challenges of reducing greenhouse gas emissions. To meet this challenge, FE/NETL is interested in evaluating advanced power cycles that will maximize system efficiency and performance, while minimizing CO_2emissions, water use, and the costs of CCS.

Solid oxide fuel cell (SOFC)-based power cycles for utility-scale applications are an advanced technology currently undergoing research and development. Utilizing electrochemical conversion of the energy contained in the fuel instead of combustion as with conventional power generation technologies, SOFCs are very efficient (with successful development, better than 60% HHV - natural gas with 90% CO_2capture) and have low pollutant emissions. A key feature of SOFCs is the ability to maintain the fuel and oxidant (air) streams physically separate, thereby facilitating carbon capture along with the capture and reuse of process water from the oxidized fuel stream (anode exhaust).

The objective of this study is to investigate the potential for solid-oxide fuel cell (SOFC)-based electric power generation to enable projected growth in electricity demand while at the same time meeting water availability constraints in regions of the country that are both vulnerable to drought and have heavy competition for available water for other needs (e.g., municipal, agriculture, industry). The approach is to select a specific river basin that is vulnerable to drought and carryout a case study to test the potential benefits that may be associated with the utilization of SOFC technologies to meet projected electricity demand while at the same time meeting projected water availability constraints in the area. The Sandia National Laboratory (SNL) Western region Water Availability, Cost, and Use database[1]is used to provide estimates for water availability and demand from the agricultural, municipal, industrial, and thermoelectric power sectors along with other environmental demands. The tasks that were carried out to develop the methodology included selection of a specific river basin; estimation of current water demand, supply, and availability through 2040; estimation of electric power demand through 2040, selection of power generation technology concepts, and their associated performance and cost, potentially used to meet the growth in electricity demand through 2040; and integration of the information into a system spreadsheet model.

The Lower and Middle Brazos River basin was selected to represent a region that already exhibits water stress and low water availability for new uses while also exhibiting projected increase in electricity demand. It is hypothesized that solid-oxide fuel cell technologies could be deployed to meet increasing electricity demand while minimizing water use in the electric power sector for the area. Three future water availability scenarios and eight electric power technology options were included in the study.

The analysis provides projections for the remaining water availability through 2040 for different water availability scenarios and electric power generation technology options. The three levels of water availability are represented by increasing levels of drought: no drought/minor drought, a Sandia reference case, and a climate change case (in which severe drought was assumed). The electric power technologies represented in the analysis vary across scenarios, with one scenario being represented by today's natural gas combined cycle (NGCC) technology (current NGCC), a scenario with current NGCC and an advanced NGCC plants available in 2030 (current and advanced NGCC), a scenario with current NGCC and a shift to utility scale SOFC plants in 2030 (current NGCC and SOFC utility), and a final scenario with current NGCC until 2025 when SOFC distributed generation (DG) technology is assumed to be available (current NGCC and SOFC DG). The results show the significant effect of the drought level and the selection of power generation technology on the projected water availability. In the severe drought scenario, DG SOFC would allow for an increase in electricity demand while at the same time meeting water availability constraints.

Cost savings due to the implementation of SOFCs were estimated compared to the current generating fleet. These savings were based on the cost of water shortfalls due to unavailable water from increased generation. Costs were incurred from the cleanup of municipal wastewater and the diversion of water from agriculture. Agriculture water diversion includes both the cost of transferring the water right from agriculture to electric power and the cost of lost agricultural sales estimated using data from the Census of Agriculture for the State of Texas.

Additionally, the impact of each scenario on CO_2 emissions was evaluated. It was found that the introduction of new electricity generation technologies provide the opportunity for significant reductions in the levels of CO_2emissions associated with new electricity generation capacity for the various scenarios.

This analysis shows the beneficial role that SOFC technologies can play in water-stressed regions. For the Middle and Lower Brazos River region, the analysis shows that use of SOFC technologies allows the 2040 water balance to remain positive under more scenarios than the use of NGCC technologies only. Further, SOFC technologies offer additional benefits: cost savings and reduced CO_2emissions.

The analysis provides a useful methodology and applied case study that identifies potential water availability issues at the watershed level. The methodology is useful for: a) assessing relative water availability across watersheds with relative degrees of availability indicating which watersheds are most likely to experience issues under specific scenarios; and b) exploring the relative impacts of different generation scenarios and technology choice combinations within a particular watershed.

---

[1] SNL Water Availability, Cost, and Use. http://energy.sandia.gov/climate-environment/water-security-program/energy-and-water-in-the-western-and-texas-interconnects/water-availability-cost-and-use/#waterForm

431407 Simulation and Validation of 15 Mwth Oxy-Coal Power Boiler

Abstract:

Advanced ultra-super critical power generation with ox-coal combustion and carbon capture appears to be a viable method for reducing coal's carbon footprint. Accurate simulation of these systems may help to accelerate the deployment of this new technology through design and optimization. In this study we are modeling Alstom's 15 MWth boiler simulation facility and validating our results with experimental measurements. These measurements are viewed as the largest-scale results with detailed experimental work - taken on an ox-coal system today. We demonstrate our ability to accurately model the system's heat flux and temperature profiles by using Arches, an in-house large eddy simulation code, and by utilizing high-performance computing to enable us to resolve the critical physics of the system.

433303 Development and Deployment of a Large Eddy Simulation Code for Simulating Full-Scale, Coal-Fired Boilers

Abstract:

The Carbon-Capture Multidisciplinary Simulation Center (CCMSC) at the University of Utah is demonstrating extreme computing with validation/uncertainty quantification to more rapidly deploy a new technology for providing low cost, low emission electric power generation to meet the growing energy needs of the U.S. We are using a hierarchal validation approach to obtain simultaneous consistency between sets of selected experimental data from different scales that embody key physics components to predict performance in a 350MWe oxy-fired boiler.

In this presentation, we discuss the development and deployment of our in-house Large-Eddy simulation capability, Arches, a key to the CCMSC program. The Arches code solves equations for the evolution of both the particle and gas phases. The phases are tightly coupled, conservatively exchanging mass, momentum and energy. Using large numbers of computing cores provides detailed data in both time and space. In addition to discussing the existing capability, we highlight the current development pathway of improving the physical models while looking towards performing coal simulations on an exa-scale platform.

The CCMSC project is funded by the DOE Predictive Science Academic Alliance Program.

415523 Thermal Characterization of Ash Deposits in a 1.5 MW Reactor

Abstract:

The University of Utah is involved in an extensive effort to develop predictive simulation capabilities for pulverized oxy-coal flames in order to facilitate the design of next-generation low-carbon-emission utility boilers. This work is being carried out by the Carbon Capture Multidisciplinary Simulation Center (CCMSC), which is a $16M research center funded by the U.S. Department of Energy. A key component of the center is a detailed Validation and Verification effort with coupled Uncertainty Quantification (V&V/UQ). The V&V/UQ work is being carried out using experimental data at multiple scales with the ultimate goal of providing confidence in the predictive capabilities of the code for simulations at a scale where no experimental data exists.

One specific endeavor is the simulation of experimental data obtained in the 1.5 MW oxy-coal furnace (L1500) located at the University of Utah. The quantity of interest (QOI) for the V&V/UQ effort is heat flux, since ability to predict this parameter accurately throughout the boiler was identified by our industrial partner as being the most critical need for designing a next-generation boiler. Predictions of heat flux are highly influenced by the thermal properties of the furnace walls; thus, the values for properties of coal ash deposits on the walls must be carefully selected for the model. Due to the highly variable nature of coal ash, however, there is a great deal of uncertainty and variability in the information available in the literature on these properties. Thus, a study was initiated to provide detailed information on the actual deposits obtained from the furnace walls.

The thermal properties examined in this work are the infrared emissivity and thermal diffusivity of ash deposits. The walls of the furnace are composed of multiple layers of insulation and refractory, which become covered over time by layers of ash deposits. These ash deposits can exhibit significant variability from location to location, especially in comparison with the values for pure refractory. Observations inside the furnace (1.1 x 1.1 m^2 internal cross section, 12 m length) showed that the deposits varied greatly in color and structure. Thus, it appeared that previous models, which have used uniform values over the entirety of the furnace, may not have been sufficient. The objective for this study was to evaluate that assumption by charting the differences in the thermal diffusivity and infrared emissivity of the deposits as a function of their location on the interior walls of the furnace.

This paper will present the results of this study and provide recommendations for a suitable approach for addressing the variability of ash deposits in simulations of large-scale boilers and furnaces.

425306 Techno-Economic Analysis of Gas Purification for CO2 Transport in Pipeline Networks and Injection for Storage

Abstract:

Techno-economic analysis of gas purification for CO_2 transport in pipeline networks and injection for storage

Clea Kolster*[1,2] , Evgenia Mechleri[1,2], Sam Krevor[3], Niall Mac Dowell[1,2]

*Presenting author's e-mail: clea.kolster10@imperial.ac.uk

[1] *Centre for Environmental Policy, Imperial College London, South Kensington Campus, SW7 1NA, UK*

[2] *Centre for Process Systems Engineering, Imperial College London, South Kensington Campus, SW7 2AZ, UK*

[3] *Department of Earth Science and Engineering, Imperial College London, South Kensington Campus, SW7 2AZ, UK*

Abstract

The IEA Energy Technology Perspectives 2014 stated that Carbon Capture and Storage (CCS) is a crucial component in capping the world's average surface temperature increase by 2 °C (IPCC, 2014) and, if applied to the power generation and heavy industry sectors, could reduce cumulative emissions by 14% in 2050 (IEA, 2014). In addition, when considering the decarbonisation of the heavy industry sector, CCS is currently the only available option. Therefore, it is imperative that the CCS process be optimised to run smoothly through each stage of operation and that its costs be minimised in order to accelerate its global financial and political acceptance. In view of these issues, this paper focuses on the operational modelling of a CO_2 capture process, modelling and quantifying the economic benefits resulting from a CO_2 transport network system and the cost reduction ensuing from CO_2 capture plant clusters.

Oxy-combustion capture comprises of two energy intensive sections: the air separation unit (ASU) and the CO_2 purification and compression unit (CO_2CPU). The CO_2CPU brings the raw flue gas from a low purity atmospheric CO_2 stream to a high purity, high pressure CO_2 stream suitable for transport. So far little has been discussed regarding the optimal design and operation of the CO_2CPU (Boot-Handford et al., 2014). This work focused on minimising the capital and operational costs involved in the CO_2CPU while prioritising an EOR suitable end product. Three CO_2 compression and purification units processing flue gas from a pulverised fuel oxy-combustion plant were modelled in Aspen HYSYS. These CO_2CPU models build on similar separation processes presented by Posch et al. (Posch & Haider, 2012) and Pipitone et al. (Pipitone & Bolland, 2009).

It was established that, as the CO_2CPU models increased in complexity and purity of the produced CO_2 stream, they increased in operational and capital costs but decreased in capture efficiency i.e. increase the amount of CO_2 vented to the atmosphere. The capital and operational costs for each process model were translated into a price for each stream of CO_2 marketed for EOR. A non-monotonic and non-linear relationship between CO_2 price and CO_2 purity was observed. A CO_2-EOR suitable product stream, requiring a very high purity, led to a price of £20 per tonne, while the least complex system provides a stream that is suitable for transport, but not for injection, at a price of £11 per tonne.

In addition, this paper discusses the possibility of having one CO_2 capture plant for a cluster of power and/or industrial plants as opposed to several smaller capture plants. Modelled in Aspen HYSYS, it was found that instead of having 4 CO_2CPUs with a 5 Mt.CO_2 yearly capacity, having one large CO_2CPU with a yearly capacity of 20Mt.CO_2 reduces capital expenditure by 16% and yearly operational expenditure by 4%.

As CCS is deployed on a large scale, we will deploy transport networks, as opposed to a large number of single point-source to sink links. It is therefore vital to understand how these complex networks will behave, and indeed contribute to the flexibility of these systems. Therefore, a second part of this work considers a hypothetical UK-based network of CO_2 sources connecting to a single sink in the North Sea (Prada, Konda, & Shah, 2010). The CO_2 sources considered include combined cycle gas turbine (CCGT) plants, coal fired power plants and industrial plants. The CO_2 transport network allows for high purity sources of CO_2 derived from post-combustion capture power plants to be combined with less expensive, lower purity sources of CO_2 using oxy-combustion capture, to produce a CO_2 stream which is suitable for injection, i.e., a CO_2 stream composed of >96 wt.% CO_2. This led to a 19% reduction in the price of CO_2.

Bibliography

Boot-Handford, M. E., Abanades, J. C., Anthony, E. J., Blunt, M. J., Brandani, S., Mac Dowell, N., ... Fennell, P. S. (2014). Carbon capture and storage update. *Energy & Environmental Science*, *7*(1), 130.

IEA. (2014). Energy Technology Perspectives 2014. http://doi.org/10.1787/energy_tech-2010-en

IPCC. (2014). Working Group 3 - Technical Summary 2.

Pipitone, G., & Bolland, O. (2009). Power generation with CO_2 capture: Technology for CO_2 purification. *International Journal of Greenhouse Gas Control*, *3*(5), 528–534. http://doi.org/10.1016/j.ijggc.2009.03.001

Posch, S., & Haider, M. (2012). Optimization of CO_2 compression and purification units (CO_2CPU) for CCS power plants. *Fuel*, *101*, 254–263. http://doi.org/10.1016/j.fuel.2011.07.039

Prada, P., Konda, M., & Shah, N. (2010). *Development of an Integrated CO_2 Capture, transportation and Storage Infrastructure for the UK and North Sea using an Optimisation Framework.*

437652 Accelerate Materials Design through Accurate Performance Prediction

Abstract:

Material challenges are at the core of many DOE energy objectives. Critical to Fossil Energy (FE) is the ability to predict performance under severe mechanical (creep and fatigue) loadings in extreme (corrosive and oxidative) environment over long time periods. Methods that can accurately predict the evolution of the mesoscale microstructural features that control a material's response to mechanical loadings and damage initiation are not currently available, let alone the tools needed to predict performance in extreme environments and service life consistent with advanced energy applications. For advanced FE systems, the critical needs are for predictive models that simultaneously describe the mechanical and chemical performance of a material component over its entire service life, at conditions consistent with the application, to include changing environments, unexpected and planned cycling, and varying feedstocks. Such a simulation tool set does not exist today. Advanced simulation in materials development for the next generation of fossil energy systems will provide the pathway to accelerate the life cycle of materials design, development, and deployment, and reduce uncertainty, risk, and cost.

436586 Democratizing Fossil Fuel Conversion

Abstract:

Do you have a billion dollars? Neither do most of the 6 million companies in the Unites States. In fact, less than 10 companies in the U.S. can reasonably finance a billion dollar investment on their own. Whether you are using coal or natural gas to make electricity or chemicals, today economic production is achieved through "economies of scale" where huge upfront capital investments are required—on the order of a billion dollars or more. This high price of entry is a financial barrier that stifles all but the most incremental technological advancements. In this talk, the long-held belief in "scale-up" to achieve economic production will be challenged and "economies of number" will be proposed as an alternative to achieve cost-reductions through the manufacturing of many small, mass-produced modules. A multitude of advantages are offered by small-scale, modular plants over traditional scaled-up, including: 1) less upfront capital, no down time, and less risk; 2) faster market response and adaptable output; 3) the creation deployable assets (no sunk capital); 4) access to remote resources/assets; 5) better integration resulting in less waste; 6) access to new capital resources and consumer markets; and 7) faster innovation through more players and a lower barrier to experimentation. However, simply reducing the size of an existing process alone will not be sufficient to realize the benefits of mass production—innovations are required. Fundamentally, there a two key technical challenges to operating processes at small-scale: 1) high surface-to-volume ratio prevents near-adiabatic operation, and 2) common gravity-based processes (separations) may not yield sufficient residence times. These challenges and overall production efficiency will require innovations in process intensification that dramatically reduce the process volume, capital, and/or environmental footprint per rate of production. Recent advances, emerging opportunities, and challenges ahead in process intensification for fossil fuel conversion will be presented.

436591 Oxy-Coal Power Boiler Simulation and Validation through Extreme Computing

Abstract:

Coal remains the low-cost but high carbon power generation option around the globe. We have studied the design of a 500 MW pulverized-coal boiler under advanced ultra-supercritical (AUSC) oxy-coal firing with pure oxygen (no flue gas recirculation) for carbon capture and high efficiency. This study was performed using high performance computing(HPC) at the extreme scale (two - twenty thousand processors per simulation) and dynamic large eddy simulation (LES). The HPC and LES in this study allowed for 1-2 cm resolution of the turbulence within the boiler and temporal resolution at the microsecond time scale. All of the scales of the particle transport and reaction are fully resolved (DNS) for the entire particle size distribution. Particle size segregation and clustering are spatially and temporal resolved.

This study targeted a formal validation and uncertainty quantification by using oxy-coal data from the 15MW testing program operated by Alstom Power under 5 years of DOE funding. For this validation study, 160 different temperature measurements, 20 different measurements of the heat flux, and 40 different O2 concentration were used as the specific quantities of interest as a function of wall and deposit thermal conductivity, thickness and emissivity, fuel feed rate, and coal reactivity. Our validation approach identified the region of simultaneous consistency between all experimental and simulation data, according to the consistency constraint, $u_e \geq [y_m(x) - y_e] \geq l_e$, where the defect between the quantity of interest as predicted by the model and as measured by the experiment is bounded by the measurement uncertainty.

Our validation approach identified the region of consistency between experimental and simulation data; thus, providing insight into the experimental measurements that could not be achieved without concurrent simulation and validation. This simulation and validation approach to predictive design provides a path to accelerating deployment of new low-cost, low-carbon technologies.

437180 Impact of US Natural Gas Abundance on Carbon Management Strategies

Abstract:

The abundant supply of natural gas provided by the fracking boom has dramatically changed the options for carbon dioxide management. It appears that much of the near-term regulatory need for carbon management will be met by switching natural gas for coal. However, long term supplies are still uncertain and may strongly affect that transition. These reductions will at most result in about 500 million tons of US energy sector reduction. The remaining 2,000 million tons will have to be obtained through advanced capture systems that work on coal and natural gas, which will be a decided challenge for existing technology. A new generation of carbon capture technologies will be necessary to meet this challenge.

Mitglied der Helmholtz-Gemeinschaft

JÜLICH
FORSCHUNGSZENTRUM

Autothermal reforming of diesel fuels on structured cordierite monoliths coated with oxide supports and Rh as active phase

AIChE Annual Meeting 2015, Salt Lake City, USA,
November 8 – November 13

20. November 2015 | Joachim Pasel, Sebastian Wohlrab, Mikhail Rotov, Katrin Löhken, Ralf Peters and Detlef Stolten

JÜLICH
FORSCHUNGSZENTRUM

Facts and Figures on Forschungszentrum Jülich

People

- Employees: 5.768
- 900+ guest scientists (>45 countries)

Budget

- Budget: 617 Mio €
- 3rd party:~ 197 Mio €
- **+ € 1,6 billion** project management

Science

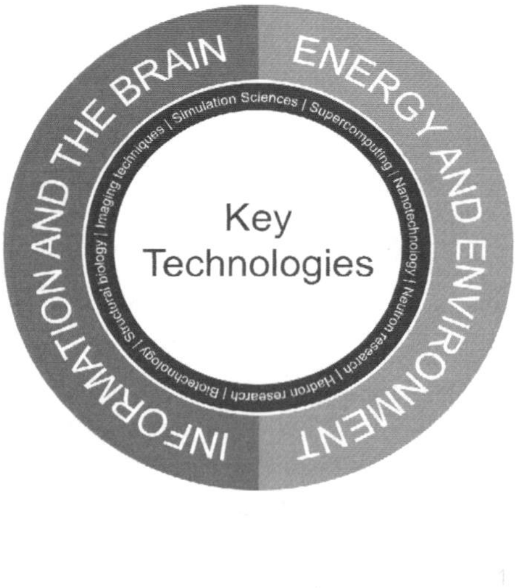

Institute of Energy and Climate Research

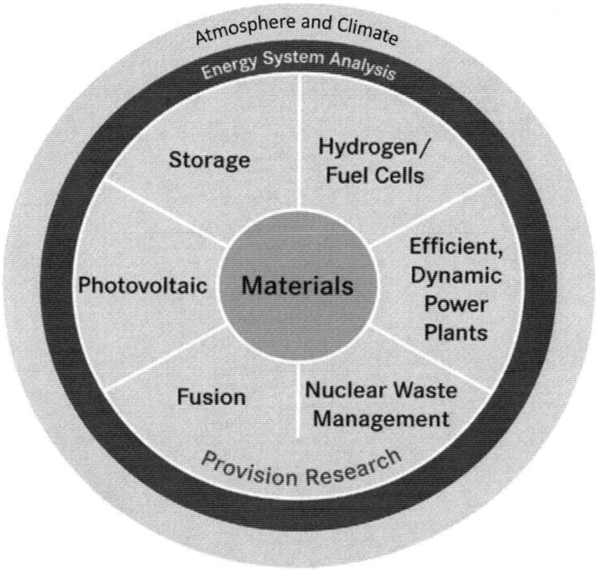

877 Staff

760 Energy Research

117 Climate Research

168 PhD Students:
Helmholtz Graduate school

Helmholtz Interdisciplinary Doctoral Training in Energy and Climate (HITEC)

Materials Research as Core Competence

IEK-3: Electrochemical Process Engineering

2

Background and motivation for research on hydrogen and fuel cells

- By 2030, global CO_2 emissions will be more than 50% higher than today (according to the current policies scenario)
- World population is growing to more than 9 billions of people in 2050 (according to United Nations, Department of Economic and Social Affairs, Population Division (2009): World Population Prospects: The 2008 Revision)
- Demand for energy increases – strongly driven by China and India (according to IEA World Energy Outlook 2009)
- Transport accounts for approx. 25% of the CO_2 emissions in Europe

→ Two-step approach at Juelich for more sustainable energy conversion and supply:

i. Development of efficient fuel cell systems

ii. Use of renewable, Bio-to-Liquid (BtL) diesel fuel

IEK-3: Electrochemical Process Engineering

Potential applications and environmental relevance of fuel cell systems

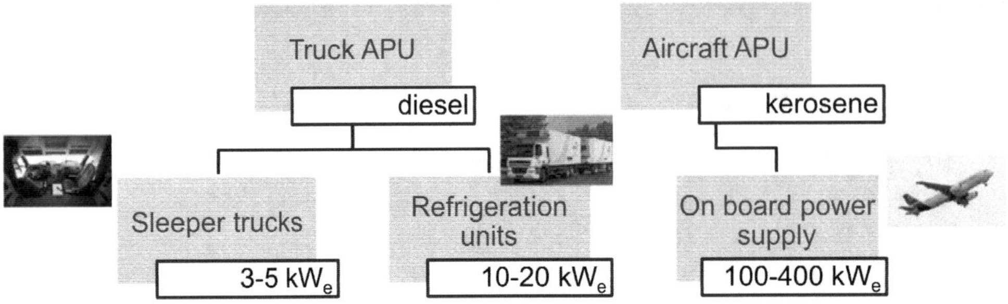

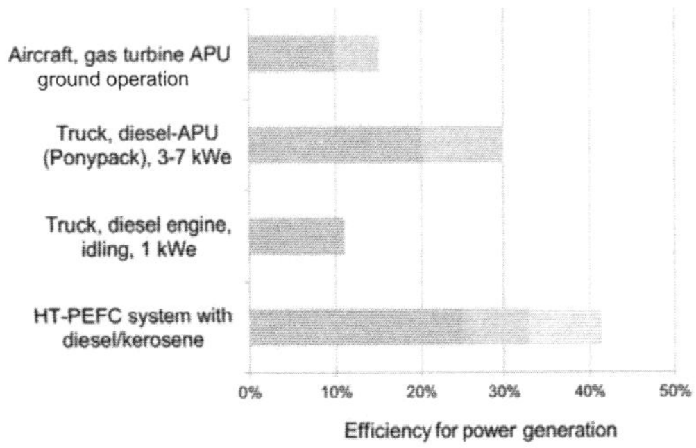

www.eberspaecher.com

IEK-3: Electrochemical Process Engineering

Fuel cell system with autothermal reforming of diesel and kerosene

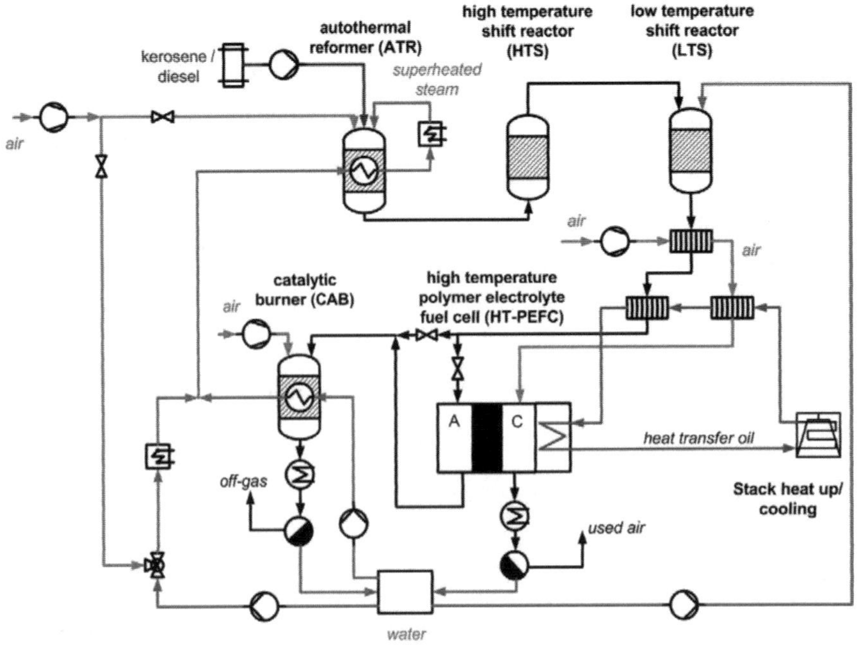

IEK-3: Electrochemical Process Engineering

Catalysts for autothermal reforming of diesel fuel

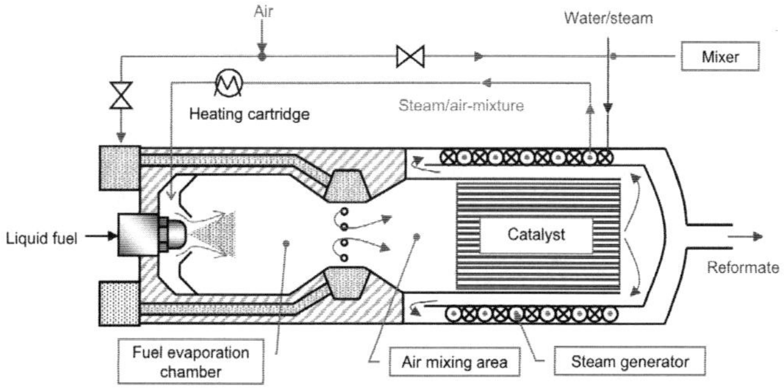

- Monolithic cordierite substrate (l = 90 mm, d = 18 mm, 600 cpsi)
- Different oxide supports with 10 ma%: γ-Al_2O_3, La-Al_2O_3, CeO_2, Gd-CeO_2, ZrO_2 or Y-ZrO_2
- Rh as active phase with 1 ma%

IEK-3: Electrochemical Process Engineering

Catalytic test reactor

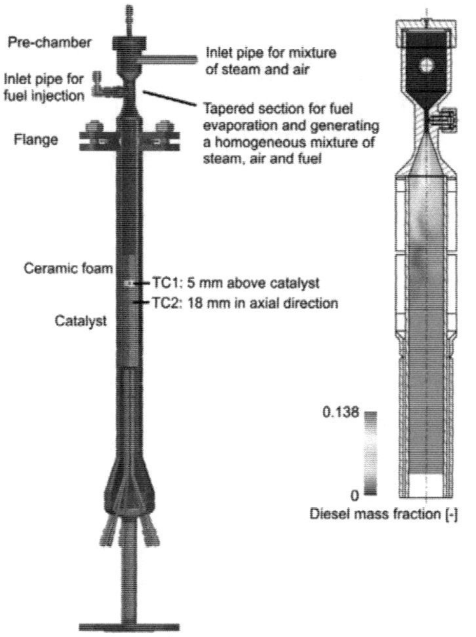

- Made of stainless steel
- Steam at 450 °C, air not pre-heated
- Venturi nozzle like configuration for fuel injection
  - Fuel injection without pulsation as fine droplets
  - Homogeneous mixing of the three reactants
- Flange for quick and easy replacement of catalysts
- Reaction zone with catalyst and ceramic foam
- Thermocouples for temperature measurement
- Reactor assembled in electrically heated oven

IEK-3: Electrochemical Process Engineering

Catalytic tests

- Evaluation pattern – **12 sequences**
  - $n(H_2O)/n(C)$ between 2.5 and 1.3 at three different gas hourly space velocities with NExBTL diesel
  - $n(O_2)/n(C)$ between 0.49 and 0.31 at three different gas hourly space velocities with NExBTL diesel
  - $n(H_2O)/n(C)$ between 2.5 and 1.3 at three different gas hourly space velocities with Ultimate diesel
  - $n(O_2)/n(C)$ between 0.49 and 0.40 at three different gas hourly space velocities with Ultimate diesel

Fuel	Supplier	Final boiling point [°C]	Mass fraction aromatics / S [wt.%] / [ppm]	Molecular formula
NExBTL diesel	Nesteoil Finland	321	< 0.1 / < 1	$C_{17}H_{36}$
Ultimate diesel	ARAL Germany	360	~ 15 / <1	$C_{19}H_{38}$

Catalyst characterization

- Temperature-programmed desorption, oxidation and reduction
- Thermogravimetric analyses
- Transmission electron microscopy

IEK-3: Electrochemical Process Engineering

H_2 concentration as measure for catalytic activity and stability

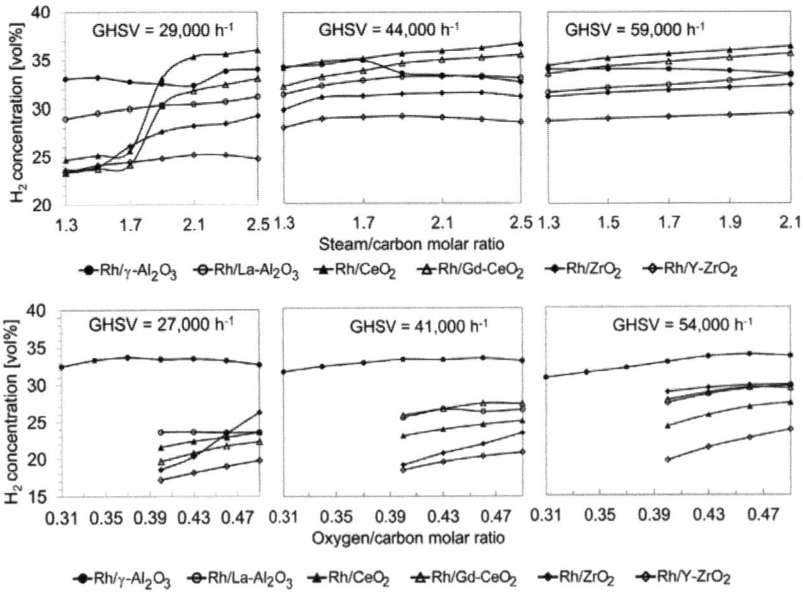

Sequences 1-3 with NExBTL diesel fuel

Sequences 4-6 with NExBTL diesel fuel

IEK-3: Electrochemical Process Engineering

H_2 concentration as measure for catalytic activity and stability

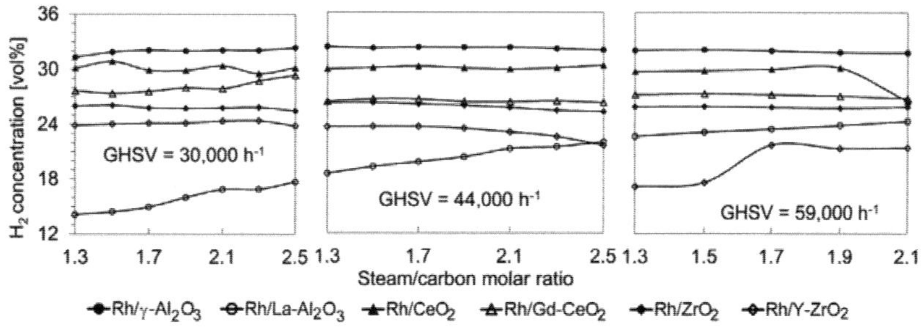

Sequences 7-9 with Ultimate diesel fuel

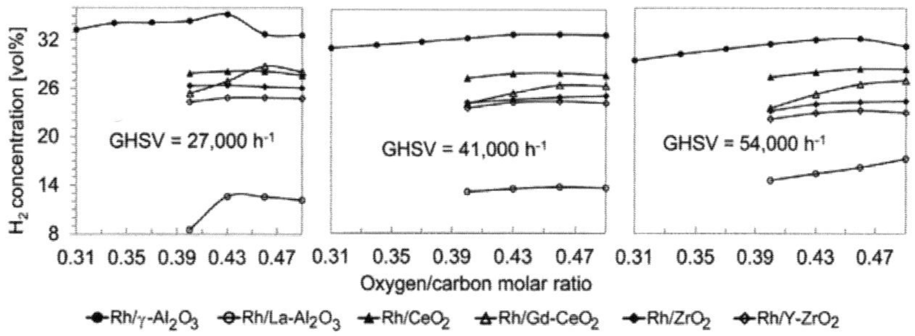

Sequences 10-12 with Ultimate diesel fuel

IEK-3: Electrochemical Process Engineering

Temperature-programmed reduction

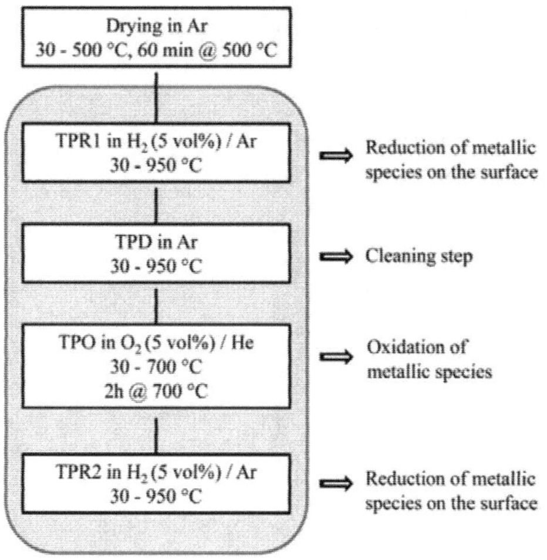

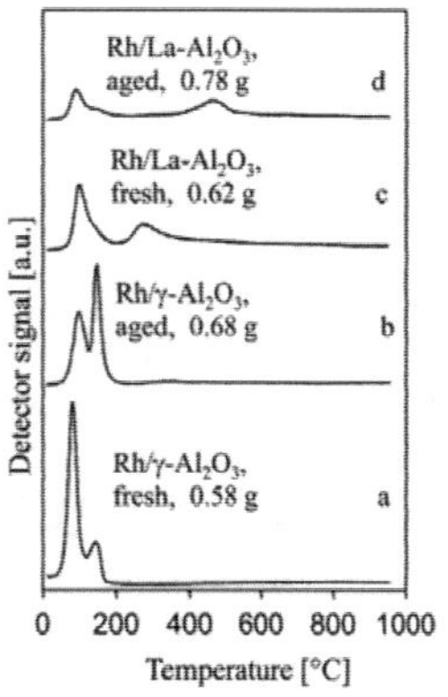

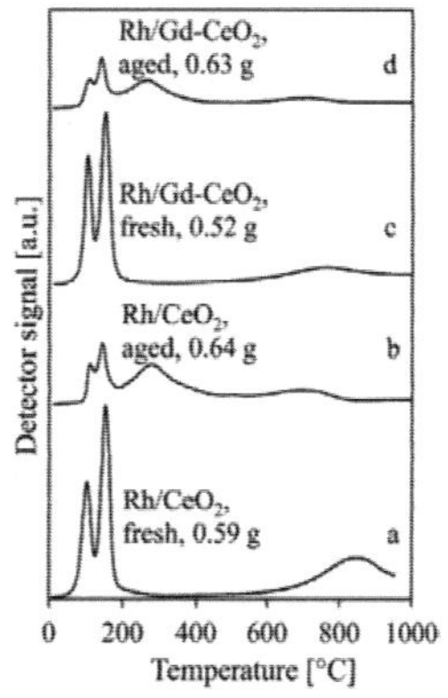

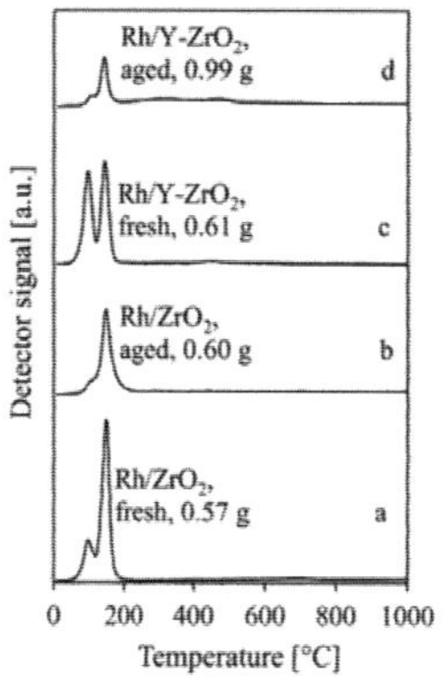

IEK-3: Electrochemical Process Engineering

Thermogravimetric analyses

Catalysts	Change in mass [%]
- $Rh/\gamma\text{-}Al_2O_3$	- 0.5
- $Rh/La\text{-}Al_2O_3$	- 2.5
- Rh/CeO_2	+ 0.2
- $Rh/Gd\text{-}CeO_2$	+ 0.2
- Rh/ZrO_2	- 0.1
- $Rh/Y\text{-}ZrO_2$	- 1.5

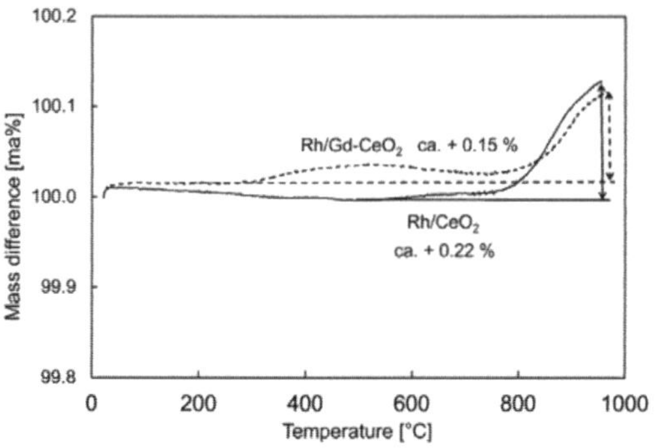

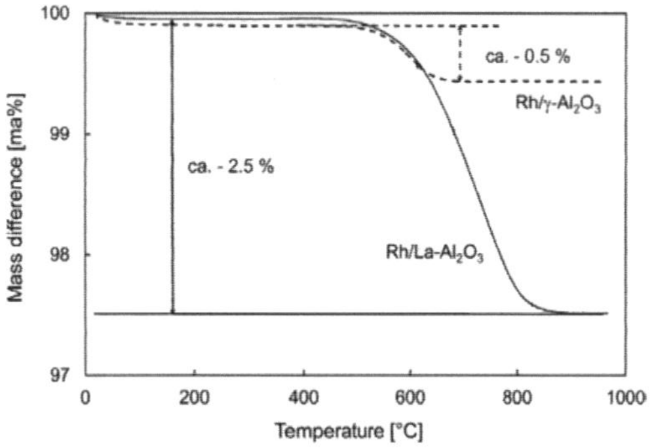

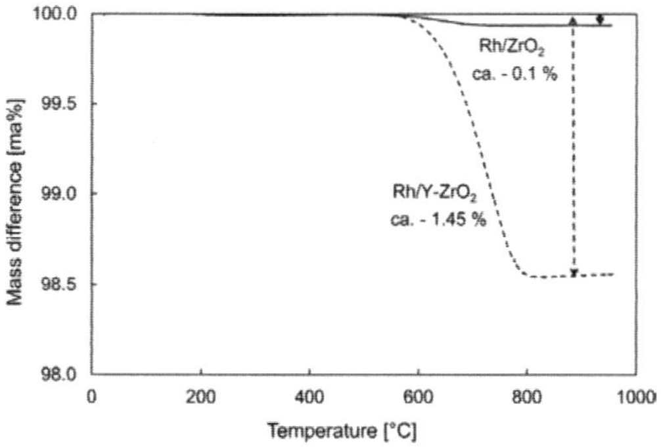

IEK-3: Electrochemical Process Engineering

Transmission electron microscopy of $Rh/\gamma\text{-}Al_2O_3$ - fresh and aged

fresh

aged

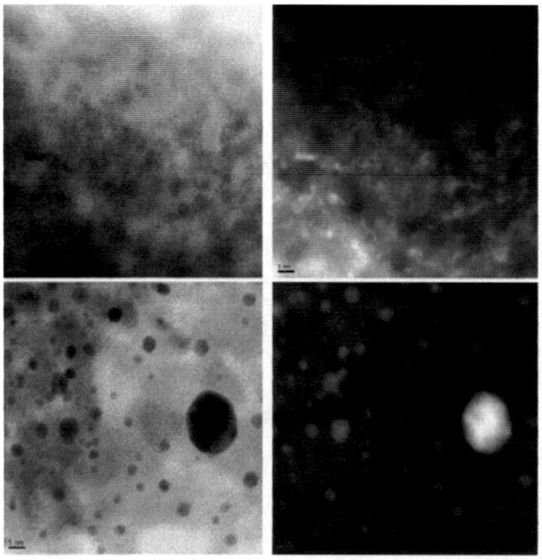

STEM mode on the left:
Al appears bright, Rh darker

HAADF-STEM mode on the right:
Rh appears brighter than Al

IEK-3: Electrochemical Process Engineering

JÜLICH
FORSCHUNGSZENTRUM

Summary and conclusions

- $Rh/\gamma\text{-}Al_2O_3$ was found to be the most active and robust catalyst.
- In contrast, $Rh/La\text{-}Al_2O_3$ showed the strongest deactivation.
- Considerable number of correlations between catalytic activity and physico-chemical properties were observed for all catalysts.
- Two different mechanisms for the deactivation of the catalysts were derived:
  - Carbonaceous deposits on the catalyst surface
  - Lowered reducibility of the active phase
- Operational measures should be taken to keep temperatures as low as possible (< 900 °C) to maintain the catalysts' redox properties.
- The order in which the reactants are arranged during start-up and shut-down procedures should be optimized to suppress the formation of byproducts, such as ethene and propene.

IEK-3: Electrochemical Process Engineering

JÜLICH
FORSCHUNGSZENTRUM

Special thanks to:

Fuel Processing team at Jülich

All project and cooperation partners

Thank you very much for your attention

Parts of the work were funded by:

ADELHEID Project

Ministerium für Klimaschutz, Umwelt, Landwirtschaft, Natur- und Verbraucherschutz des Landes Nordrhein-Westfalen

progres.nrw

EUROPEAN UNION
EUROPEAN REGIONAL DEVELOPMENT FUND

IEK-3: Electrochemical Process Engineering

JÜLICH
FORSCHUNGSZENTRUM

JÜLICH
FORSCHUNGSZENTRUM

FINDING TOMORROW TODAY

427443 Performance of Monolithic Catalyst for Autothermal Reforming of Diesel for APU (Auxiliary Power Unit) Applications

Abstract:

Fuel cell application is considered as one of the potential technologies for auxiliary power unit (APU) since fuel cell has higher efficiency compared to competing technologies. However, high quality hydrogen feed is needed for stable fuel cell operation of an APU system.

Using autothermal reforming (ATR) of diesel, hydrogen-rich reformate gas can be obtained with high efficiency. Monolithic structure of catalysts is very important for APU applications due to its mechanical strength and ability to maintain high flow rate with low-presure drop. However, there are concerns about catalytic performances to use long time because it has not been varified for long-term performances.

In this study, stability test for monolithic catalyst was carried out using diesel ATR conditions. Catalyst powders are coated on cordierite monolith support and micro-reactor testing was used to evaluate catalytic performances under ATR operating conditions. Over 2,000 hours of continuous operation was achieved with monolithic ATR catalyst. Also, start-up tests for 1kWe-class diesel reformer were optimized and achieved with the coated monolithic catalyst. Initial heat-up process and stabilization strategies were investigated, and optimization processes of those procedures were studied. Detailed start-up processes will be discussed in this study. Flow conditions were controlled to maintain stable range of kW-class reforming reactor. With those results, strategies for fast start-up and steady operation of ATR were investigated with monolithic ATR catalyst at kW-class reactor.

432784 Highly Efficient Fuel Processor and System for Hydrogen Production

Abstract:

The conversion of hydrocarbon fuels, including natural gas, to H_2 can be carried out by several reaction processes, including steam reforming (SR), partial oxidation (POX), and autothermal reforming (ATR). Steam reforming is generally the preferred approach to achieve high purity hydrogen, having higher system efficiency in comparison to POX and ATR. Since steam reforming is endothermic, some amount of the fuel must be burned and the heat must be transferred to the reformer via heat exchangers. Thus, effective energy utilization from waste heat and poor quality fuels is of importance to an integrated H_2 generation unit for improving the overall system efficiency. Precision Combustion, Inc. (PCI) has been developing a novel heat-integrated steam reforming reactor for efficient syngas (i.e., H_2 and CO) generation based on its patented Microlith® technology, providing higher heat flux compared to other state-of-the-art reactors. The Microlith based reactor design results in a remarkably compact, lightweight, and efficient steam reforming reactor, which is otherwise widely known to be limited by the heat transfer resistance from the exothermic (heat source) to the endothermic reforming side. The burner component comprises of catalytic oxidizer instead of flame-stabilized combustion, resulting in improved temperature control and uniformity. In addition to the steam reformer, PCI has also developed an oxygen-blown, natural gas ATR that was designed for producing a syngas stream with a specific hydrogen-to-carbon monoxide ratio. This ATR system was developed and demonstrated to supply syngas with a specific H_2-to-CO ratio to meet customer requirements. This system was integrated by an industrial fluid and gas handling firm under PCI's supervision after performance shakedown and validation performed at PCI's facility. The system was delivered to the customer, accepted and intermittently used over a period of two years to support its technology development and testing needs. The system was used for a total of 4,000 hours and produced 50,000 Nm^3 of syngas.

In this paper, results from testing the steam reformer with a distillate fuel (containing 2-5 ppm_w sulfur) and natural gas will be presented, where near equilibrium product distribution was observed. The effect of steam-to-carbon ratio, operating pressure (i.e., up to 10 atm), and thermal inputs on the catalyst performance, fuel conversion and H_2 production for the steam reformer will be discussed. The sulfur tolerance studies of the burner catalyst will also be presented. Catalytic sulfur tolerance allows effective utilization of poor quality or waste fuels in the burner, thus significantly improving the overall process efficiency. Results from hundreds of hours of testing of the steam reformer unit showed minimal catalyst degradation, thus providing a measure of the reactor performance under realistic conditions and demonstrating the durability of the reformer. Separately, results from testing the oxygen-blown ATR for converting natural gas to syngas will be presented. The operational specifications and product stream quality will also be highlighted.

432347 Contaminant-Tolerant Fuel Reforming/Upgrading Catalyst for Fuel Cell Systems

Abstract:

Fuel cells can efficiently produce electricity using hydrogen. Fuels proposed for use in fuel cell processors include gaseous fuels (e.g. natural gas, liquefied petroleum gas, landfill gas, and digester gas), liquid fuels (e.g. gasoline, jet fuel, and diesel) and solid fuels (e.g. coal, biomass, and municipal solid waste). All fuels are either reformed or gasified to produce syngas ($CO + H_2$) and the syngas is then subjected to water gas shift (WGS) to produce hydrogen for the fuel cell. Typically, solid fuels are thermally (non-catalytically) gasified whereas gaseous and liquid fuels are catalytically reformed using steam and air. The contaminants in the solid fuels, particularly sulfur and nitrogen, appear in the syngas as H_2S and NH_3. In addition, gasification invariably leads to light gaseous hydrocarbon and tar formation, unless carried out at temperatures in excess of 1300°C. The raw syngas from solid fuel gasification containing light hydrocarbons, H_2S, NH_3, and tar must be cleaned prior to its use in the fuel cell to prevent the poisoning of the fuel cell anode catalyst. Also, to reduce cost, the light hydrocarbons need to be reformed to produce additional syngas. Gaseous and liquid fuels on the other hand need to be desulfurized prior to the catalytic reformer to avoid the poisoning of the conventional nickel-based reforming catalyst. This paper describes the development and laboratory-scale test results of cost-effective contaminant-tolerant reforming/upgrading catalysts suitable for reforming of sulfur containing gaseous and liquid fuels as well as for cleaning/upgrading of raw syngas from solid fuel gasification. Experiments were carried out using raw simulated gasifier syngas. Results showed that the best performing catalysts maintained high methane reforming, tar reforming, and ammonia decomposition activity for the duration of the testing (200 hours) in the presence of about 100 ppmv H_2S. Furthermore, the catalysts demonstrated significant tolerance to H_2S even at H_2S concentration levels of 500 ppmv. Techno-economic evaluation showed that the use of the sulfur- and contaminant-tolerant catalysts developed in this work for an auto-thermal reformer, as opposed to using a desulfurizer prior to the auto-thermal reformer employing a conventional nickel reforming catalyst, can result in a more compact and economical overall fuel processor for gaseous and liquid fuels.

427656 Syngas Production By Biogas Steam and Oxy Steam Reforming Processes on Rh/CeO2 Catalyst Coated on Ceramics Monolith and Open Foams

Abstract:

Today there is a worldwide interest in the use of biofuel as an environmental friendly renewable energy source [1]. Biogas can be considered as one of the most widespread renewable fuel obtained from biomass produced from a variety of organic raw materials in various sectors ranging from zootechnical to agro-industrial [2-3]. Biogas, basically, it is constituted by 50-75% CH_4, 25-45% CO_2, 2-7% H_2O (at 20-40 °C), 2% N_2, less than 1% H_2/H_2S and traces of O_2, NH_3, halides and siloxanes [4]. The utilization of biogas to produce syngas/hydrogen by reforming processes [5] to feed fuel cells represents an alternative route to the traditional utilization of this renewable fuel in less efficient and polluting engines to produce energy and heat. Among the possible reforming reactions, steam reforming (SR) is widely used for hydrogen production. It is the largest and generally most economical route especially for industrial applications [6]. However, SR is strongly endothermic, and large amount of heat supply is required. Another option can be represented by the oxy steam reforming (OSR) process. The simultaneous occurrence of the reactions due to the presence of O_2 and H_2O can help to overcome critical factors as the deactivation by coke deposition reducing at the same time the overall endothermicity of the process [7]. One of the central issue of reforming processes is the development of efficient, stable and low cost catalysts. The most used catalysts in reforming processes are based on transition metals supported on oxides. Noble metals (Pt, Rh, Ru, Pd, Ir) are normally, more active and tolerant to coke formation [8] with respect to other transition metals as Ni. However, an excessive use of these elements increases the total cost of the catalyst. Another important challenge is the development of more compact and lightweight fuel processors, especially to achieve the targets for small-scale applications such as for distributed hydrogen production [9]. In this respect different catalyst configurations (foams, honeycombs, gauze, microchannels) were proposed as alternative to traditional packed-bed reactors for the realization of compact reactors [10]. The advantages of structured catalysts are related especially to the large surface-to-volume ratio that leads to good heat and mass transfer properties and low pressure drop. In this work, the performances of structured catalysts, based on Rh (1.5 wt.%), supported on CeO_2, coated on cordierite monoliths (400 cpsi, diameter 1 cm, length 1.5 cm) and alumina foams (20, 30, 40 ppi) were investigated and compared. The structured support was lined by combining the solution combustion synthesis with the impregnation technique [11-13]. The deposition process was repeated several times until reaching the desired amount of catalytic layer over the supports (6.5 mg cm^{-2} referred as weight of active metal plus oxide carrier on the bare support geometric surface). In addition to the preparation of the coated monoliths, the corresponding catalyst at powder level was prepared and used for the chemical-physical characterization of the supported catalyst (XRD, TEM, SEM and chemisorption analyses). The pressure drop and the porosity across the bare ceramics supports and the structured catalysts were determined at different superficial velocities and compared. In addition the adherence between the catalytic layer and the support walls was evaluated via sonication treatment [12,13]. Biogas OSR and SR experiments were conducted in a fixed-bed quartz reactor with inner diameter of 1 cm at atmospheric pressure. The structured catalysts were placed between two quartz wool plugs in the center of the quartz tube, inserted into a furnace heated at the reaction temperature T_{SET} and controlled through a PID temperature controller. Instead, the influence of temperature and of the WSV was assessed on all the prepared structured catalysts varying the temperature (T_{SET}) between 700–900 °C and the WSV between 34,000 and 18,000,000 Nml g_{cat}^{-1} h^{-1} at fixed S/C =1 for the SR reaction, and the temperature (T_{SET}) between 700–900 °C and the WSV between 36,000 and 24,0000,00 Nml g_{cat}^{-1} h^{-1} at fixed S/C =1 and O/C = 0.2 for the OSR reaction. Model biogas containing 60% CH_4 and 40% CO_2, was prepared from high purity CH_4, O_2 and N_2 gas cylinders. Steam was added to the feed by using an isocratic pump and a specially designed evaporator. The feed and reaction products were analyzed using an Agilent 6890 Plus gas chromatograph equipped with thermal conductivity (TCD) and flame ionization (FID) detectors. Before testing the catalytic performance of the prepared monoliths, each coated support was reduced under hydrogen-rich atmosphere directly in the quartz reactor of the test rig. All the structured catalysts showed a uniform thin coating with thickness between 20–30 μm, high mechanical strength and low pressure drop. In detail the monolith based catalyst has shown a negligible weight loss (0.4%) of the catalytic layers after two sanction baths (1 h overall) in a mixture of water and acetone (50–50 wt.%), instead the foam based samples have shown a higher loss of catalytic layer. This different behaviour is probably due to the different surface morphology of bare supports, the bare monolith is characterized by the presence of a more intense macro-porosity over the walls respect the surface of the foams supports, this can help to obtain a more stable layer of CeO_2 during the first step of coating procedure. Also, regarding the pressure drop measurement there are differences between the foam and monoliths. The monolith based catalysts show a lower pressure drop respect that of the foam catalysts, in addition the values increase increasing the ppi. Both monolith an foam catalysts have shown good performances during SR and OSR processes at all the WSV investigated in terms of high CH_4 conversion (99.9–98% for SR and 99.9–96% for OSR), high H_2 concentration (64-62% for SR and ≈58% for OSR, in dry basis and nitrogen free), the H_2/CO ratio changed from about 2.8 for the SR to 1.9 for the OSR. The utilization of CO_2 contained in the biogas in both reforming processes was also registered, the results have shown that the CO_2 conversion can be regulated principally tuning the S/C and the O/C molar ratio. In SR reforming condition, increasing the WSV, the CO_2 conversion decreased from 10 to 3%, instead in OSR condition it decreased from 49 to 42%. The presence of O_2 and lower amount of steam in the OSR process probably leads to convert more CO_2 trough the dry reforming reaction that is disadvantaged in SR condition by the large presence of steam.

[1] E. Shafiei, B. Davidsdottir, J. Leaver, H. Stefansson, E. I. Asgeirsson. Comparative analysis of hydrogen, biofuels and electricity transitional pathways to sustainable transport in a renewable-based energy system. Energy 83,2015,614–627.

[2] W. Edelmann Biogas production and usage. In: Kaltschmitt M, Hartmann H, editors. Energy from biomass: basic principles, technologies and processes. Leipzig, Germany: Springer; 2001.

[3] G. Alkanok, B. Demirel, T.T. Onay. Determination of biogas generation potential as a renewable energy source from supermarket wastes. Waste Manag. 34,2014,134-140.

[4] I. Wheeldon, C. Caners, K. Karan, Brant Peppleyd, Utilization of biogas generated from Ontario wastewater treatment plants in solid oxide fuel cell systems: A process modeling study. Int. J. Green Energy 4,2007,221–231.

[5] J. Xuan, M.K.H. Leung, D.Y.C. Leung, M. Ni, A review of biomass-derived fuel processors for fuel cell systems, Renew. Sust. Energy Reviews 13,2009,1301–1313.

[6] M. Ashrafi, T. Proll, C. Pfeifer, H. Hofbauer. Experimental study of model biogas catalytic steam reforming: 1. thermodynamic optimization. Energy Fuels 22,2008,4182–4189.

[7] U. Amjad, A.Vita, C. Galletti, L. Pino, S. Specchia. Comparative study on steam and oxidative steam reforming of methane with noble metal catalysts. Ind. Eng. Chem. Res. 52,2013,15428−15436.

[8] L. Pino, A.Vita, F. Cipitì, M. Laganà, V. Recupero, Catalytic performance of $Ce_{1-x}Ni_xO_2$ catalysts for propane oxidative steam reforming. Catal. Lett. 122, 2008, 121–130.

[9] S. Specchia, Fuel processing activities at European level: A panoramic overview. Int. J. Hydrogen Energy 39,2014,17953–17968.

[10] G. Kolb, T. Baier, J. Schürer, D.Tiemanna, A. Ziogas, H. Ehwald, P. Alphonse, A micro-structured 5 kW complete fuel processor for iso-octane as hydrogen supply system for mobile auxiliary power units: Part I. Development of autothermal reforming catalyst and reactor. Chem. Eng. J. 137,2008,653–663.

[11] S. Specchia, E. Finocchio, G. Busca, V. Specchia, Combustion synthesis, Lackner M, Winter F and Agarwal AK, Handobook of combustion, Wiley-VCH Verlag GmbH & Co. KGaA, Volume 5, Chapter 17, Weinheim (Germany), 2011, 439–472.

[12] A.Vita, G. Cristiano, C. Italiano, S. Specchia, F. Cipitì, V. Specchia, Methane oxy-steam reforming reaction: performances of $Ru/\gamma\text{-}Al_2O_3$ catalysts loaded on structured cordierite monoliths. Int. J. Hydrogen Energy 39,2014,18592–18603.

[13] A.Vita, G. Cristiano, C. Italiano, L. Pino, S. Specchia, Syngas production by methane oxy-steam reforming on Me/CeO_2 (Me = Rh,Pt,Ni) catalyst lined on cordierite monoliths. Appl. Catal. B: Environ. 162,2015,551–563.

Acknowledgements

This work was funded by the Italian Ministry of Education, University and Research (MIUR, PRIN 2010–2011) within the project IFOAMS ("Intensification of catalytic processes for clean energy, low-emission transport and sustainable chemistry using open-cell FOAMS as novel advanced structured materials", protocol no.2010XFT2BB).

422186 Hydrogen Production By Steam Reforming of Dimethyl Ether - Development of Catalysts for the Steam Reforming

Abstract:

Dimethyl ether (DME) is expected as a clean fuel for the 21st century [1, 2]. DME has also recently become a hydrogen carrier, storage, and potential fuel for hydrogen production to be used in fuel cells. I have been studying on steam reforming of DME for the hydrogen production. The new type catalysts prepared using a sol-gel method have been developed [2-4]. These catalysts produce hydrogen more effectively at lower reaction temperature than mixed catalysts with DME hydrolysis catalysts and methanol steam reforming catalysts. Cu-Zn(29-1wt%)/Al_2O_3 is the most effective catalyst for hydrogen production in my recent study. Aiming at improving the catalysis activity for hydrogen production, the preparation method and additive effect have been investigated. In this paper and presentation, I will show my latest development of the sol-gel catalyst for DME steam reforming.

Two Cu(29wt%)-Zn(1wt%)/Al_2O_3 and two Cu(29wt%)-Zn(1wt%)-CeO_2(1wt%)/Al_2O_3 powder catalysts were prepared using the two different preparation procedures of sol-gel method, respectively. One is prepared in one-step (one-step method) [2-4], and the other is made in two steps (consecutive method), respectively.

BET specific surface area of Cu-Zn/Al_2O_3 (one-step), Cu-Zn/Al_2O_3 (consecutive), Cu- Zn-CeO_2/Al_2O_3 (one-step), and Cu-Zn-CeO_2/Al_2O_3 (consecutive) were 209, 190, 208, and 202 $m^2\ g^{-1}$, respectively. Specific surface area of catalysts prepared by one-step method is larger than that of catalysts prepared by consecutive method. Amount of adsorbed CO is related to Cu sites of catalyst surface. The values of the four catalysts were 0.18, 0.14, 0.19, and 0.15 mmol g^{-1}, in the same order above-mentioned. Adsorbed CO amount of catalysts prepared by one-step method is larger than that of catalysts prepared by consecutive method like surface area. The physical property data of the catalysts prepared by the consecutive method are lower than those of the catalysts prepared by the one-step method, however, hydrogen production rate of the four catalysts order is Cu-Zn-CeO_2/Al_2O_3 (consecutive) > Cu-Zn/Al_2O_3 (consecutive) > Cu-Zn/Al_2O_3 (one-step) > Cu-Zn-CeO_2/Al_2O_3 (one-step). The acidity of the catalysts was analyzed using pyridine adsorption by FT-IR. Lewis acid was detected, but Brønsted acid was not detected. Acidity of the catalysts prepared by the consecutive method is stronger than that of the catalysts prepared by the one-step method. Steam reforming of DME ($CH_3OCH_3 + 3H_2O \rightarrow 6H_2 + 2CO_2$) consists of two reaction steps; the first step is hydrolysis of DME to methanol ($CH_3OCH_3 + H_2O \rightarrow 2CH_3OH$), and the second step is methanol steam reforming to hydrogen ($CH_3OH + H_2O \rightarrow 3H_2 + CO_2$). The DME hydrolysis reaction is the rate-determining step. The stronger acidity of Lewis acid sites works for the hydrolysis well and well-produced methanol reforms to hydrogen on the copper sites, therefore copper alumina catalysts prepared by consecutive method produce hydrogen more than copper alumina catalysts prepared by one-step method. CeO_2 accelerates methanol steam reforming in case of the catalyst prepared by the consecutive method, but in case of the catalyst prepared by the one-step method CeO_2 weakens the acidity of the Lewis acid sites of the catalyst. Therefore, Cu-Zn-CeO_2/Al_2O_3 (one-step) was not work most for hydrogen production among the four catalysts.

I have developed DME steam reforming catalysts. Sol-gel method has been developed, and the consecutive sol-gel method is optimum for preparation of DME steam reforming catalysts. CeO_2 improves methanol steam reforming and DME steam reforming.

REFERENCES

[1] T. H. Fleisch, A. Basu, M. J. Gradassi, J. G. Masin, Studies in Surface Science and Catalysis, 107 (1997) 117.

[2] K. Takeishi, Biofuels, 1(2010) 217.

[3] K. Takeishi, H. Suzuki, Applied Catalysis A: General, 260 (2004) 111.

[4] K. Takeishi, K. Yamamoto, Japan Patent No. 3951127; US Patent No. 7,241,718 B2; etc.

434438 Conversion of Alcohols, Ethers and Light Hydrocarbons - on the Way to Multifuel Reformer

Abstract:

In recent years, the leading research centers in the world have been focusing considerable efforts on the extended R&D studies of compact, efficient and reliable fuel reformers generating synthesis gas and hydrogen. The fuel reformers are assumed to become a key element of fuel cell based power units for stationary and portable (decentralized) applications. Its operation is based on catalytic reforming of a fuel into synthesis gas (hydrogen-rich gas mixtures) followed by removal/conversion of the components inhibiting efficient oxidation of hydrogen in the fuel cell.

Among promising feedctocks for hydrogen-rich gas production are alcohols (methanol, ethanol), ethers, light hydrocarbons (methane, LPG), gasoline and diesel fractions, biodiesel. The main task to be solved is to specify the optimum catalyst composition and develop a reformer which capable to convert of more than one fuel under similar reaction conditions. This approach allows the development of universal power generating units operating with various fuel types.

As continuation of our previous R&D on multi-fuel concept realization and production of hydrogen-rich gas from synthetic, renewable and conventional fuels [1], we report in the present work the results on steam reforming of C1-based oxygen containing compounds (methanol, dimethyl ether, dimethoxymethane) over the same type catalysts under similar reaction conditions. The other part of the study was performed using the natural gas, LPG, methane-propane, and flare (associated petroleum) gas, which are widely used for fueling power generating units. This approach will facilitate the using of available infrastructure for gradual introduction of high-temperature fuel cell based power generating units.

The considered in this work multi-fuel concept – production of hydrogen-rich gas from various hydrocarbon feedstocks using the same type catalyst and under similar reaction conditions – has been proved promising. The selected nickel oxide based catalyst provides efficient reforming of various hydrocarbon fuels into synthesis gas and methane-hydrogen mixtures suitable for feeding power generating units on the base of high-temperature solid oxide fuel cells and spark, diesel and gas-diesel internal combustion engines. The selected copper based catalyst provides efficient reforming of methanol, dimethyl ether, dimethoxymethane into hydrogen-rich gas with low CO content suitable for feeding power generating units on the base of high-temperature polimer electrolyte membrane fuel cells.

Acknowledgments

The work was partially supported by MES (Russia). AM wishes also to acknowledge the NYSERDA for financial support.

1. P.V. Snytnikov, S.D. Badmaev, G.G. Volkova, D.I. Potemkin, M.M. Zyryanova, V.D. Belyaev, V.A. Sobyanin, Catalysts for hydrogen production in a multifuel processor by methanol, dimethyl ether and bioethanol steam reforming for fuel cell applications // Int. J. Hydrogen Energy, **37** (2012), pp. 16388-16396.

440988 Fuel Processing for Fuel Cells in Mobile and Stationary Applications

Abstract:

Fuel cell systems based on diesel steam reforming and polymer electrolyte fuel cells (PEFC) offer a great potential for auxiliary power units (APU) in mobile and stationary applications. In a joint research project with partners from industry, Oel-Waerme-Institut GmbH has developed an integrated fuel cell system for mobile power generation that can also be used as a stationary cogeneration unit. The system is based on a diesel steam reformer and allows the operation of low-temperature and high-temperature PEFC. In preceding investigations of the author's group, coupled operation of a steam reformer with an LT-PEFC as well as fuel processor optimization regarding start-up and system integration have been reported [1, 2, 3].

The focus of this work is the system integration of the fuel processor and a fuel cell. Results from the stand-alone operation including balance-of-plant components are presented. In addition to an LT-PEFC, the steam reformer was coupled to an HT-PEFC using logistic diesel from a gas station as a fuel. The effect of sulfur and residual hydrocarbon concentration on the PEFC performance was investigated. The system design was optimized for fast system start-up and high system efficiency. This research and development project was funded by the German Federal Ministry of Economics and Technology (BMWi, FKZ 03ET2052A).

One of the research areas of the Institute for Building Technology is the development of a bidirectional real-time energy management system enabling the user to actively control energy fluxes within a building. The basis for the energy management system is a recently constructed residential building that incorporates a fuel cell cogeneration unit as well as a solar power and battery storage system. Key design parameters of the energy management system will be presented.

References

[1] Engelhardt, P.; Maximini, M.; Beckmann, F.; Brenner, M.: Integrated fuel cell APU based on a compact steam reformer for diesel and a PEMFC. In: International Journal of Hydrogen Energy 37 (2012), pp. 13470-13477 - ISSN 0360-3199

[2] Maximini, M.; Engelhardt, P.; Brenner, M.; Beckmann, F. ; Moritz, O.: Fast start-up of a diesel fuel processor for PEM fuel cells. In: International Journal of Hydrogen Energy, Volume 39 (2014), pp. 18154–18163

[3] Engelhardt, P.; Maximini, M.; Beckmann, F.; Brenner, M.; Moritz, O.: Coupled operation of a diesel steam reformer and an LT- and HT-PEFC. In: International Journal of Hydrogen Energy 39 (2014), pp. 18146-18153 - ISSN 0360-3199

429178 Sustainability Assessment of Direct and Indirect Mechanical Energy Recovery Devices

Abstract:

Sustainable development has become a key concern in industries, largely due to natural resource depletion, global competition, environmental pressure, etc. Despite the efforts for sustainability improvement, still over a half of energy consumption is wasted in manufacturing sectors, where the chemical industry is responsible for the energy efficiency lower than it should be. Many attempts have been made to recover the thermal energy using heat integration techniques. Although work is more expensive than process heat, no efficient solution has been studied to recover mechanical energy yet. In chemical plants, many process streams need to be pressurized or depressurized in different operational stages. Therefore the energy of these streams can be recovered by a new class of exchange, which is called work exchange (Huang and Fan, 1996). Two possible ways to exchange mechanical energy between two streams are direct and indirect recovery devices.

Some studies related to indirect recovery devices, such as single-shaft-turbine-compressor (SSTC) units, have been reported, where a superstructure-based Work Exchange Network (WEN) synthesis has been introduced (Razib et al., 2012). The direct recovery of mechanical using Work Exchanger as a key device is still at the stage of analysis. In this paper, the technical feasibility and thermodynamic efficiency of these two types of the work exchange units are discussed in details. Sustainability assessment with the main focus on the energy recovery efficiency and cost analysis, beside the safety of these two type of WE's, is introduced. Indirect recovery devices like SSTC are traditional compressors and turbines with a mature technology and known capital cost data. Direct recovery devices are pair of combined operated piston pumps without any accurate capital cost data which needs to be estimated in the presented work. The possibility of work and heat integration through direct work exchange units for non-isothermal and adiabatic operating conditions is the other parameter which is analyzed. Mechanical energy recovery using different types of units are compared through a case study.

References:

Huang, Y.L. and L.T. Fan, "Analysis of a Work Exchanger Network," *Ind. Eng. Chem. Res.*, 35(10), 3528-3538, 1996.

Razib, M.S., M.M.F. Hasan, and I.A. Karimi, "Preliminary Synthesis of Work Exchange Networks," *Comp. & Chem. Eng.*, 37, 262-277, 2012.

Design of CHP Systems for Housing Complexes under Uncertainty

Luis Fabián Fuentes-Cortés, José Ezequiel Santibañez-Aguilar and José María Ponce-Ortega.
Chemical Engineering Department, Universidad Michoacana de San Nicolás de Hidalgo, Morelia, Michoacán, México.

Abstract

In the residential sector of Mexico, the most important energy demands are related to the supply of electricity and hot water. These demands can be met efficiently using cogeneration systems. However, the design of a combined heat and power (CHP) system involves problems related to technology selection, sizing and operation of the central CHP unit. To solve this problem, there have been used optimization algorithms that consider environmental parameters, usually related to ambient temperature, economic, as fuel costs or prices of electricity and operating, maintenance and capital costs of equipment and the energy demands of the end user. Traditionally, it was considered that these elements behave regularly over time. However, the reality is that these parameters fluctuate and the designer has no control over them. Therefore, this paper presents a methodology based on a mixed-integer nonlinear programming model that considers the selection of technologies, a thermal storage tank, an auxiliary water heating system using a conventional boiler, the optimal sizing of the system and the interactions of purchase - sale of energy with the grid of the local company to meet the energy demands of a housing complex. As design parameters under uncertainty there have been considered the ambient temperature, energy demand related to the floating population and local energy market prices. As objective function there is considered a multi-objective scheme that allows to minimize the total annual cost of the system and secondly to minimize the generation of greenhouse gas emissions associated to fuel consumption. As a case study is considered a housing complex located in the center of Mexico with 1440 households. The results obtained under different scenarios show economic and environmental benefits in implementing CHP systems.

Introduction

Combined heat and power (CHP) systems are efficient alternatives for supplying the needed power and heat in residential complexes. This is because CHP systems offer several advantages in terms of efficiency (1), environmental impact (2) and economic cost (3) compared with conventional systems. The design of cogeneration systems is determined by several factors, as the availability of natural resources that can feed the system (4), weather conditions (5), energy demand of end users, available technology and the conditions and policies of the local energy market (6). Usually designing CHP systems is based on average values of the parameters that represent significant uncertainty (7). Nevertheless, this approach is not the best way to account for the involved uncertainty, since the ambient temperature, the energy market prices and the energy demands are factors that have involved significant uncertainty in designing residential CHP systems.

To overcome previous limitations, this paper presents an optimization approach for designing domestic CHP systems, for accounting for the involved uncertainty in the ambient temperature, energy market conditions, demand for electricity and demand for hot water, based on stochastic forecasting methods. The proposed model is based on a superstructure that allows determining the optimal structure for the CHP system, and the optimal selection for the central CHP units, which can be internal combustion engine (ICE), fuel cell (FC), Stirling engine (SE) and microturbine (MT), coupled to a thermal insulated storage tank (ST) and interacting with the grid of the electric company to smooth the gaps between the demands for electricity and hot water. Weather conditions (especially temperature) modify the energy demand of domestic users. In this case, the uncertainty was taken into account through normal distributions based on the historical behavior of ambient temperature, floating population and changes in energy market prices are used for creating different scenarios. The objective functions are used to measure the financial risk, which is associated with the total annual cost (TAC) and the environmental risk, which in turn is associated with CO_2 emissions.

Problem Statement

The problem addressed in this paper is schematically described in **Figure 1**. Given are the hourly seasonal thermal consumptions associated to hot water for sanitary use (HWS), as well as the power demands of a housing complex, the average hourly seasonal temperature and the prices for fuel and hourly electrical rates. The energy demands are subjected to the ambient temperature and demographic changes; especially for the floating population. The ambient temperature is subjected to the seasonal and climate change conditions, which can be predicted by analyzing the historical behavior for a given location. The prices in the energy market also fluctuate and these can be predicted considering inflation and historical factors. Therefore, based on historical data, stochastic sampling techniques can be used to predict the uncertain data generating multiple scenarios with different conditions on hourly energy demand, hourly rates purchase – sale of electricity, fuel costs and hourly temperature, and with this information stochastic optimization approaches can be used for designing domestic CHP systems accounting for the uncertain data. Then, the problem consists in determining the optimal configuration of the cogeneration system that meets the energy demands considering factors such as sizing and selection of the prime mover, sizing of the ST as well as the sizing of an auxiliary thermal system, which consists of a boiler (B) to provide the extra hot water needed, and the scheme purchase-sale of power with the grid of the local company, while accounting for the simultaneous assessment of the financial risk through the values of the total annual cost (TAC) and the analysis of the environmental risk using the greenhouse gas emissions (GHGE) associated to the consumption of the fuels.

This work proposes a superstructure (see **Figure 2**) that operates cooperatively with the grid of the local utility. The proposed superstructure is basically formed by four integrated technologies, an auxiliary water heater and a ST, and CHP technologies are used to meet the electricity and hot water demands. First, the produced electricity is sent to the housing complex, and the excess of produced electricity is then sent to the grid of the electric company (i.e., when the CHP system produces surplus energy, this is sold to the local electric company). Furthermore, when the CHP system produces less electricity than the one needed in the housing complex, and then the needed electricity is purchased to the local electric company. Because the hot water demand behavior in the housing complex is different than the electricity demand, there is included a ST for hot water from the CHP system in the proposed scheme when there is a surplus, and also there is an auxiliary equipment for providing extra hot water to the housing complex. Finally, before to send the water to the housing complex, the hot stream is mixed with cold water to reach the needed temperature for domestic use. Therefore, the algorithm must determine the type and size of the CHP system, the size of the ST and the needed auxiliary heating system to meet the hot water demands, as well as the interactions with the local electric company. The optimization formulation accounts for the minimization of the TAC of the system as well as the simultaneous minimization of the associated GHGE accounting for the involved uncertainty of the system.

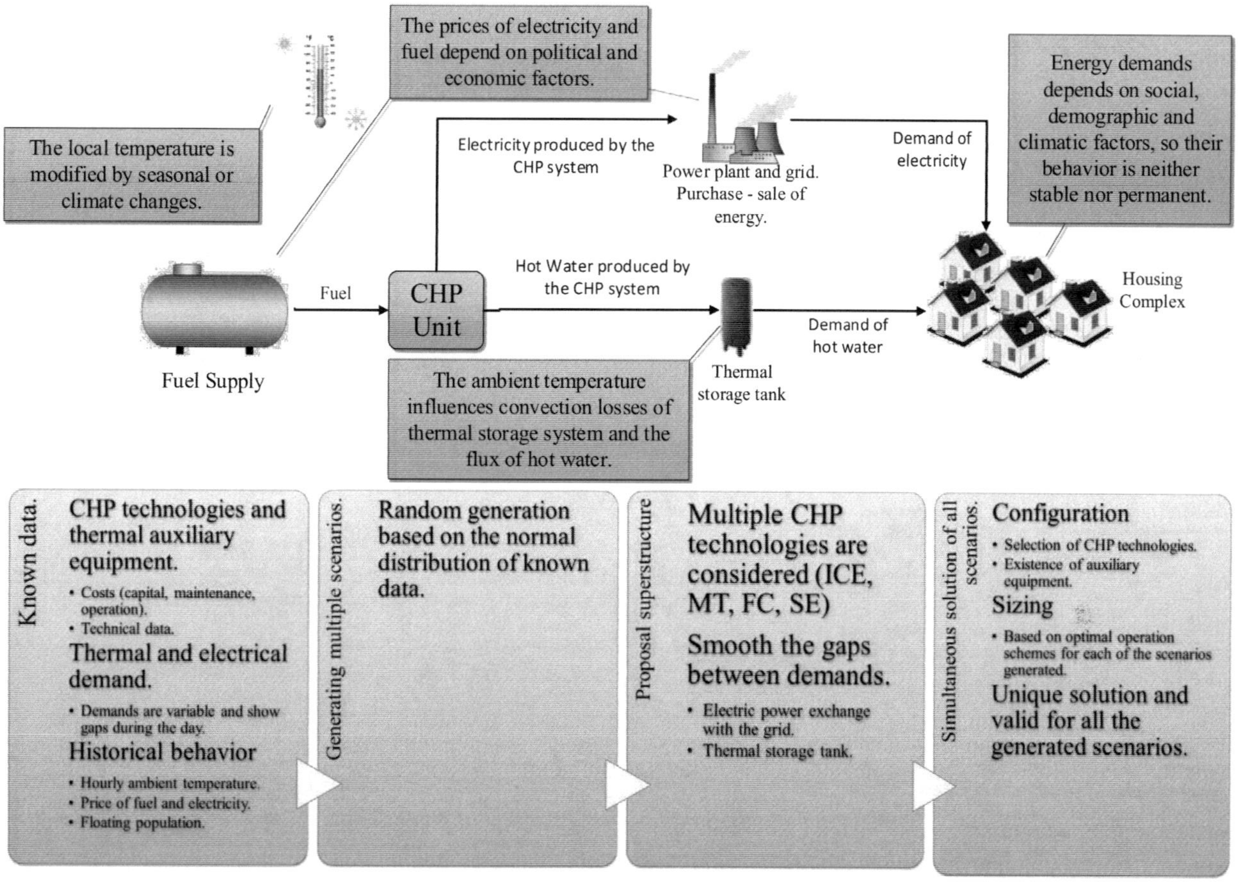

Figure 1. Problem Statement

Model Formulation

The proposed model formulation is a mixed-integer nonlinear programming problem (MINLP), based on the proposed superstructure (**Figure 2**). The mentioned uncertain parameters are considered; the uncertain scenarios are generated in a random way considering the normal distribution of the historical data. The objective functions that determine the optimal design of the system are, in economic terms, to minimize of the expected TAC for all scenarios, and in environmental terms, to minimize the mean direct GHGE associated with the consumption of fuels to meet the energy demands of the housing complex, which are described as follows. The total annual cost for each scenario is determined by the sum of the annualized capital costs, purchasing power from the grid of the local electric company, operating costs and maintenance costs, minus the incomes generated by the sale of electricity and heat. In this case, each scenario generates different operating costs, and this way the TAC is different for each scenario, even that the capital costs are the same.

$$TAC_{sc} = CostCap^{T} + CostOp_{sc}^{T} + CostOM_{sc}^{T} + CostPower_{sc}^{GRID} - PowerSale_{sc}^{T} - HeatSale_{sc}^{H}, \ \forall sc \in SC$$

Then, the objective function to minimize is the expected for all the analyzed uncertain scenarios generated stochastically.

$$\overline{TAC} = \sum_{sc=1}^{n} TAC_{sc} \cdot \left(\frac{1}{n}\right)$$

Moreover, the worst economic scenario is defined by the worst TAC for all the analyzed scenarios (WTAC). This represents the greatest TAC for all scenarios.

$$WTAC \geq TAC_{sc}, \quad \forall sc \in SC$$

The greenhouse gas emissions generated for each scenario are calculated based on the generation of CO_2 of each of the elements of the proposed system.

$$GHGE_{sc}^{T} = GHGE_{sc}^{\mathrm{CHP}} + GHGE_{sc}^{GRID} + GHGE_{sc}^{B}, \quad \forall sc \in SC$$

Then, the average of the GHGE for the generated scenarios is determined as follows:

$$\overline{GHGE} = \sum_{sc=1}^{n} GHGE_{sc}^{T}\left(\frac{1}{n}\right)$$

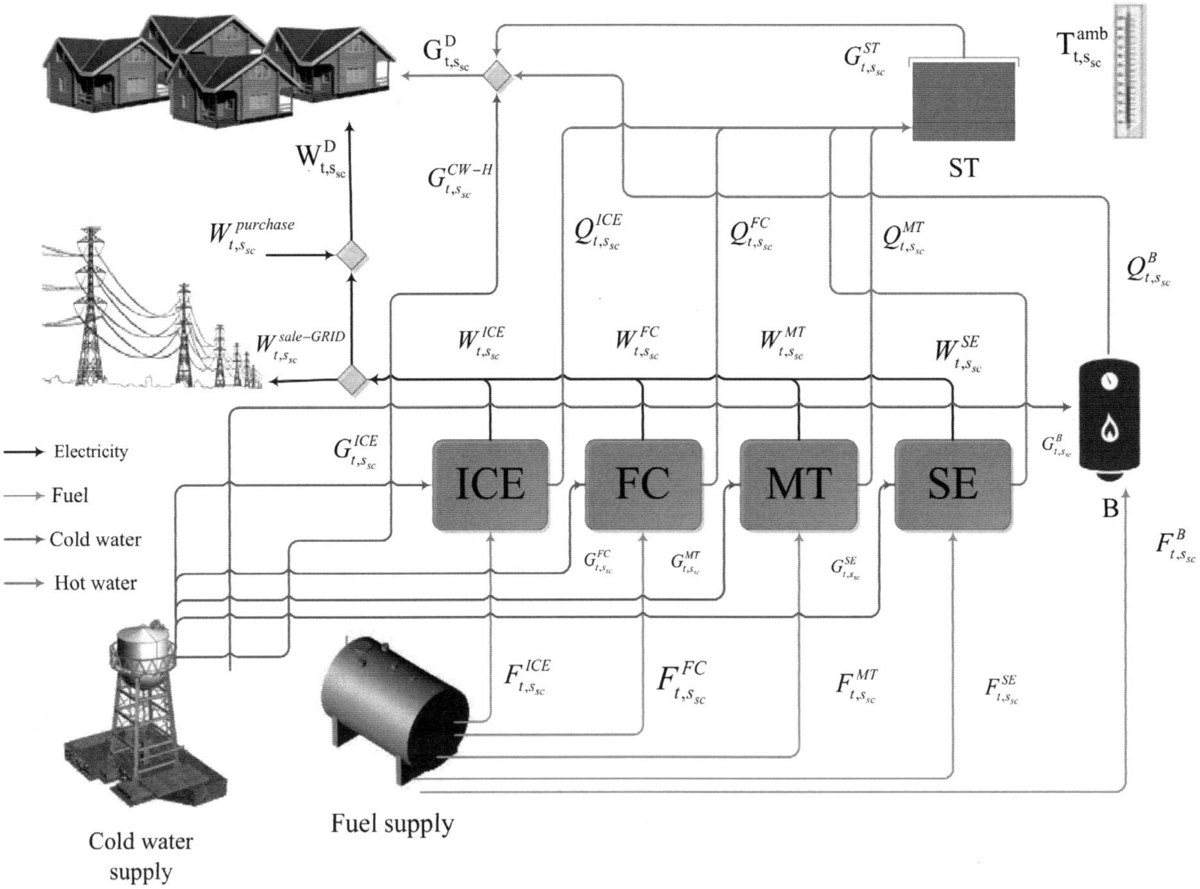

Figure 2. Proposal Structure.

The worst scenario for the greenhouse gas emissions is determined as follows:

$$WGHGE \geq GHGE_{sc}^{T}, \quad \forall sc \in SC$$

The electricity demanded by the housing complex is satisfied by the sum of the energy purchased to the grid and energy produced by the CHP system considering each technology.

$$W^{D}_{t,s_{sc}} = W^{purchase}_{t,s_{sc}} + W^{ICE-H}_{t,s_{sc}} + W^{MT-H}_{t,s_{sc}} + W^{FC-H}_{t,s_{sc}} + W^{SE-H}_{t,s_{sc}}, \forall t \in T, \forall s \in S, \forall sc \in SC$$

The electricity produced by each CHP unit can be distributed according to the needs as energy for domestic consumption, the one sent to the housing complex, and the energy sold to the grid as surplus production.

$$W^{\text{CHP-Technology}}_{t,s_{sc}} = W^{\text{CHP-Technology}-H}_{t,s_{sc}} + W^{\text{CHP-Technology}-sale}_{t,s_{sc}}, \forall t \in T, \forall s \in S, \forall sc \in SC, \forall \text{CHP-Technology}$$

A design problem in CHP systems is the one associated with synchronizing the operation of the system with the energy demands; this way, an appropriate thermal storage device is needed and in this case a ST is used. The hot water stored in the tank over the time period *t*, over the season *s* and for scenario *sc* is the result of the addition of hot water stored over the time period *t – 1* plus the water sent to the ST from the CHP system minus the water sent to meet the demand of the housing complex. The convective loss through the ST wall is included. The water temperature inside the ST is a decision variable.

$$S^{water}_{t,s_{sc}} Cp^{W\text{-}ST}_{t,s_{sc}} T^{ST}_{t,s_{sc}} = S^{water}_{t-1,s_{sc}} Cp^{W\text{-}ST}_{t-1,s_{sc}} T^{ST}_{t-1} + G^{ICE}_{t,s_{sc}} Cp^{ICE}_{t,s_{sc}} T^{ICE}_{t,s_{sc}} + G^{MT}_{t,s_{sc}} Cp^{MT}_{t,s_{sc}} T^{MT}_{t,s_{sc}} + G^{FC}_{t,s_{sc}} Cp^{FC}_{t,s_{sc}} T^{FC}_{t,s_{sc}} + G^{SE}_{t,s_{sc}} Cp^{SE}_{t,s_{sc}} T^{SE}_{t,s_{sc}}$$
$$-G^{ST}_{t,s_{sc}} Cp^{ST}_{t,s_{sc}} T^{ST}_{t,s_{sc}} - Q^{loss}_{t,s_{sc}}, \forall t \in T, \forall s \in S, \forall sc \in SC$$

The parameters that define the operation of the CHP system are the thermal and electrical efficiencies. Both represent the relationship between the products, either electricity and heat, and fuel required to produce them. A parameter, which becomes a decision variable, is the partial load. None unit can operate below a certain value of partial load because of their design conditions. In these equations, *y* is a binary variable used to determine the existence of the CHP technology in the optimal solution.

$$\eta^{W^{\text{CHP-Technology}}} = \frac{W^{\text{CHP-Technology}}_{t,s_{sc}}}{F^{\text{CHP-Technology}}_{t,s_{sc}}}, \forall t \in T, \forall s \in S, \forall sc \in SC, \forall \text{CHP-Technology}$$

$$\eta^{Q^{\text{CHP-Technology}}} = \frac{Q^{\text{CHP-Technology}}_{t,s_{sc}}}{F^{\text{CHP-Technology}}_{t,s_{sc}}}, \forall t \in T, \forall s \in S, \forall sc \in SC, \forall \text{CHP-Technology}$$

$$PL^{\text{CHP-Technology}}_{t,s_{sc}} = \frac{W^{\text{CHP-Technology}}_{t,s_{sc}}}{W^{MAX-\text{CHP-Technology}}}, \forall t \in T, \forall s \in S, \forall sc \in SC, \forall \text{CHP-Technology}$$

The dimensioning of the CHP units is determined by the highest capacity available in the market and the highest load required for operating the system during any time period for each season and for all the uncertain scenarios generated.

$$W^{MAX-\text{CHP-Technology}} \leq W^{UB-\text{CHP-Technology}} y^{\text{CHP-Technology}}, \forall \text{CHP-Technology}$$

$$W^{MAX-\text{CHP-Technology}} \geq W^{\text{CHP-Technology}}_{t,s_{sc}}, \forall \text{CHP-Technology}$$

Case study

As case study is presented a residential zone from Mexico, which consists of 1,440 homes (the overage demands are shown in **Figure 3**). The considered technical and operating parameters in the case study are presented in **Table 1**.

Table 1. Technical and economic parameters for the considered CHP technologies and auxiliary equipment.

Parameter	RICE	MTG	FC	SE	Boiler	ST
Electrical efficiency (η^W - %)	37.25	26	38	30		
Thermal efficiency (η^Q - %)	47.5	47.5	50	60	70	
Maximum size (W^{UB}, G^{UB} and S^{UB} – kW, l/h and m^3)	15,800	12,500	2,000	1,500	15,000	200
Minimum partial load (PL^{MIN} - %)	35	60	50	42.5		
Unitary fuel cost (UFC - \$/kWh)	0.02	0.02	0.02	0.02	0.04	
Unitary cost of cold water (UCCW - \$/l)	0.001	0.001	0.001	0.001	0.001	
Fixed cost (FC - \$)	100	100	100	100	100	120
Variable cost (VC - \$)	400	450	2500	450	70	125
Scale factor (β)	1	1	1	1	1	1
Annualization factor (k_F)	0.23	0.23	0.23	0.23	0.23	0.23
Maintenance cost (UCOM - \$/kWh)	0.015	0.065	0.015	0.015	0.156	
Outlet temperature (°C)	90	90	90	90	70	

Figure 3. Case Study

Energy demand for the housing complex

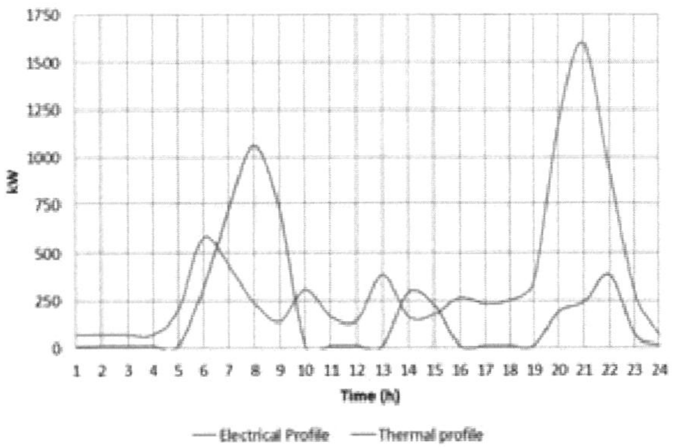

Seasonal environmental hourly temperature

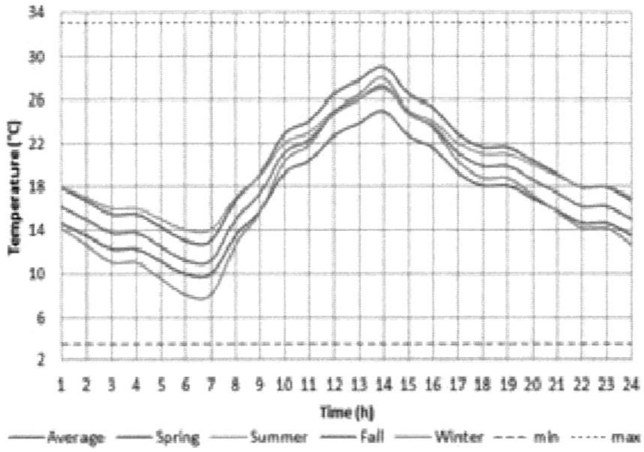

Results

In the economic analysis, the worst scenario always remains below the mean case for the same probability. For the analysis of emissions, there is a small area of risk, where the worst case is greater than the mean value for the same probability (see **Figure 4**). However, this area is minimal compared with respect to the rest of the area generated below the crossing, where the mean TAC is above the worst TAC for the same probability. The Pareto front (**Figure 5**) shows a fixed WTAC, while other values change due to the operating conditions. In this case, Solutions 2, 3 and 4 (see **Table 2**) remain with a fixed technological structure and capital costs. The predominant technology from an economic standpoint is the ICE. The analysis for the emissions includes the FC and SE technologies (see **Figure 5** and **Table 3**). The last solutions share technological configuration but have different operating schemes. A comparative analysis of emissions between the two groups of solutions shows that the differences between the two groups in that area is not as substantial as the economic side. Therefore, an appropriate solution could be Solution 1, which has low emissions and a cost considerably lower than those presented in this group.

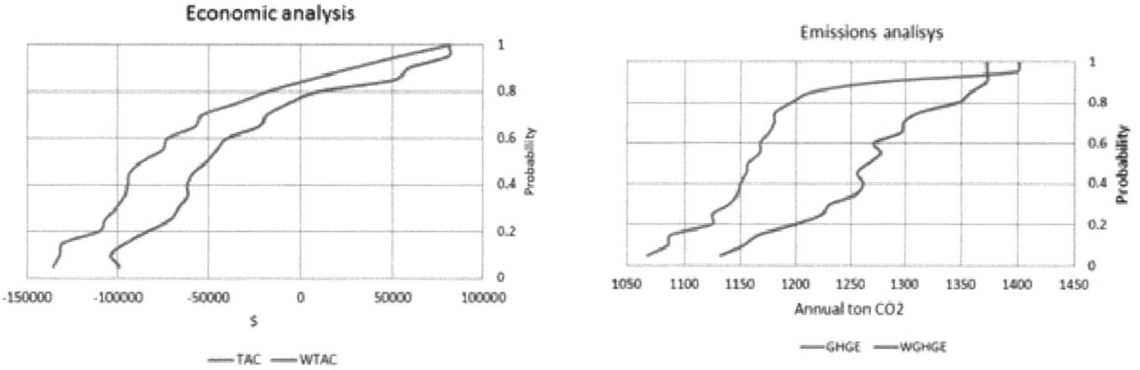

Figure 4. Sensitivity Analysis

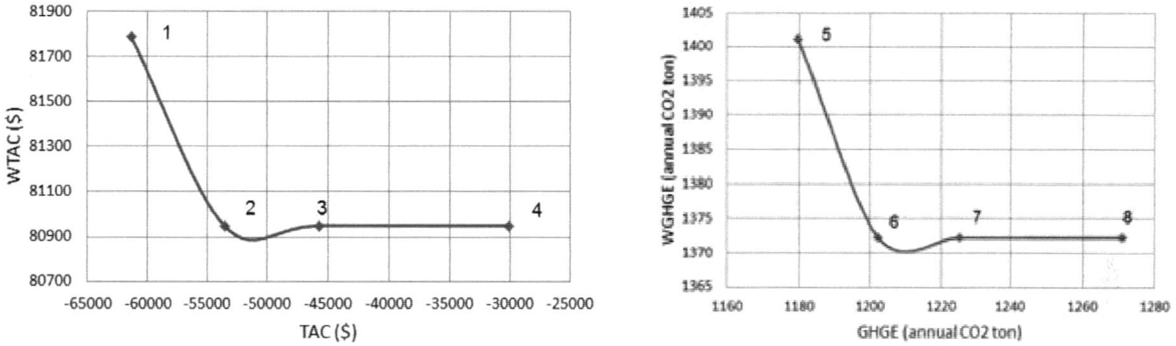

Figure 5. Pareto Fronts

Table 2. Solutions for Pareto front $\overline{TAC}$ - *WTAC*.

	Solution 1	Solution 2	Solution 3	Solution 4
$\overline{TAC}$ ($)	- 61,380	-53,560	-45,740	- 30,100
WTAC ($)	81,788	80,948	80,948	80,948
$\overline{GHGE}$ (ton of CO_2)	1,314	1,351	1,412	1,451
WGHGE (ton of CO_2)	1,577	1,612	1,702	1,740
CHP technologies	ICE	ICE	ICE	ICE
Sizing of CHP technologies (kW_e)	124.5	107.9	107.9	107.9
Sizing ST (m^3)	21.7	22.5	22.5	22.5
Capital cost (annual $)	12,127	10,625	10,625	10,625

Table 3. Solutions for Pareto front $\overline{GHGE}$ - *WGHGE*.

	Solution 5	Solution 6	Solution 7	Solution 8
$\overline{TAC}$ ($)	211,540	223,610	231,780	239,330
WTAC ($)	342,060	347,120	356,460	366,430
$\overline{GHGE}$ (ton of CO_2)	1,179	1,202	1,225	1,270
WGHGE (ton of CO_2)	1,401	1,372	1,372	1,372
CHP Technologies	FC SE	FC SE	FC SE	FC SE
Sizing of CHP technologies (kW_e)	3.9 65.3	17.4 65.3	17.4 65.3	17.4 65.3
Sizing ST (m^3)	23.2	22.4	22.4	22.4
Sizing of thermal auxiliary equipment (l/h)	15,000	15,000	15,000	15,000
Capital cost (annual $)	251,300	259,010	259,010	259,010

Conclusions

This paper has presented a superstructure for the optimal design of domestic CHP systems involving the optimal selection of multiple technologies to satisfy electricity and hot water demands under the uncertainty associated to the ambient temperature, energy market conditions and energy demands considering demographic factors. Based on this superstructure, a mathematical programming formulation has been developed for the simultaneous minimization of the total annual cost and the associated greenhouse gas emissions. A proper solution approach based on the epsilon constraint method has been implemented to show compensated solutions that trade off the economic and environmental objectives. The proposed model also considers the possibilities of implementing auxiliary thermal systems, the sizing of equipment and thermal storage system. The scenarios used to model the uncertainty generated randomly considering a normal distribution based on historical data of temperature, floating population, fuel prices and purchase of electric energy. Simultaneously, the worst scenarios for the objective functions are minimized to generate a second set of target functions. The results show low economic and environmental vulnerability for the proposed CHP systems, which is designed to meet the demand of the case study, and involving changes in weather, economic and/or demographic conditions.

References

1. Maghanki M.M., Ghobadian B., Najafi G. and Galogah R.J. (2013) Micro combined heat and power (MCHP) technologies and applications. Renewable and Sustainable Energy Reviews 28 510–524.
2. Peacock A.D. and Newborough M. (2005) Impact of micro-CHP systems on domestic sector CO2 emissions. Applied Thermal Engineering 25 2653–2676.
3. De Paepe M., D'Herdt P. and Mertens D. (2006) Micro-CHP systems for residential applications. Energy Conversion and Management 47 3435–3446.
4. Tchanche B.F., Pétrissans M. and Papadakis G. (2014) Heat resources and organic Rankine cycle machines. Renewable and Sustainable Energy Reviews 39 1185–1199.
5. Lazos D., Sproul A.B. and Kay M. (2014) Optimisation of energy management in commercial buildings with weather forecasting inputs: A review. Renewable and Sustainable Energy Reviews 39 587–603.
6. Fuentes-Cortés L.F., Ávila-Hernández A., Serna-González M., Ponce-Ortega J.M. (2015) Optimal design of CHP systems for housing complexes involving weather and electric market variations. Applied Thermal Engineering 90 895-906.
7. Fuentes-Cortés L.F., Ponce-Ortega J.M., Nápoles-Rivera F., Serna-González M., El-Halwagi M.M. (2015) Optimal design of integrated CHP systems for housing complexes. Energy Conversion and Management 99 252-263.

435598 The Use of Dimethyl Carbonate for Etherification

Wednesday, November 11, 2015: 9:20 AM
258 (Salt Palace Convention Center)

Rodrigo Cella, Chemical Engineering, Centro Universitário da FEI, São Bernado do Campo, Brazil and Alex S. T. Z. Tabu, Chemical Engineering, Centro Universitário da FEI, São Bernardo do Campo, Brazil

Introduction

In this work, we present the use of dimethyl carbonate (DMC) as an *O*-alkylating agent to prepare ethers from the reaction of alcohol and DMC (Sheme 1).[1] DMC is considered a green alternative for alkylation's reactions, once the final sub products are carbon dioxide and methanol, which can be latter, recovered to DMC.[2]

Scheme 1

Usually, those reactions undergo over high temperature and high pressure, to overcome these hard conditions we have used ultrasound as energy source. Ultrasound has been utilized recently to accelerate a number of synthetically useful reactions.[3],[4],[5],[6] The use of ultrasound in chemistry is called sonochemistry and the ultrasound effects observed on organic reactions are due to cavitation, a physical process that create, enlarge, and implode gaseous and vaporous cavities in an irradiated liquid. Cavitation induces very high local temperatures and pressure inside the bubbles (cavities), leading to a turbulent flow in the liquid and enhanced mass transfer.

Methodology

In a conical flask phenol (5 mmol), DMC (50 mmol), base and additive (see table 1) were added in quickly agitated for homogenization. After 1 minute, an ultrasound horn was introduced into the flask and ultrasound wave was irradiated for 2 hours in the reaction medium (Amplitude= 40; Pulse-on=15 seconds and pulse-off= 10 seconds). The reaction was monitored by GC-MS.

Results and Discussion

The *O*-alkylation reaction between phenol and dimethyl carbonate was chosen as the model reaction and a variety of conditions were screened (Table 1). Inorganic base such as KOH, *t*-BuOK and K_2CO_3 were tested in the solvent dimethylformamide (DMF) (Table 1, entries 6-8), however no products were observed. Potassium carbonate was also tested in the presence of phase transfer catalyst such as tetrabutylammonium chloride (TBACl) and tetrabutylammonium bromide (TBABr) (Table 1, entries 9 and10) and the product was observed in 8 and 15% GC yield, respectively.

Table 1: Screening condition of the *O*-alkylation of phenol using DMC and ultrasound:

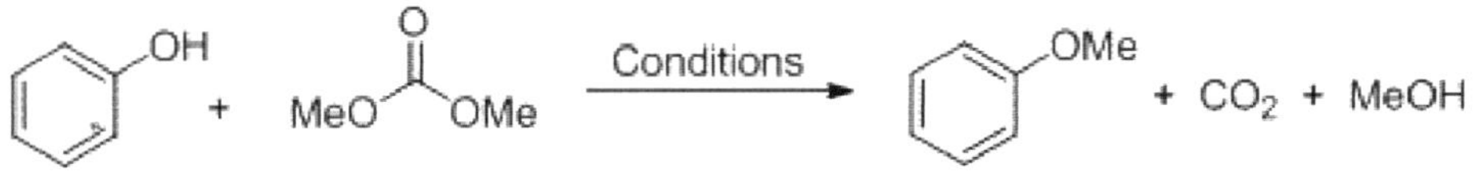

Entry	Phenol (mmol)	DMC (mmol)	Base (mmol)	Additive (mmol)	Solvent (ml)	GC Yield (%)
1	5	60	DBU (0.5)	-	-	ND
2	5	10	DBU (0.5)	-	DMF (5)	9
3	5	60	DABCO (5)	-	-	46
4	5	10	DABCO (5)	-	DMF (5)	ND
5	5	10	TMEDA (5)	-	DMF (5)	ND
6	5	60	KOH (5)	-	-	ND
7	5	10	K_2CO_3 (5)	-	DMF (5)	ND
8	5	10	*t*-BuOK (5)	-	DMF (5)	ND
9	5	10	K_2CO_3 (5)	TBABr (1)	DMF (3)	8
10	5	10	K_2CO_3 (5)	TBACl (1)	DMF (3)	15
11	5	10	*t*-BuOK (5)	TBACl (1)	DMF (3)	ND

Better results were observed when organic base were used (Table 1, entries 1-5). We have tested 1,8-Diazabicyclo[5.4.0]undec-7-ene (DBU), 1,4-diazabicyclo[2.2.2]octane (DABCO) and tetramethylethylenediamine (TMEDA). The best condition was using DMC as the solvent and DABCO in 1 equimolar quantity, the GC yield achieve was 46% (Table 1, entry 3).

Conclusion

In summary, we have showed our initial results in the *O*-alkylation of phenol using DMC as an alkylating agent using sonochemistry as a new tool to overcome hard conditions. Despite the fact that the yields are still low, the reaction is showing potential. The screening for better conditions is ongoing in our lab.

Acknowledgments

The authors are grateful to FAPESP (Grant 2013/11822-1 and Fellowship 2014/21897-1) for financial support.

[1] TUNDO, P.; ROSSI, L.; LORIS, A. *J. Org. Chem.*, v. 70, p. 2219, **2005**.

[2] DELLEDONNE, D.; RIVETTI, F.; ROMANO, U. *J. Organomet. Chem. v. 488*, p. C15, **1995**.

[3] MARGULIS, M. A. *High Energ. Chem.*, v. *38*, p. 135, **2004**.

[4] Mason, T. J. *Chem. Soc. Rev.*, v. *26*, p. 443, **1997**.

[5] Cravotto, G.; Cintas, P. *Chem. Soc. Rev.*, v. *35*, p. 180, **2006**.

[6] Suslick, K. S. *Science*, v. *247*, p.1439, **1990**.

435424 Comparative Life Cycle Analysis of Different Light Bulbs Used for Street Lighting

Abstract:

Tremendous technological and sustainable development has forced the society to adopt modern and more energy efficient lighting devices instead of old fashioned less efficient incandescent lamps. Recently, incandescent lamps has been replaced with compact fluorescent lamps (CFLs) and Light Emitting Diode (LED) lamps. In comparison to incandescent lamps, these devices are so energy efficient that they provide similar light output at the expense of only $1/5^{th}$ part of electricity consumption. This is due to the fact that less energy is lost in the form of heat during luminance phenomenon.

The ecological footprint evaluation for lighting system network in Veszprem county, Hungary has been carried out utilising Sustainable Process Index (SPI) methodology (Sandholzer et. al, 2005; Narodoslawsky and Krotscheck, 1995). It is an aggregate method resulting in an ecological footprint, calculating the area required to embed the whole life cycle in an ecosphere to sustainably provide products or service. The analysis is carried out considering three different light bulbs i.e. conventional or old fashioned less energy efficient incandescent lamps and high tech more energy efficient CFLs and LED lamps. The analysis results reveals that there is a potential to decrease environmental impact by 2 to 4 times by changing lamps from conventional incandescent to CFL and LEDs.

436643 One Year Operation of a Salinity Gradient Solar Pond in Northern Cyprus

Abstract:

One Year Operation of a Salinity Gradient Solar Pond in Northern Cyprus- Experimental Investigations and CFD Simulation

[1]Soudabeh Gorjinezhad, [1]Sultan Kadyrov, [2]Mohammad Askari, [1]Negar Zare Pakzad, [1]Mehdi Amouei Torkmahalleh, [3]Goodarz Ahmadi

1- *Chemical Engineering Program, Middle East Technical University Northern Cyprus Campus, Guzelyurt, Mersin 10, Turkey*

2- *Mechanical Engineering Program, Middle East Technical University Northern Cyprus Campus, Guzelyurt, Mersin 10, Turkey*

3- *Mechanical Engineering Department, Clarkson University, Potsdam, NY, 13699- 5725*

1. Introduction

Today, renewable energy sources gain importance day by day. It is crucial to develop devices and processes to supply energy from non-polluting and renewable energy sources for sustainable development of the world. Solar pond is an example of such devices that basically collects solar energy and stores it as thermal energy for a long period. Studies indicate that the temperature in solar ponds may in general reach up to 70-80°C implying that the thermal energy from solar ponds can be useful to applications with low grade energy demand [1]. One of the most important advantages of solar pond over other renewable energy sources, such as solar collectors, is its lower cost of investment [1]. Solar pond is environmentally friendly in particular when it is used for electricity generation. The heat obtained from solar pond can be converted into electric power even at low temperatures [2]. In this regard, organic Rankine cycle engines are generally run using the temperature difference in a solar pond. For applications where the organic fluid fails to operate due to low temperature difference, thermoelectric generators can be a good candidate to replace organic Rankine cycle engines for power generation [3].

Solar Ponds normally consist of three different salinity layers. The first layer, known as the upper convective zone (UCZ), is located at the top of the pond, and contains the least salinity level. The second layer, whose salinity level increases with depth, is called non-convective zone (NCZ). This layer is responsible to act as an insulator to prevent heat from escaping to the UCZ, maintaining higher temperature at deeper zones. The last layer made of a saturated salt solution, is responsible for energy storage, and is known as the lower convective zone (LCZ) [5]. The working principles of solar ponds are quite simple. In the absence of salinity gradient, the solar radiations reaching a pond are mostly absorbed by the lower water levels and will cause the water to heat up. The heated water is then risen up to the pond surface due to buoyancy effect and loses its thermal energy to the atmosphere. Thus, the main reason behind imposing salinity gradient to solar ponds is to create a density gradient restricting such buoyancy driven natural convection and as a result to trap the thermal energy at the bottom of the pond [4].

Northern Cyprus is enriched in solar energy in particular during summer. The performance of a solar pond in this region has not been evaluated yet. To take advantage of solar energy in Northern Cyprus, a salinity gradient solar pond has been constructed and operated at Middle East Technical University Northern Cyprus Campus (METU NCC) located at Guzelyurt, Northern Cyprus (***Figure 1***). This study is primarily concerned with describing the development, validation and use of CFD for computer modeling of flow and thermal analysis of the experimental salinity gradient solar pond at METU NCC. The pond has started operating since Oct. 2014 and experimental results are being recorded accordingly. The CFD commercial software considered to be utilized is ANSYS FLUENT and the challenge is to visualize the transient effect of temperature distribution, salinity concentration, velocity field and the degree of influence of wind speed on the overall stability and performance of the solar pond over time.

Figure SEQ Figure * ARABIC 1. Constructed Solar Pond at Middle East Technical University (Northern Cyprus)

2. Materials and Experiment

A cylindrical salinity gradient solar pond of 61 cm diameter, 55 cm height and 1.2 cm thickness installed and operated since October 8th, 2014 at Middle East Technical University Northern Cyprus Campus (METUNCC) located at Guzelyurt, Northern Cyprus. Three inlet ports were installed to carefully add the solutions to the pond to create the three salinity layers. Also, these inlets were used to add proper amount of solutions to each zone during the experiments to compensate for evaporation and sampling losses as well as surface washing. The pond has been divided in three zones. The bottom zone (LCZ) consisted of 75 liters of saturated salt solution (concentration C). The middle zone (NCZ) consisted of three equally divided sub-layers, each 15 liters in volume, with 0.75C, 0.5C and 0.25C concentrations. The top zone (UCZ) consisted of 15 liters of fresh water. The rest of the pond was left empty. The pond was equipped with three sampling valves to withdraw samples from each layer to monitor the salt concentrations using conductivity measurements. Six thermometers (2 in each zone) were installed to monitor the temperature variations at 9 a.m., 1 p.m., 5 p.m., and 10 p.m. every day. Temperature recording at other time of the day was performed as needed. The bottom and sides of the pond were insulated with a 2.5cm thick thermal insulator. The inner surfaces of the pond were painted black. Excess (undissolved) salt was added to the bottom of the pond, to ensure saturation of the bottom layer. The ambient temperature and solar radiation were monitored using a sun tracker (Kipp and Zonen) placed next to the solar pond. Wind velocity and relative humidity were also monitored throughout the experiment.

3. Computational Scheme and Modeling

The geometry and meshing of the solar pond were constructed by ANSYS Workbench Design Modeler and Meshing environment. FLUENT version 14.5 was used as the CFD solver. A two-dimensional computational domain is considered for investigating the transient behavior of the pond representing a cross-section of the solar pond.

Accordingly, the rectangular domain of consideration is that of the existing pond dimensions with 61 (cm) in width and 50 (cm) in height. The lower wall of the rectangle is assumed to be the ground and upper boundary to be the top water level open to atmosphere. Vertical walls and the bottom surface of the pond are set to be impermeable and thermally insulated with the corresponding thermal conductivity value of the insulation material. Also the top surface, due to contact with ambient air, is taken to be at ambient temperature with an average wind speed according to local meteorological data. The salinity concentration of the LCZ is kept at saturated level, while the top surface is set to zero concentration, initially.

The modeling of the pond is divided into two phases. Firstly, a simple model is going to be developed in which it is assumed that the concentration level remains the same due to the present regular injection of brine solution to the system. Accordingly, the thermofluid properties will be assumed to be primarily a function of temperature and will be evaluated for each layer at its initial concentration. Thus, in the first phase, the concentration will not be considered as a modeling parameter and it is aimed to primarily study the variation of the temperature using the most simplified model possible. In the second phase, the model will be enhanced by defining the concentration as a varying parameter to the system. In this regard, the variation of properties with concentration can be taken into account and the significance of salt diffusion and its influence on the stability of the pond can be investigated. In addition, the results between both cases and the experimental results will be compared to justify the computational complexity added to the model.

4. Results and Discussion

Figure 2 shows the average monthly solar pond temperature variations from October 2014 to February 2015) as well as the average monthly ambient temperature and irradiance. As can be seen, the pond has been functioning as the temperature differences among the three layers have been established. The pond temperature variations are well correlated with the ambient temperature variations. The temperature at all layers decreased from October 2014 to end of December 2015, but then beginning January 2015, it increased as solar irradiance increased.

Figure 1. Average monthly temperature variations of the solar pond

References

[1.] Sukhatme, S., & Nayak, J. (2008). *Solar energy: Principles of thermal collection and storage* (3rd ed.). New Delhi: Tata McGraw-Hill.

[2.] Akbarzadeh, A., Bernad, F., Casas, S., Gibert, O., Cortina, J. L., & Valderrama, C. (2013). Salinity gradient solar pond: Validation and simulation model. *Solar Energy*, *98*, 366-374.

[3.] Singh, B., Gomes, J., Tan, L., Date, A., & Akbarzadeh, A. (2012). Small Scale Power Generation using Low Grade Heat from Solar Pond. *Procedia Engineering,* 50-56

[4.] Mehdizadeh, M., & Ahmadi, G. (2014). Two-dimensional Computer Simulation of Salinity Gradient Solar Pond Operation. *Proceedings of ASME 2014 Fluids Engineering Summer Meeting*, *FEDSM2014*.

[5.] Jaefarzadeh, M. (2005). Thermal behavior of a large salinity-gradient solar pond in the city of mashhad. *Salinity Gradient Solar Pond*. Retrieved October 14, 2014, from http://profdoc.um.ac.ir/articles/a/203017.pdf

424818 Evaluation of Electric Vehicle Energy Consumption for Traveling and Air-Conditioning

Abstract:

“Smart Photovoltaic power generation (hereinafter referred to as PV) and Electric Vehicle (hereinafter referred to as EV) combined system" has been proposed and the effect has been clarified. In order to promote “SMART PV and EV combined system”, it’s necessary to spread the EV. In order to spread the EV, it’s necessary to solve some problems which are high energy consumption of car air-conditioner, high cost of storage battery and risk aversion of a dead battery.

In order to use effectively utilize an EV battery, it is necessary to evaluate the remaining battery capacity and predicting the power consumption. Evaluation method of the remaining battery capacity with high accuracy has been proposed. Moreover, the prediction method of the power consumption for traveling has been proposed. However, the power consumption for a vehicle-mounted air-conditioner ocuupy 30% of all energy consumption of a vehicle. Therefore, it is crucial to evaluate the power consumption for both traveling and air-conditioning, in order to acculately predict the really usable power that is left in the battery.

In this report, the authors have constructed a theoretical model which calculates the EV power consumption for both traveling and air-conditioning. The calculated results of this model have been compared to the actual traveling datas that are obtained by the experiment of a converted type EV and i-MiEV which is made by Mitsubishi Motors Corporation. In addition, the factors which have a impact on the EV power consumption have been studied.

As a result, the EV power consumption is greately varied by acceleration resistance, slope resistance, and the required air-conditioning load. Moreover, it's clarified that the air-conditioner consumes about 25 to 50 % of the total power consumption in a hot summer day. Additionally, the acceleration and the air-conditioning load differ depending on each vehicle driver. Therefore, in order to acculately predict the EV power consumption practically, it's necessary to reflect the characteristics of each vehicle driver based on the theoretical model.

416663 Dry Reforming of Methane on Ni in a Fixed-Bed Reactor: Spatial Reactor Profiles and Detailed CFD Simulations

Abstract:

Dry Reforming of Methane on Ni in a Fixed-Bed Reactor: Spatial Reactor Profiles and Detailed CFD Simulations

Gregor D. Wehinger[1], Matthias Kraume[1], Viktor Berg[2], Katharina Mette[3], Malte Behrens[4], Robert Schlögl[3], Oliver Korup[2], Raimund Horn[2]

[1] Chemical and Process Engineering, Technische Universität Berlin, Fraunhoferstr. 33-36, 10587 Berlin, Germany

[2] Institute of Chemical Reaction Engineering, Hamburg University of Technology, Eißendorfer Str. 38, 21073 Hamburg, Germany

[3] Department of Inorganic Chemistry, Fritz Haber Insitute of the Max Planck Society, Faradayweg 4-6, 14195 Berlin, Germany

[4] Faculty of Chemistry, Inorganic Chemistry, University of Duisburg-Essen, Universitätsstr. 7, 45141 Essen, Germany

Fixed-bed reactors are the workhorses of industrial catalysis. They are characterized by a complex interplay between physical transport processes and chemical reactions.

Current design of fixed-bed reactors is based on pseudo-homogeneous or heterogeneous reactor models in which the structure of the solid phase is not explicitly accounted for. This approach leads to unsatisfactory results in particular for small tube-to-particle-diameter ratios.

The discrete element method[1] allows generating a random packing of catalyst particles in a single tube. Industrial multi-tubular reactors contain thousands of tubes, hence this randomly filled tube can be considered representative. For the geometry of the packing, it is possible to simulate the flow field and the transport processes of mass and heat between the bulk fluid and each individual catalyst particle. In this way the usage of lumped transport correlations can be avoided. The catalytic chemistry is implemented by detailed microkinetics.

In the present contribution we demonstrate this simulation approach for dry reforming of methane on nickel. To validate the simulations spatial reactor profiles of species concentrations and temperature were measured in a dedicated reactor[2]. The nickel catalyst is applied as a washcoat of a Ni, Mg, Al hydrotalcite-like precursor[3] on α-Al_2O_3 spheres of 1 mm diameter.

The conservation equations of total mass, momentum, energy, and the species balances were solved in 3D assuming laminar flow. The local temperature of the catalyst particles was set to the experimental value and the particles were considered to be isothermal. Further details on the simulation procedure can be found in Ref. [1].

For heat transfer only, simulation and experiment are in excellent agreement. If chemistry is included, the experimental species profiles are qualitatively reproduced by the model, but some quantitative deviations occur. This can be due to catalyst deactivation during the measurements or to inaccuracies in the microkinetics. The presented approach gives more insight into catalytic fixed-bed reactors than conventional reactor simulations treating the phases as a continuum.

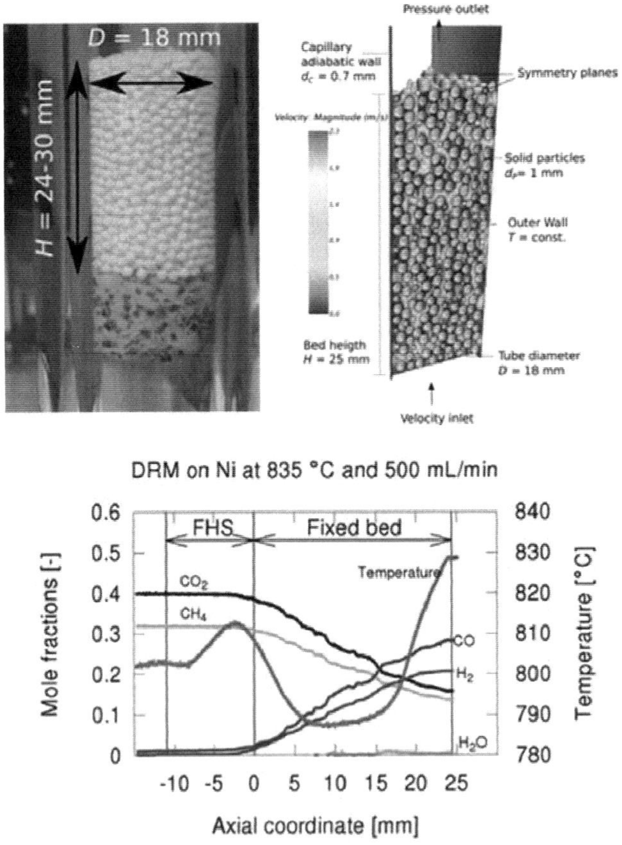

Fig. 1: Top left: Reactor tube with catalyst packing. Top right: Velocity field in the bed. Bottom: Experimental species and temperature profiles.

[1] G. D. Wehinger, T. Eppinger, M. Kraume *Chem. Eng. Sci.* 122 (2015) 197.

[2] R. Horn, O. Korup, M. Geske, et al., *Rev. Sci. Inst.* 81 (2010) 064102.

[3] K. Mette, S. Kühl, H. Düdder, K. Kähler, A. Tarasov, M. Muhler, M. Behrens *ChemCatChem* 6 (2014) 100.

423534 Nickel Supported on Ceria-Zirconia for Syngas Production Via Dry Reforming of Methane

Abstract:

1.Introduction

Dry reforming of methane (DRM) produces synthesis gas with higher purity and lower H_2/CO ratio than either partial oxidation or steam reforming. Lower H_2/CO ratio is a preferable feedstock for the Fischer-Tropsch synthesis of long-chain hydrocarbons [1,2]. Another important advantage of this method is the use of harmful for the environment greenhouse gases (CO_2 and CH_4) to produce hydrogen and syngas [1-3].

The typical catalysts for DRM are nickel-containing materials and noble metals (Pt, Rh and Ru. Ni-based catalysts supported on Al_2O_3, MgO and hydrotalcite are used more often in comparison with noble metals due to low price and satisfactory activity. However, such nickel-based catalysts exhibit certain disadvantages such as formation of carbon deposits and sintering at high temperatures. The use of oxide supports e.g. CeO_2, CeO_2-ZrO_2, which contain mobile lattice oxygen active in removing carbon deposits and in regeneration of active sites, can improve the lifetime of Ni-based catalysts [3].

Temperature is another important factor that affects the activity and stability of Ni-based catalysts. Since DRM is an endothermic process, higher temperature improves the conversion of both CH_4 and CO_2, but can also cause deactivation of the catalyst due to the formation of carbon deposits. Hence, it is favourable to carry out the DRM process at low temperature (≤550 °C), which also lowers the cost of energy [3]. Until now, only few papers concerning low-temperature DRM have been published [3].

The goal of this work is to study the activity of DRM catalysts containing nickel deposited on the ceria-zirconia support in the low-temperature DRM (550°C). A particular attention has been devoted to the influence of variable nickel loading and the temperature on the amount of hydrogen produced and the value of the molar ratio H_2/CO.

2. Experimental

The catalysts containing 2,4 and 10 wt. % of nickel were prepared by the impregnation of commercial ceria-zirconia support ($Ce_{0.62}Zr_{0.38}O_2$, Rhodia Electronics & Catalysts) with aqueous solutions of nickel nitrate. The obtained catalysts were dried and calcined in air at 550 °C for 2 h and were denoted as Ni(2)/CZ, Ni(4)/CZ and Ni(10)/CZ.

Catalysts were characterized by X-ray diffraction, low temperature N_2 sorption (S_{BET}), temperature programmed desorption of CO_2 (CO_2-TPD) and reduction by H_2 (H_2-TPR). The activity of Ni/CZ catalysts in low temperature dry reforming of methane (550 °C) was studied. Before the experiments the catalyst was reduced *in situ* at 800 °C for 2 h with 3 % vol. H_2 in Ar. The composition of the reaction mixture was $CO_2/CH_4/Ar$= 1/1/8 and the GHSV was 20 000 h^{-1}. Catalytic tests were also performed at 650 and 750 °C.

After DRM tests the catalysts were characterized by XRD and thermogravimetric analysis (TG) to examine the formation of carbon deposits.

3.Results and discussion

The XRD patterns of the support and catalysts exhibit planes characteristic for the fluorite cubic structure of ceria-zirconia mixed oxides. Planes characteristic of the regular structure of NiO were observed only for the catalysts with the highest Ni loading. Catalysts with low loading of the Ni probably contain well dispersed NiO crystallites, so the respective signals were not recorded during XRD measurements. Because of the active sites in the DRM process are metallic centres, the Ni/CZ catalysts and the support were reduced in 3 % H_2/Ar at 800 °C for 2 hours. X-ray diffraction patterns of Ni/CZ catalysts after reduction contained no planes characteristic of the regular NiO structure. For the Ni(10)/CZ catalyst, reflexes characteristic of metallic Ni were visible. It proves that NiO crystallites were completely reduced to Ni^0.

The SSA of the calcined CZ, Ni(2)/CZ, Ni(4)/CZ and Ni(10)/CZ catalysts were 125, 92, 87, 83 m^2/g, respectively. The decrease of SSA determined for Ni/CZ catalysts in comparison with the support can be explained by blocking pores of the support by the active phase. The same phenomenon was observed for reduced catalysts Ni/CZ (SSA: Ni(2)/CZ=85, Ni(4)/CZ= 79, Ni(10)/CZ=73 m^2/g).

H_2-TPR results indicate higher hydrogen consumption after the introduction of the Ni to the support (CZ=0.75, Ni(2)/CZ=1.14, Ni(4)/CZ=1.64, Ni(10)/CZ=2.6 mmol/g_{kat}). It is caused by strong interaction between nickel and the support, which improves the mobility of lattice oxygen and redox properties.

The results of CO_2-TPD suggest that calcined Ni/CZ catalysts and the support exhibit basic properties (CZ=0.28, Ni(2)/CZ=0.28, Ni(4)/CZ=0.29, Ni(10)/CZ=0.38 mmol/g_{kat}). The presence of weak (OH groups), medium (metal-oxygen couples) and strong (O^{2-} ions) basic centres has been observed [3]. During the reduction of Ni/CZ catalysts the weak basic centres disappear (elimination of lattice oxygen) (CZ=0.18, Ni(2)/CZ=0.20, Ni(4)/CZ=0.21, Ni(10)/CZ=0.23 mmol/g_{kat}). It has also been found that introduction of nickel creates new basic centres (the amount of desorbed CO_2 increases). These phenomena can be explained by the strong interaction between Ni and ceria, which can enhance the basicity of Ni/CZ catalysts. According to H_2-TPR results, the addition of Ni to the ceria-zirconia improves the redox properties of the material, favouring oxygen transfer required for CO_2 activation [3].

The highest conversion of methane and carbon dioxide at 550 °C was observed for the Ni(10)/CZ catalyst (average conversion of %CH_4=38.6, %CO_2=37.5). High activity of Ni(10)/CZ, in comparison with other samples, may be caused by the high loading of the active phase and a greater number of active sites. A similar tendency for the conversion of CO_2 was also observed. It can be explained by the increasing catalyst basicity with the Ni loading. According to our results the best H_2 production was observed for the Ni(10)/CZ catalyst. For other Ni/CZ catalysts more carbon dioxide than methane was used in the reaction (which lowers the value of the H_2/CO molar ratio). This phenomenon was caused by a competitive reaction of reverse water-gas shift ($CO_2+H_2=CO+H_2O$) giving H_2/CO molar ratio of <1. After DRM experiments, the catalysts were characterized using XRD and thermogravimetric analysis to identify and estimate the amount of formed carbon material. The results of TG analyses show that the highest amount of carbon deposits was observed for Ni(10)/CZ catalyst 40 % of the mass and 5 % wt. for. Ni(2)/CZ and 10 % wt. for Ni(4)/CZ. The carbon deposition can lead to CO by CO_2 in the reverse Boudouard reaction ($C+CO_2=2CO$) or to CO_2 by the surface oxygen from the support ($C+O_2$àCO_2) [3]. These reactions have an influence on the molar ratio H_2/CO. Therefore, the lower value of the molar ratio H_2/CO <1 for Ni(2)/CZ and Ni(4)/CZ catalysts, can be caused by RWGS reaction and oxidation of carbon deposits by CO_2 [3]. The reverse Boudouard reaction can also occur over the catalyst Ni(10)/CZ.

Studies on the influence of higher temperature on the activity of the catalyst Ni(10)/CZ in DRM show the increase in conversion of both methane (at 650 °C $\%CH_4=67.9$, 750 °C $\%CH_4=91.8$) and carbon dioxide (at 650 °C $\%CO_2=67.2$, 750 °C $\%CO_2=87.7$). It seems to be a favourable effect, but the value of the molar ratio H_2/CO deteriorates as a result of side reactions (at 650 °C $H_2/CO=0.9$, 750 °C $H_2/CO=1.03$).

4.Conclusions

The activity of a series of Ni/CZ catalysts has been studied in the DRM process. Physico-chemical characterization has revealed that the catalysts obtained exhibit redox and basic properties, which depend on the loading with the active phase. Catalytic DRM tests show that the Ni(10)/CZ catalyst gives the highest conversion of CH_4 and CO_2 and the highest H_2 production as well as the optimum ratio $H_2/CO=$ comparison to Ni(2)/CZ and Ni(4)/CZ. The influence of side reactions such as RWGS and reverse Boudouard reaction on the production of hydrogen and the value of the molar ratio H_2/CO has been shown. The formation of carbon deposits, which can be oxidized to CO, has been demonstrated. This reaction can change the molar ratio H_2/CO obtained in the DRM process over Ni/CZ. It has been found that increasing temperature has a negative influence on the value of the H_2/CO ratio.

5.References

[1] S. Sokolov et. al, Int. J. Hydr. En. 38 (2013) 16121

[2] D. Harshini et.al., Catal. Lett. 144 (2014) 656

[3] M. Radlik et. al., C. R. Chim., in press DOI:10.1016/j.crci.2015.03.008

436800 Dry Reforming of Methane over Ni-Based Pyrochlore Catalysts Using Transient Pulsing

Abstract:

Dry reforming of methane requires temperatures of 750 – 950 °C to achieve near equilibrium conversions. A thermally robust and active catalyst, such as Ni-substituted $La_2Zr_2O_7$ pyrochlore is desirable to achieve long term synthesis gas production.

Here, we report pulsing experiments in which 10% CH_4/He flows over a pre-reduced catalyst continuously while pulses of 10% CO_2/He was switched with pure helium gas. The products were quantified using an in-line quadrupole mass spectrometer.

The results at 750 °C show the deactivation of the catalyst during the first pulse, but in the subsequent pulses the catalyst is found to be stable. At 800 °C and 850 °C, the CO concentration decreases during each CO_2 pulse. However, the catalyst activity was found to be increasing from its initial value before start of each pulse, which corresponds to an increase in the reaction rate of the CO_2-consuming reaction(s). Carbon, formed during the times only CH_4 flows, reacts with CO_2 to produce CO. At the higher temperatures, the reaction is faster, producing more CO and consuming the carbon faster.

423506 Catalytic Production of Hydrogen and Carbon Monoxide Via Methane Dry Reforming Reaction over Hydrotalcite-Derived Ni/Mg/Al Mixed Oxides Promoted with Zr and/or Ce

Abstract:

1. Introduction

The reaction of dry reforming of methane (DRM) might become of a great importance in the near future since it is a method of CO_2 utilization, which allows to highly valorize natural gas and biogas [1]. The commercialization of the DRM process is dependent on the offer of new stable and active catalysts. The nickel-based catalysts are commonly used in DRM reaction as they are effective in hydrogen production and nickel is a much cheaper material than noble metals. However, Ni-based catalysts undergo fast deactivation due to the formation of coke deposits on the catalysts surface via direct methane decomposition and Boudouard reactions [2]. The stability of nickel catalyst in DRM reaction may be enhanced by introduction of support or promoters with basic and/or redox properties [1]. Thus in this work the hydrotalcite-derived mixed oxides modified with Ce or Zr species were tested in DRM reaction.

2. Experimental

The starting materials – hydrotalcites were synthesized by the co-precipitation method at constant pH from the aqueous solution of metal nitrates (Mg/Ni=3, $M^{2+}/M^{3+}=3$). In this way H-Ni sample was prepared. Zirconium was introduced into the brucite-like layers at the co-precipitation stage ($Al^{3+}/Zr^{4+}=9/1$), resulting in H-NiZr sample [3]. Cerium species were introduced into the catalytic system of both hydrotalcites via adsorption form the solution of Ce-EDTA complexes (3 wt.% Ce^{3+} aqueous solution), resulting in samples H-Ni-aCe and H-NiZr-aCe. Materials underwent calcination at 550°C for 4h under flow of air and were subsequently characterized by XRD, H_2-TPR, elemental analysis, N_2 sorption, FTIR and CO_2-TPD experiments.

Catalytic tests were carried out in a fixed-bed reactor at 550, 650 and 750°C. The total flow of feed gases was equal to 100cm^3/min ($CH_4/CO_2/Ar$=1/1/8; GHSV=20,000h^{-1}). The products of the reaction were analyzed by Varian GC-490 micro-GC. Prior to each catalytic test, the samples were activated by reducing with H_2 (3%H_2 in Ar, 1h at 900°C). In order to establish the influence of direct methane decomposition reaction on the overall process, additional catalytic tests with only methane in the feed (CH_4/Ar=2/8) were carried out at 550°C.

3. Results and discussion

The physicochemical characterization of the catalysts precursors and products of their thermal decomposition confirmed successful synthesis of hydrotalcite-like materials modified with Ce or Zr. The addition of cerium and zirconium species modified the basicity of tested samples and thus most probably their catalytic properties [4]. The calcined materials exhibited specific surface areas in a range from 100-160 m^2/g. The highest values of textural parameters were observed for Zr-containing samples.

The results of catalytic tests revealed that all catalysts were active in the DRM reaction at 550, 650 and 750°C. The obtained conversions of methane and carbon dioxide for the prepared materials at 550°C followed the sequence: HNi > HNi-aCe > HNiZr-aCe > HNiZr. However, the obtained distribution of products of the DRM reaction, suggests that activity of each catalyst is quite different in obtained values of H_2/CO molar ratio and the overall process is influenced by the occurrence of side reactions. The un-promoted catalyst showed excess of H_2 in the products of the reaction, pointing to the occurrence of direct CH_4 decomposition. The addition of promoters (especially Zr) inhibited this side reaction, which was confirmed in additional catalytic test with only methane in the feed. The differences in conversions of methane and carbon dioxide between the studied samples decrease with increasing temperature. At 750°C all tested catalyst showed CH_4 conversion of ca. 86% and CO_2 conversion of ca. 90%. The obtained values of H_2/CO molar ratio at 750°C were also similar and closed to unity. The difference in catalysts performance was shown by the measurement of coke deposition by TG experiments after reaction, the lowest shown for zirconia promoted samples.

4. Conclusions

The obtained results showed that the catalytic activity of Ni-catalyst depends strongly on the promoter used in low-temperature dry reforming. The addition of zirconia prevents the occurrence of direct methane decomposition, but, on the other hand, RWGS reaction is well developed over Zr-promoted catalyst. The addition of ceria results in increased CO_2 conversion and H_2/CO close to unity, pointing to the increased oxidation of carbon deposits. The promotion of Ni-based catalyst with Ce or Zr results in increase in catalyst stability.

ACKNOWLEDGMENTS

The work was financed by AGH grant 11.11.210.213.

References

[1] K. Asami et. al, Catal. Today 84, 27 (2003)

[2] Y.H. Hu wt al., Adv. Catal. 48, 297 (2004)

[3] D. Tichtit et. al, Chem. Mater. 14, 1530 (2002)

[4] R. Dębek et. Al, Catal. Today, in press DOI: 10.1016/j.cattod.2015.03.017

425798 ATR-Based Compact Reactor System for the Distributed Hydrogen Production

Wednesday, November 11, 2015: 9:54 AM
250E (Salt Palace Convention Center)
Vincenzo Palma[1], **Antonio Ricca**[1], Biagio Addeo[1], Maurizio Rea[2], Gaetano Paolillo[2] and Paolo Ciambelli[3], (1)Department of Industrial Engineering, University of Salerno, Fisciano, Italy, (2)Sol S.p.a., Monza, Italy, (3)University of Salerno, Fisciano, Italy

ATR-based compact reactor system for the distributed hydrogen production

Vincenzo Palma[1], Antonio Ricca[1*], Biagio Addeo[1], Maurizio Rea[2], Gaetano Paolillo[2], Paolo Ciambelli[1]

[1*]Dep. of Industrial Engineering, University of Salerno, via Giovanni Paolo II 132, 84084 Fisciano, ITALY, aricca@unisa.it

[2]R&D - SOL s.p.a, Via Borgazzi 27, 20900 Monza (MB), ITALY

INTRODUCTION

The well-known problems linked to the depletion of fossil fuels and to the growing energy demand address the industrial research toward the development of effective innovative routes for alternative energy resources. If in one hand hydrogen is pointed as the best clean energy vector (to employ in fuel cell systems), in the other hand its production through hydrocarbon reforming still appears as the most viable solution in a transition period towards a hydrogen based economy. One of the main limitation connected to the hydrogen employment diffusion is linked to economic issues related to the H_2 transport and storage. In this aim, distributed hydrogen production appeared an optimal solution, both for the stationary and mobile installations. From this point of view, hydrocarbons catalytic Auto-Thermal Reforming (ATR) appears the best candidate process[1]: it combines the partial oxidation and the steam reforming processes, since hydrocarbons react with both oxygen and steam to obtain a self-sustained and very fast process. Anyway the high carbon monoxide content of the produced stream require a further catalytic stage, the Water-Gas Shift (WGS) in which CO reacts with steam to produce further hydrogen. Anyway, the very different operating temperature of ATR and WGS stages resulted in the necessity of a heat exchange system, with a consequent increasing in plant size and operating costs.

Aim of this work was to design and test a thermally integrated fuel processor based on natural gas ATR, aimed to the hydrogen production intensification and to the external duty minimization.

EXPERIMENTAL

A compact fuel processor based on Auto-Thermal Reforming process for natural gas conversion was designed and configured to produce up to 10 Nm^3/h of hydrogen. Water and air were delivered to the system, together with the hydrocarbon. The system was provided to 2 catalytic stage: in the ATR module, the hydrocarbons conversion to syngas was obtained, in the WGS module the CO present in the syngas was converted increasing hydrogen content. In the ATR module a 0.5 L commercial (Johnson Matthey) honeycomb catalyst noble metals based was loaded, while in the WGS 2 L of commercial (JM) pellets catalysts were employed. Due to the very different temperature in the two stages, a compact and optimized heat exchange module were designed and placed between the two reaction volumes. In this way, sensible heat of effluent gas from ATR volume was exploited to pre-heat reactants before feeding to it; moreover, effluent gas were cooled to a temperature consistent to the WGS stage. Such configuration allowed to feed all reactants at room temperature, without external heat exchangers; moreover, the thermal integration was able to reduce system start-up transient times. Preliminary studies were carried out in order to evaluate the reactants rate and the feed ratios effect on system performances.

RESULTS AND DISCUSSION

Preliminary tests confirmed that the designed system was characterized by very short start-up times, being able to produce up to 5 Nm^3/h of hydrogen (27%vol) in less than 3 minutes. The transitory phases were strictly related to the overall gas rate, since higher stream rate resulted in faster system response. Despite thermodynamic predictions, due to the thermal integration the system seemed not to be greatly affected by steam content, since too high steam-to-carbon ratio resulted in a decreasing in system temperature. In Fig. 1, the system performances are summarized in a selected operating condition. As reported, the ATR module well approached the thermodynamic equilibrium, reaching a total methane conversion. On the other hand, WGS module was able to convert only the 51% of CO, mainly due to the thermodynamic limitations, so also affecting the thermal efficiency that anyway reached very promising values.

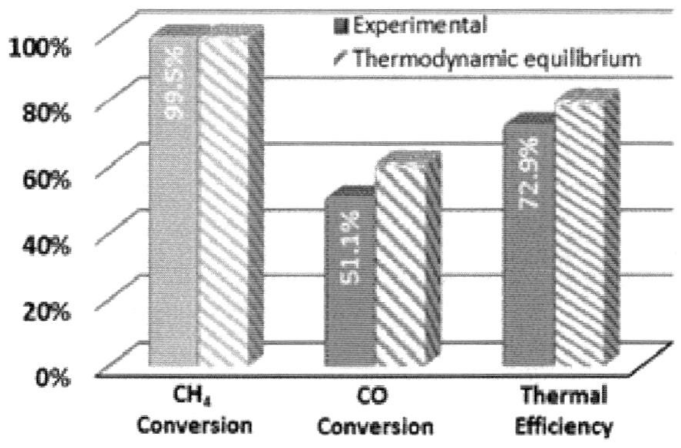

Fig. 1 System performances (H_2O:O_2:CH_4 = 0.65:0.65:1)

CONCLUSION

An ATR-based fuel processor thermally integrated was designed and realized. Preliminary tests evidenced the quick start-up and response of the system, that showed very good performances by producing up to 7.4 Nm^3/h of H_2. WGS module seemed to suffer for thermodynamic limitation, suggesting to feed more steam to the system, as well as to modify catalytic formulation. As a future enhancement, a heat recovery module may be placed downstream the WGS module for a further heat recovery.

427325 Sensitivity and Economical Analysis of Fuel Processors Based on SR Integrated with WGS and PSA for Pure Hydrogen Production from Natural Gas

Abstract:

The vision of an "hydrogen economy" with the use of hydrogen as an energy carrier will require increasing hydrogen production by more than an order of magnitude of the current production levels [1]. Hydrogen can be produced from a variety of sources. Fossil fuels may be used to produce hydrogen by first converting their fuel value to gas phase by reaction with steam, oxygen or air (gasification/reforming) followed by hydrogen enrichment and separation through the use of fuel processors [2–4].

The choice of an hydrogen separation technology often depends on the hydrocarbon feedstock used and the resulting synthesis gas composition. PSA(pressure swing adsorption) has been the most commonly used technology for producing high-purity hydrogen as product from steam reforming processes. PSA is based on an adsorbent bed that captures the impurities in the syngas stream at high pressure and then releases the impurities at low pressure. For this purpose multiple beds are used simultaneously so that a continuous stream of hydrogen at purity up to 99.9% may be produced recovering up to the 95% of the hydrogen present in the reformer syngas [5].

Our work focuses on an hydrogen production system starting from natural gas for refilling stations. The overall process consists in an hydrogen production unit (reforming and WGS stage), a purification unit (PSA stage), and a compression and refrigeration unit with the recirculation of PSA exhausts in the burner. In our previous works a comparison between two clean-up methods (CO preferential oxidation, CO-PROX, and COselective methanation, CO-SMET) for a refilling station (not integrated system) and for an Auxiliary Power Unit (APU, FC integrated system) were investigated [6,7]. Now, the final purpose is to replace CO-PROX and CO-SMET units with a PSA unit to obtain pure hydrogen. Besides, we want investigate in the thermal conditions of the system varing the PSA hydrogen recovery level from 70% to 95% for different reforming pressure ranging from 5 to 30 bar. The PSA exhaust contains a noticeable part of the produced hydrogen, that can be sent to the integrated burner (I-BR) to sustain the reforming reaction. Passing from 70% to 95% of PSA hydrogen recovery level, there is a thermal neutral point (TNP) above that that the system is not self-sustained. In these configurations part of the total natural gas feed to the reforming is shifted to the I-BR to supply the reforming reaction. Whereas below the TNP not all the recirculated fuel is required from the I-BR. Therefore the remaining part of the fuel is send to another burner (BR) that produces waste gases to preheat the air required to the I-BR. The TNP moves in the range of hydrogen recovery with the refroming pressure variation.

Moreover, we are completing the simulations with an hydrogen compression and refrigeration stage (ammonia plant) in order to give a full overview of a refilling station from the production to the storage of the hydrogen. The sensitivity analysis of the reforming process as a fuction of the working pressure, coupled with an economical analysis of the different possible configurations will be presented in order to estabilish the best configuration in terms of reforming pressure and hydrogen recovery.

[1] K. Liu, C. Song, V. Subramani. Hydrogen and syngas production and purification technologies. AIChE, Wiley & Sons Inc., 2010.

[2] M. O'Connell, G. Kolb, K.-P. Schelhaas, M. Wichert, D. Tiemann, H. Pennemann, et al. Towards mass production of microstructured fuel processors for application in future distributed energy generation systems: a review of recent progress at IMM. Chem. Eng. Res. Des. 90 (2012) 11–18.

[3] G. Kolb. Review: microstructured reactors for distributed and renewable production of fuels and electrical energy. Chem. Eng. Process Process Intensif. 65 (2013) 1–44.

[4] S. Specchia. Fuel processing activities at European level: A panoramic overview. Int. J. Hydrogen Energy 39 (2014) 17953–17968.

[5] S. Sicar. Pressure swing adsorption: Commentaries. Ind. Eng. Chem. Res. 41 (2002) 1389–1392.

[6] M.A. Ashraf, G. Ercolino, S. Specchia, V. Specchia. Final step for CO syngas clean-up: comparison between CO-PROX and CO-SMET processes. Int. J. Hydrogen Energy 39 (2014) 18109–18119.

[7] G. Ercolino, M.A. Ashraf, V. Specchia, S. Specchia. Performance evaluation and comparison of fuel processors integrated with PEM fuel cell based on steam or autothermal reforming and on CO preferential oxidation or selective methanation. Applied Energy 143 (2015) 138–153.

Acknowledgements

This work was funded by the Italian Ministry of Education, University and Research (MIUR, PRIN 2010–2011) within the project IFOAMS ("Intensification of catalytic processes for clean energy, low-emission transport and sustainable chemistry using open-cell FOAMS as novel advanced structured materials", protocol no.2010XFT2BB).

419193 Modeling Transport and Reaction in Porous Catalyst Washcoat for Steam Methane Reforming in a Microchannel Reactor By CFD with Elementary Kinetics

Abstract:

Modeling transport and reaction in porous catalyst washcoat for steam methane reforming in a microchannel reactor by CFD with elementary kinetics

Chenxi Cao, Nian Zhang, Yi Cheng*

Department of Chemical Engineering, Tsinghua University, Beijing 100084, P.R.China

Email: yicheng@tsinghua.edu.cn

Towards an upcoming era of hydrogen economy, there is an increasing need in distributed hydrogen production for hydrogen fuelling stations, fuel cell systems and so forth. Currently, steam methane reforming (SMR) is the most widely used and cost-effective route for world-wide hydrogen production. The conventional fixed-bed based process for hydrogen production in centralized plants is unsuitable for distributed use for its drawbacks such as heat transfer and diffusional limitations, catalyst deactivation and large footprint, which has led to large research interests in the intensification of SMR using microchannel reactors. Micro-channel reactors have the advantages of highly efficient enhancement of heat and mass transfer, easy scale-up and fast startup times [1-2]. Thus, SMR in microchannel reactors have great potential in reducing capital cost and energy consumption and to realize a small-scale and compact application as required by distributed hydrogen production.

CFD simulation is a crucial tool for reactor design and process optimization, especially in the case of microchannel reactors in which experimental measurements of profile data within the reaction channel is difficult. CFD simulation has been extensively used by researchers [3-6] for SMR in microchannel reactors with Ni and Rh as reforming catalyst, but the majority of which employ global kinetics and/or neglect the diffusional effects in the catalyst washcoat, For the millisecond SMR in microchannel reactors, effective usage of the interior surface of the washcoat may be important for Ni catalyst over which the intrinsic reaction rates of SMR is slower. Also, elementary reaction kinetics are preferred to global kinetics for ultra-fast reactions and can provide key information of the reaction path, rate-controlling step and surface coverage of key species on the catalyst throughout the washcoat. This work presents a comprehensive CFD modeling and simulation of a microchannel reactor, incorporating the elementary reaction kinetics of SMR over Ni and the modeling of the catalyst washcoat on the channel wall. Special focus is put on the transport and reaction in porous washcoat of the reforming channel.

The modeling and simulation is conducted in FLUENT 14.0 software linked with an external in-house stiff ODE solver for the elementary reaction kinetics network. An isotropic porous medium model is used to describe the momentum and heat transfer in the catalyst washcoat. Diffusion in the porous washcoat is described by an effective diffusivity model that takes into account both molecular diffusion and Knudsen diffusion. All properties used in the simulation are temperature dependent and no lumped model for heat and transfer is used. The model is quantitatively validated by our experimental data. The influence of catalyst thickness, porosity, pore diameter and operational factors such as WHSV on the reactor performance is investigated. Effectiveness factors and Damköhler number of the catalyst are determined under various conditions and allow for discussions of control regime and catalyst utilization ratio. It is revealed that the control regime shifts from reaction to diffusion when the washcoat thickness increases from 10 μm to 100 μm with the total catalyst loading kept constant under normal operating conditions. Porosity and pore diameter have minor effects on the reactor performance compared with catalyst thickness. Increasing the catalyst washcoat thickness from 25 μm to 50 μm with the catalyst loading per washcoat volume kept constant causes a more than 50 % drop in WHSV to achieve the same methane conversion of 90 %. It is also demonstrated that serious consideration should be taken on the diffusional effects in the catalyst washcoat when simulating heat coupling, as internal diffusion changes the overall reaction rates profile along the reaction channel and thus leads to different results in heat matching. Detailed reaction path analysis is further performed for on the basis of the presented simulation framework.

Keywords: Hydrogen production; Steam methane reforming; Micro-channel reactor; CFD with elementary reaction kinetics; porous catalyst washcoat

References

[1] Tonkovich AY, Perry S, Wang Y, Qiu D, LaPlante T, Rogers WA. Microchannel process technology for compact methane steam reforming. Chem Eng Sci 2004;59:4819-24.

[2] Johnson BR, Canfield NL, Tran DN, Dagle RA, Li XS, Holladay JD, et al. Engineered SMR catalysts based on hydrothermally stable, porous, ceramic supports for microchannel reactors. Catal Today 2007;120:54-62.

[3] Stefanidis GD, Vlachos DG, Kaisare NS, Maestri M. Methane Steam Reforming at Microscales: Operation Strategies for Variable Power Output at Millisecond Contact Times. AlChE J 2009;55:180-91.

[4] Zanfir M, Baldea M, Daoutidis P. Optimizing the Catalyst Distribution for Countercurrent Methane Steam Reforming in Plate Reactors. AlChE J 2011;57:2518-28.

[5] Wang F, Zhou J, Wang GQ. Transport characteristic study of methane steam reforming coupling methane catalytic combustion for hydrogen production. Int J Hydrogen Energy 2012;37:13013-21.

[6] Jeon SW, Yoon WJ, Baek C, Kim Y. Minimization of hot spot in a microchannel reactor for steam reforming of methane with the stripe combustion catalyst layer. Int J Hydrogen Energy 2013;38:13982-90.

419187 Pervaporative Enrichment of 1,3-Propanediol from Model Fermentation Broths By Hydrophobic Specialty Polymers

Abstract:

While 1,3 PD can be easily separated from the other by products of bacterial metabolism, the real challenge lies in enriching it from dilute concentrations in aqueous mixtures which amounts to almost 50 – 70% of the total production cost in fermentative 1,3-PD production. The hydrophilic character of 1,3 PD compounds the difficulty of purification. Conventional and reported separation techniques have had their limitations in due cost and energy consumption and hence applicability. Membrane separation processes offer a cheap and energy efficient method of upgrading the value of specific components typically produced in dilute mixtures. Although pervaporation is an energetically advantageous process. 1,3-PD cannot be enriched by materials conventionally used for pervaporation. The development of novel materials with a good cost performance balance is essential for the establishment of industrial separation procedures for 1,3-PD enrichment from dilute broths. The present work describes a study on the pervaporation enrichment of 1,3-propanediol from model fermentation broths using three hydrophobic polymer families – functionalized siloxanes, cyclohexylamine base methacrylates and imidazolium ionic liquid based methacrylates with progressively increasing separation efficiency. While the highest 1,3-PD permeabilities were seen with the siloxane polymers, the imidazolium IL based methacrylates exhibited the best separation factors both with respect to water and glycerol. The effect of glycerol on transport coupling was evaluated. The permeability and separation factors from the pervaporation experiments with these polymers were siloxanes were collated to arrive at a Robeson's upper bound of pervaporation of 1,3-propanediol from aqueous solution by polymeric membranes. The development of novel materials with a good cost/performance balance is central to the establishment of industrial separation procedures for 1,3-PD enrichment from fermentation broths. The removal of this process bottleneck may increase the feasibility of using waste industrial glycerol as a sustainable resource for producing 1,3-PD as a biorenewable alternative to a petroleum platform.

IMPLEMENTATION OF A CUSTOMIZED GAS-SEPARATION MEMBRANE MODEL INTO COMMERCIAL FLOWSHEETING SOFTWARE TO SIMULATE A HYBRID CO_2 REMOVAL PROCESS FOR OXIDATIVE COUPLING OF METHANE

Alberto Penteado[1,2], Erik Esche[1], Günter Wozny[1]
(1) Process Dynamics and Operations Group, Technische Universität Berlin, Berlin, Germany
(2) CAPES Foundation, Ministry of Education of Brazil, Brasilia, DF, Brazil

Introduction

Ethene is one of the most produced basic chemical world-wide with a global production of 141 million tons in 2011 [1]. The oxidative coupling of methane (OCM) is a promising path for the conversion of methane into ethene, offering the possibility for utilizing natural gas, shale gas or biogas instead of oil as feedstock. It is a more direct route from methane to valuable chemicals compared to the Fischer-Tropsch process, which has synthesis gas as an intermediate [2]. The selective oxidation of methane to ethene can be seen below:

$$2CH_4 + O_2 \rightarrow C_2H_4 + 2H_2O$$

Due to the high temperatures required for catalyst activation in the OCM reaction (750 to 950 °C), products of non-selective oxidation such as carbon monoxide (CO) and carbon dioxide (CO_2) are also formed and should be removed in the separation section. The C_2 yield in the OCM reactor is not expected to exceed 28% [3], which highlights the need for an efficient downstream separation for an OCM process to become economically feasible.

The OCM process structure typically includes a reaction section, consisting of a reactor or reactor network; a compression section based on a multi-stage compressor with intermediate cooling and knock-out drums for water removal; a carbon dioxide removal section, commonly via an absorption/desorption system; and a distillation section, consisting of a demethanizer column and a C_2 splitter. This structure is reproduced in the block diagram in Figure 1 below.

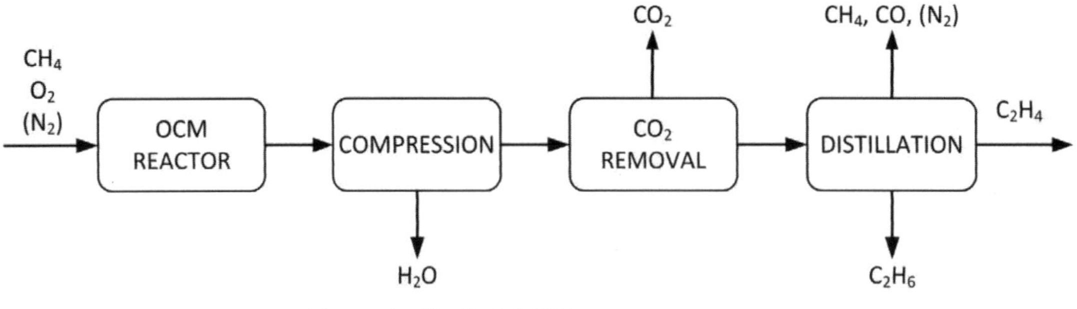

Figure 1 - Typical OCM process structure

The carbon dioxide removal is usually an energy-intensive step, since the amine solution used for the absorption needs to be regenerated in a stripping column by adding heat through the reboiler. Alternatively, a hybrid process can be employed. This consists of an upstream gas separation membrane section and a downstream absorption section. This concept has been successfully applied in a mini-plant in the Technische Universtiät Berlin, showing significant reduction in the specific energy required per kilogram of captured CO_2 [4]. In this contribution, the feasibility of applying this process at an industrial scale is assessed through simulations and preliminary cost estimations.

Process Concept

Amine solutions are often utilized for the sequestration of CO_2 in power plants or for natural gas sweetening. A similar approach is used for the OCM process. The OCM reactor outlet passes through the compression and water removal section and enters the CO_2 removal section, where it is fed to a packed column in counter-current flow with a 30 wt% Monoethanolamine (MEA) acqueous solution. The carbon dioxide is dissolved and chemically bound, producing a CO_2-free gas stream at the top and a CO_2-rich amine solution at the bottom. The rich amine solution is than flashed, pumped, and pre-heated before entering the stripper column, within which it is regenerated. A gas stream consisting mainly of CO_2 and water vapor is produced at the top of the stripper column, whilst a lean amine solution is produced at the bottom. The latter is then pumped, cooled, and recycled back to the absorption column.

In the proposed hybrid process, a gas separation membrane unit or unit cascade is added upstream of the absorption section. The carbon dioxide is partially removed in the membrane section, reducing the amine flow rate and reboiler duty in the absorption section for a fixed degree of CO_2 capture. This concept is represented in Figure 2, wherein a two-stage stripping cascade is employed.

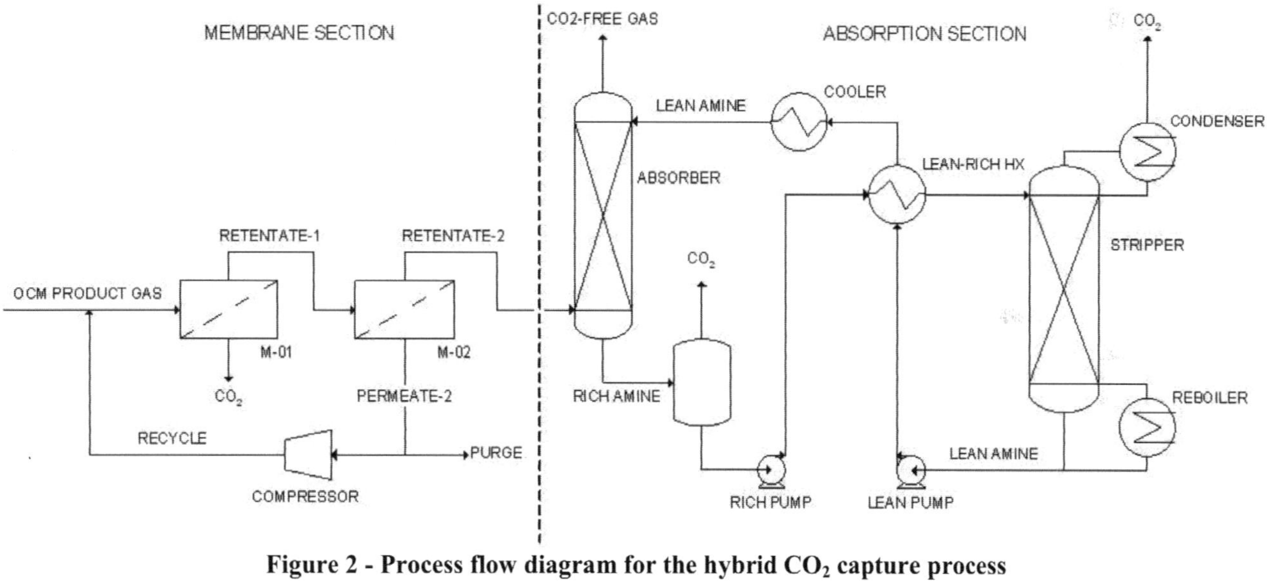

Figure 2 - Process flow diagram for the hybrid CO_2 capture process

Membrane units consist of a series of circular flat sheet membrane envelopes with segment cuts disposed in series inside a metal casting containing an inner tube for the permeate flow as seen in Figure 3. Both Polyimide (PI/Matrimid) and Poly-(ethylene oxide) (PEO) membranes are considered in this contribution. The first has higher selectivity to CO_2, whilst the second allows for a larger flow through the membrane. Membrane materials and modules are produced by coworkers at Helmholz-Zentrum Geesthacht Centre for Materials and Coastal Research, Germany, as described in [5].

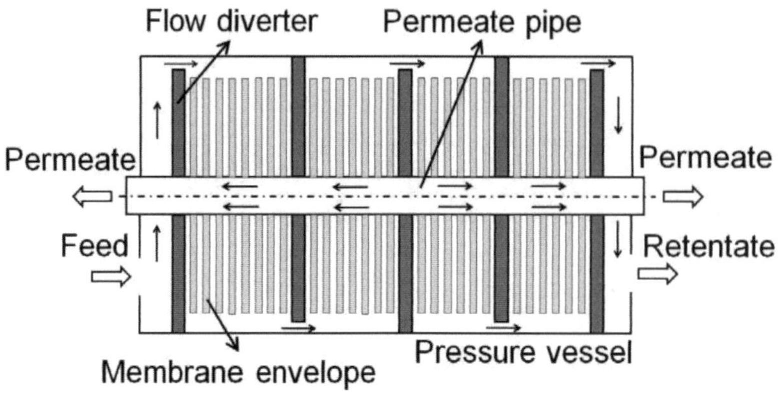

Figure 3 – Representation of the envelope type membrane unit

Modeling

The global model is divided into a membrane section and an absorption section. The model for the membrane section is implemented using the web-based modeling environment MOSAIC [6]. MOSAIC uses a document-based modeling approach to facilitate the documentation step and the exchange and reuse of models across platforms. It allows for automatic code generation for the model solution in different programming languages and environments. A code for the solution of the membrane model in Aspen Custom Modeler® (ACM) is generated and the ACM model is later exported to Aspen Plus® as a unit operation for the simulations. The absorption section is modeled in Aspen Plus® using available models from the library and from the example file [7].

Membrane Section

Equations for modeling gas permeation membranes are presented in [8]. A one-dimensional solution-diffusion model is applied and discretized using orthogonal collocation on finite elements with fourth order Lagrangian polynomials on Radau roots. Basic modeling equations are presented below: molar balances on permeate and retentate sides (Equation 1), permeance calculation for the PI membrane according to the Arrhenius law (Equation 2) and permeance calculation for the PEO membranes according to the free-volume theory (Equation 3). Further detail on modeling equations and implementation is given in [4].

$$\frac{dn_i^P}{dz} = -\frac{dn_i^R}{dz} = L_i \cdot \left(f_i^R - f_i^P\right) \cdot h \tag{1}$$

$$L_i^{PI} = L_i^{0,PI} \cdot exp\left(-\frac{Ea_i^{PI}}{R \cdot T}\right) \tag{2}$$

$$L_i^{PEO} = L_i^{0,PEO} \cdot exp\left[-\frac{Ea_i^{PEO}}{R \cdot T} + \sigma_i^2 \sum_{i=1}^{NC}\left(\frac{1}{\sigma_i^2} \cdot m_i^0 \cdot \frac{f_i^R + f_i^P}{2} \cdot exp\left(m_i^T \cdot T\right)\right)\right] \tag{3}$$

Permeabilities are obtained from co-workers at Helmholz-Zentrum Geesthacht Centre for Materials and Coastal Research, Germany [5]. Fugacitites are calculated using the Peng-Robinson equation of state on the retentate side, while ideal gas is assumed on permeate side.

Experiments carried within the mini-plant at the Technische Universität Berlin showed nearly isothermal conditions on both retentate and permeate sides, as well as negligible pressure drop and concentration polarization, therefore these effects are not taken into consideration. The model is initially used for the simulation and optimization of the mini-plant [4]. The model is then extended for the simulation of industrial size plants by using a scale-up by numbering-up approach. The inlet flow rate is divided by the design flow of the membrane module to obtain the required number of parallel modules. A single module is simulated and the resulting flows are then multiplied to the number of parallel modules to obtain the total retentate and permeate outlet flows.

Absorption Section

An unsymmetric Electrolyte NRTL model is used for the liquid phase and PC-SAFT equation of state is used for the vapor phase. Gaseous components, H_2, N_2, CH_4, C_2H_4, C_2H_6, CO, and CO_2 are selected as Henry components. The example file [7] is used as a starting point for the modeling of the absorption section. Implemented parameters from various sources are kept and the Henry constant for ethene in Monoethanolamine (MEA) is additionally estimated based on data from [9] to avoid an underestimation of ethene losses in the absorber. Columns are simulated using rate-based calculations with built-in correlations by [10] for mass transfer, by Chilton and Colburn [11] for heat-transfer and by [12] for the packing hold-ups. Kinetic parameters for the CO_2 absorption are taken from [13] and [14]. Plug flow is assumed for the vapor, while liquid is assumed to be well mixed. Column packing is Sulzer Mellapak 350X for the absorption column and 350Y for the stripping column. Other equipment (flash, pumps and exchangers) are simulated using the traditional MESH approach.

Simulations and Optimizations

Simulations are performed using Aspen Plus® version 8.6. All models are solved using the available numeric methods in the software. A sparse solver is used for the membrane modules, Broyden method is applied for converging tears in recycle calculations and SQP is used for sequential optimization of the membrane section.

Different reactor design and feeding policies for OCM are currently under investigation as described in [15] and [16]. As they impact the downstream separation processes drastically, an integrated approach is required to synthesize an efficient OCM process. In this contribution, multiple feed scenarios are considered to reflect possible reactor outlets of a plant producing 100 kton/year of ethene. All gas flows are assumed to be compressed, cooled, and dry, as if they were coming from the compression section. The feed scenario details are shown in Table 1 below.

Table 1 - Inlet gas conditions for the simulations

Feed Scenario	Temperature (°C)	Pressure (bar)	Molar Composition (%)			
			CO_2	C_2H_4	N_2	CH_4
I	35	10	27.4	15.4	34.4	22.8
II	35	10	9.0	9.0	16.7	65.3
III	35	10	24.5	4.5	8.0	63.0
IV	35	10	38.0	5.0	10.0	47.0

Scenario I reflects the use of nitrogen diluted flow in the reactor. Scenario II considers no dilution, but the use of a high methane to oxygen ratio instead. Scenarios III and IV consider carbon dioxide dilution, which has a positive impact on the C_2 selectivity and helps to control the high exothermicity of the OCM reaction network. For scenarios II, III, and IV, in which no nitrogen is present, other components (H_2, CO, and C_2H_6) are simulated as N_2, given that no permeances are available for those.

All simulations aim at the removal of 97% of the CO_2 content in the inlet gas. A base case consisting of a standalone absorption process is simulated for each scenario for the latter comparison with hybrid processes. Due to the size and complexity of the system, no actual optimization could be performed in the absorption section. Therefore, optimal design is carried in the form of extensive simulations and sensitivity analyses. A specific cost per kilogram of captured CO_2 is then calculated.

The membrane section is simulated separately. An objective function consisting of a simplified operational cost is used for its optimization. It comprises the compression cost, cost attributed to ethene losses and the cost for removing the remaining CO_2 through absorption at a fixed price calculated in the base case. Flowsheets containing a single PI membrane and a stripping cascade of PI and PEO membranes are considered. The decision variables are the membrane materials, the membrane areas and the recycle/purge ratio in the case of membrane cascades. Bad convergence for the recycle calculations limits the steps of the optimization solver. Many starting values are tested leading to different local optima. The downstream absorption section is then simulated only for the cases presenting a simplified operational cost lower than the base case.

Utility consumption and cost rates are calculated inside Aspen Plus®, allowing for the operational cost to be quickly assessed during the flowsheeting stage. Utility costs are set to their default values. Apart from the utilities, costs are also attributed to the ethene losses during the process, based on current market prices published on [17]. These costs are shown in Table 2 below.

Table 2 – Base costs for economic calculations

Type	Specific Cost	Unit
Cooling Water	2.12×10^{-7}	$/kJ
Electricity	0.0775	$/kWh
Steam	2.2×10^{-6}	$/kJ
Ethene Loss	1.00	$/kg

Cooling water is used in the condenser of the stripping column, amine solution cooler, and interstage coolers of compressors. Electricity is used in the amine circulation pumps and gas compressor. Steam is used in the reboiler of the stripping column. Ethene is lost via the permeate stream, the purge stream, and the absorber's bottom stream. The total operational cost (OPEX) is calculated as the sum of all applied utilities and the ethene losses.

Results

Table 3 shows the results of the cost estimations for the standalone absorption and for the hybrid processes for each feed scenario. It consists of the operational cost and the specific operational cost per ton of removed CO_2 and the membrane areas per module.

Table 3 - Operational costs for different feed scenarios

Scenario		OPEX (US$/h)	Specific OPEX (US$/ton$CO_2$)	PI area per module (m^2)	PEO area per module (m^2)
I	Standalone absorption	483.56	14.78		
	Single PI module	517.24	15.81	0.1	
	PI + PEO cascade	695.65	21.26	0.5	0.1
II	Standalone absorption	241.35	13.17		
	Single PI module	301.29	16.44	0.5	
	PI + PEO cascade	454.03	24.78	0.5	0.1
III	Standalone absorption	978.37	9.81		
	Single PI module	**924.75**	**9.27**	**1.0**	
	PI + PEO cascade	1132.32	11.35	2.3	0.1
IV	Standalone absorption	1456.66	10.46		
	Single PI module	**1353.27**	**9.72**	**2.5**	
	PI + PEO cascade	1548.33	11.12	2.0	0.1

For feed scenarios I and II, the cost for operating the hybrid process with a single PI module is higher than that of the standalone absorption. Even though the specific energy per kilogram of CO_2 removed and total utility costs are reduced, the total operational cost of the hybrid process is still higher because ethene losses are higher in the membrane process compared to the absorption. When utilizing CO_2 diluted streams, as in scenarios III and IV, the hybrid process with a single PI module becomes economically attractive. The higher CO_2 content in the feed stream increases the driving force for the separation in the membranes. Furthermore, the greater amount of CO_2 to be captured requires a larger amine recirculation rate, which also increases ethene loss in the absorption section up to the point that they become comparable to those in the membrane section. Economically feasible scenarios were found by applying a single PI module to remove carbon dioxide from feed scenarios III and IV.

The utilization of stripping membrane cascades intends to increase carbon dioxide removal for a fixed ethene loss, or reduce ethene losses for a fixed carbon dioxide removal. However, optimizations showed that the use of such membrane cascades is costlier than the use of single membrane modules or even the standalone absorption for the given application. This can be explained by the high cost of recompressing the permeate gas stream for recycle, or purging the gas stream containing ethene.

The diameters of the absorber and stripper do not change significantly between the standalone absorption and hybrid process with a single PI module for cases I and II. For cases III and IV the columns for the hybrid process have a smaller diameter as the total gas flow is significantly reduced by partially removing the CO_2 in the membrane section. Investment cost for the membrane material, which is to be replaced over time, is relatively small compared to the cost of the carbon steel casting. These are both relatively small if compared to the cost of the absorption and stripping columns with structured packing in stainless steel. Thus, the hybrid process also has the potential to offer a reduction in the capital investment costs.

Conclusions

MOSAIC is a convenient platform for documentation-based modeling of unit operations for subsequent export and integration into flowsheets. It allows the user to concentrate fully on the modeling activity, avoiding typographic mistakes and difficulties in model reuse, which are typical for coding in different environments. In this contribution, a gas separation membrane module is successfully exported from MOSAIC and integrated into Aspen for simulations and optimizations.

While very attractive for applications like natural gas sweetening or biogas enrichment, the utilization of gas permeation membranes for the OCM process is highly sensitive to product (ethene) losses. Hybrid processes employing a single PI membrane offer an economic advantage for OCM when CO_2 dilution is used in the reactor. Applying gas permeation membranes in this case leads to a significant reduction in the operational costs. For scenario IV, this reduction amounts to 103.39 US$/h. The use of CO_2 dilution is especially interesting for combining OCM with Dry Reforming of Methane as described in [18].

Outlook

A more effective and efficient approach for process optimization needs to be developed. The use of simplified models in an initial stage, especially in the absorption section, could ease the optimization and pave the way for the subsequent implementation of more rigorous models. Despite being a convenient platform for flowsheeting and preliminary cost analysis, Aspen Plus® is far from being a powerful optimization tool. The use of other tools needs to be considered for carrying a proper process optimization, notably simultaneous optimization, for the entire system. To this end, MOSAIC offers a built-in connection with the NEOS Server [19], which is a web-based service for optimization using state-of-the-art solvers.

Capital investment estimations need to be performed for the most promising scenarios in order to properly assess the advantages of the hybrid process. Cost of membrane units and membrane material replacement are yet to be estimated, and both negatively impact the hybrid process. More detailed economic analyses need to be performed, including the determination of the return on investment and payback times.

Finally, it is necessary to evaluate the operational and capital investment costs of the OCM process as a whole to assess its feasibility. Alternative unit operations like membrane reactors and pressure and temperature swing adsorption need to be considered in the OCM process superstructure. The integration of OCM with other processes as ethane cracking and methane dry reforming also need to be properly assessed.

Acknowledgement

Present work is developed with the support from CAPES, Coordination for the Improvement of Higher Education Personnel – Brazil (Grant Number BEX 11946/13-0). Financial support by the Cluster of Excellence “Unifying Concepts in Catalysis” (DFG EXC 314) is gratefully acknowledged.

Nomenclature

Latin letters

Ea	activation energy	kJ/mol
f	fugacity	Pa
h	membrane height	m
m^T	parameter in free-volume theory	K^{-1}
m^0	parameter in free-volume theory	Pa^{-1}
n	molar flow	kmol/s
z	membrane length	m
L	permeance	kmol/(m^2 s Pa)
R	universal gas constant	kJ/kmolK
T	temperature	K
y	molar fraction	kmol/kmol

Greek letters

σ	Lennard-Jones molecular diameter	Å

Superscripts

P	permeate
R	retentate
0	reference condition
PI	polyimid membrane
PEO	poly-(ethylene oxide) membrane

Subscripts

i	component index

References

[1] W. True, "Global ethylene capacity continues advance in 2011," *Oil & Gas Journal,* vol. 110, no. 7, 2012.

[2] A. T. Ashcroft, A. K. Cheetham, J. S. Foord, M. L. H. Green, C. P. Grey, A. J. Murrell and P. D. F. Vernon, "Selective oxidation of methane to synthesis gas using transition metal catalysts," *Nature,* vol. 344, pp. 319-321, 1990.

[3] Y. S. Su, J. Y. Ying and W. H. Green Jr., "Upper bound on the yield for oxidative coupling of methane," *Journal of Catalysis,* vol. 218, no. 2, p. 321–333, 2003.

[4] E. Esche, D. Müller, S. Song and G. Wozny, "Optimization during the process synthesis: enabling the oxidative coupling of methane by minimizing the energy required for the carbon dioxide removal," *Journal of Cleaner Production,* vol. 91, pp. 100-108, 2015.

[5] T. Brinkmann, C. Naderipour, J. Pohlmann, J. Wind, T. Wolff, E. Esche, D. Müller, G. Wozny and B. Hoting, "Pilot scale investigations of the removal of carbon dioxide from hydrocarbon gas streams using poly(ethyleneoxide)–poly(butyleneterephthalate) PolyActiveTM) thin film composite membranes," *Journal of Membrane Science,* vol. 489, p. 237–247, 2015.

[6] S. Kuntsche, H. Arellano-Garcia and G. Wozny, "MOSAIC, an environment for web-based modeling in the documentation level," *Computer Aided Chemical Engineering,* vol. 29, pp. 1140-1144, 2011.

[7] Aspen Technology, Inc., "Rate-Based Model of the CO2 Capture Process by MEA using Aspen Plus," Burlington, MA, USA, 2014.

[8] K. Ohlrogge und K. Ebert, Membranen: Grundlagen, Verfahren und industrielle Anwendungen, Weinheim, Germany: WILEY-VCH Verlag, 2006.

[9] J. J. Carroll and A. E. Mather, "A model for the solubility of light hydrocarbons in water and aqueous solutions of alkanolamines," *Chemical Engineering Science,* vol. 52, no. 4, pp. 545-552, 1997.

[10] B. Hanley and C.-C. Chen, "New Mass-Transfer Correlations for Packed Towers," *AIChE Journal,* vol. 58, no. 1, pp. 132-152, 2012.

[11] R. Taylor and R. Krishna, Multicomponent Mass Transfer, New York: Wiley, 1993.

[12] J. L. Bravo, J. A. Rocha and J. R. Fair, "A comprehensive model for the performance of columns containing structured packings," *I. ChemE. Symp,* vol. 129, no. A439, 1992.

[13] H. Hikita, S. Asai, H. Ishikawa and M. Honda, "The Kinetics of Reactions of Carbon Dioxide with Monoethanolamine, Diethanolamine, and Triethanolamine by a Rapid Mixing Method," *Chem. Eng. J.,* vol. 13, pp. 7-12, 1977.

[14] B. R. Pinsent, L. Pearson and F. J. W. Roughton, "The Kinetics of Combination of Carbon Dioxide with Hydroxide Ions," *Trans. Faraday Soc.,* vol. 52, pp. 1512-1520, 1956.

[15] E. Esche, H. Arellano-Garcia, G. Wozny and L. T. Biegler, "Optimal Operation of a Membrane Reactor Network," *Computer Aided Chemical Engineering,* vol. 31, p. 1321–1325, 2012.

[16] H. R. Godini, S. Jašo, S. Xiao, H. Arellano-Garcia, M. Omidkhah and G. Wozny, "Methane Oxidative Coupling: Synthesis of Membrane Reactor Networks," *Ind. Eng. Chem. Res.,* vol. 51, no. 22, pp. 7747-7761, 2012.

[17] A. Platts.com news & Analysis Feature, "www.platts.com," PLATTS McGraw Hill Financial, [Online]. Available: http://www.platts.com/news-feature/2014/petrochemicals/pgpi/ethylene. [Accessed 2015 07 07].

[18] H. R. Godini, S. Xiao, M. Kim, O. Görke, S. Song and G. Wozny, "Dual-membrane reactor for methane oxidative coupling and drymethane reforming: Reactor integration and process intensification," vol. 74, pp. 153-164, 2013.

[19] Wisconsin Institutes for Discovery at the University of Wisconsin in Madison, "NEOS Server," [Online]. Available: http://www.neos-server.org/neos/. [Accessed 17 09 2015].

434107 Membrane-Based Process for the Continuous Enzymatic Saccharification of Lignocellulosic Biomass

Abstract:

We are currently evaluating the feasibility of performing continuous enzymatic hydrolysis of lignocellulosic biomass to product sugars using a membrane-assisted reaction/separation process. The overarching technical goals are to continuously remove the sugars—this lowers product feedback inhibition—retain and recycle active enzyme, and continuously recover the co-product of lignin. Experimental data is being collected to support preliminary technoeconomic analysis of the speculative process designs versus the current state-of-the-art batch enzymatic hydrolysis. Prospective savings are anticipated due to higher time-averaged productivity from capital equipment investments, and lower enzyme usage. Our initial studies included characterizing cake-layer resistances for varying slurry solids loadings during crossflow filtration. These results highlighted the fact that the membrane's nominal permselective properties are less important than its module configuration and ease-of-cleaning because the cake formation controls the flux and fine tunes the selectivity. Thus, our current studies include two aspects. Firstly, we are coupling tubular ultrafiltration membranes to a continuous stirred-tank reactor and continuously removing the sugars and recycling the bulk of the water back to the reactor after concentrating them. Overall reaction kinetics and yield are being assessed and compared with classical batch hydrolysis results. Secondly, we are identifying sufficient protocols for in-situ membrane cleaning and duty cycling.

435012 Recovery of Nutrients from Swine Wastewater Using Ultrafiltration: Applications for Microalgae Cultivation in Photobioreactors

Abstract:

Algal biomass has been identified as a promising feedstock that could be used in a number of industrial applications, including biofuel production, aquaculture, and pharmaceutical production. Through the use of sunlight and CO_2, microalgae species are capable of producing biomass rich in lipids and carbohydrates, which can be extracted from the plant material and used in the production of commodities such as biofuels. Algae have been shown to have higher rates of productivity and higher lipid contents than traditional bioenergy crops, and do not require high-quality land for cultivation, and would not compete with current agricultural products for space.

While the high productivity and lipid content of microalgae make it a promising feedstock for a number of industrial applications, there are several challenges associated with their cultivation in large-scale production systems. The production of microalgae can require significant amounts of nitrogen and phosphorus fertilizer. In order to make the production of microalgae economical, it is important that low cost sources of nitrogen and phosphorus be available to producers. Considering the relatively high nitrogen content of microalgae (4 to 8 percent), the nitrogen inputs required for microalgae cultivation can be significant. In addition, the synthesis of nitrogen fertilizer produces around 2 kg of CO_2 kg^{-1}. The additional carbon production, if added to the overall lifecycle, undermines the favorable carbon balance for microalgae feedstocks.

Membrane separations technology constitutes an alternative method for the removal of biological contaminants from the nutrient-rich wastewater prior to use in microalgae cultivation. A membrane consists of a thin film that separates two phases, and can be used in liquid-liquid separations. Microfiltration and ultrafiltration involve the use of porous membranes to remove micro- or macro-particles from a solution. In this study we used ultrafiltration membranes for removing contaminants from swine wastewater for use in algal growth. Through several trials we have found that the algal wastewater is completely contaminant free and the growth rate of the algae is greater than optimal conditions.

433601 Use of Forward Osmosis in Treatment of Hyper-Saline Produced Water

Abstract:

Oilfield produced water forms the largest waste stream in the production of petroleum. The quality of produced water varies greatly, with some waters being close to potable while others are oily or extremely saline. Hyper-saline produced water is an exceptionally challenging problem as there are few treatment options available. Desalination of such hyper-saline water is extraordinarily expensive because we are limited to evaporative processes which have enormous energy requirements. Conventional desalination technologies, like reverse osmosis are unusable due to the high osmotic pressure which require exceptionally high hydraulic pressure to overcome.

Forward osmosis circumvents this issue by using osmotic pressure instead of hydraulic pressure. Using a high concentration draw solution, osmotic potential can force osmotic flow from even the highest salinity solutions. The draw solution can then be subsequently regenerated (or used) by another process chosen specifically for that draw solute. While forward osmosis will not use less energy than other technologies, it does have the capability of using lower quality energy (low temperature thermal) or operating with other advantages (low fouling propensity, for instance). FO achieves this while relying simply on an osmotic potential gradient generated by a concentrated draw solution across a semi-permeable membrane. In this work, the feasibility of using FO in treatment of produced water with TDS of about 240,000 ppm was studied. Both an asymmetric cellulose triacetate (CTA) and thin film composite (TFC) forward osmosis (FO) membrane from Hydration Technologies Innovation (HTI) were used for this study. Synthetic produced water was used as a feed solution to simulate the composition of the oilfield produced water. We evaluated the individual effect of the different salts in produced water (NaCl, $CaCl_2$, $MgCl_2$, $MgSO_4$, $NaHCO_3$, and $FeCl_3$) on the flux performance. Two types of draw solutions were used in this study: 6M ammonia-carbon dioxide draw solution and 4.8M magnesium chloride draw solution. Results show that although NH_3-CO_2 has the advantage of being economically separated and recycled to the FO process, it has high reverse solute flux across the membrane. This reverse solute flux can also have unforeseen effects such as the formation of low solubility salts that cause scale or a pH change (in this case, a rise), which can also impact the solubility of salts. While ammonia is a key solute that crosses the membrane, carbonate can also precipitate readily if it diffuses. The use of $MgCl_2$ showed promise as it gave higher water flux with less scaling, although the recovery method (membrane distillation or evaporation) would have greater energy demands and higher cost.

425116 Membrane Distillation for Separating Water from Ionic Liquid Solutions

Abstract:

Ionic liquids, salts molten at temperatures below 100 °C, have exciting potential as "green" recyclable solvents for industrial processes. Ionic liquids have been found that can dissolve the lignin from biomass at relatively low temperatures. The pretreated biomass is then ready for hydrolysis and fermentation to produce biofuels. Water is added to the ionic liquid solution to surround the ions so that the dissolved lignin precipitates out. This water must be removed prior to recycling the ionic liquid for pretreatment. Two ionic liquids known to be effective in biomass pretreatment are 1-ethyl-3-methylimidazolium acetate and 1-ethyl-3-methylimidazolium formate. These were diluted with water and their conductivities and densities measured at varying concentrations. Membrane distillation of the ionic liquid – water mixtures was performed using two different well-characterized PTFE membranes. The feed was maintained at 55 °C and the distillate at 20 °C. The feed and distillate were circulated counter-currently on opposite sides of the membrane at 1.5 L/min. Water flux obtained was ~10 $L/m^2 \cdot h$. Concentration of IL in the feed concentrate and distillate was determined by conductivity and density measurements. The membrane distillation process provided a concentration factor of 10 (from 5% IL to 50% IL) with near 100% rejection; as no ionic liquid was detected in the final distillate. Membrane distillation is a promising technique for re-concentrating ionic liquids for recycling.

433229 The Research on a Polyamidoamine-Based Demulsifier for Oil/Water Separation

Abstract:

Oil-in-water emulsions widely exist in the waste water resulted from polymer flooding oil recovery process. Because there are large amounts of interfacial active substance and ultrafine oil droplets in this kind of emulsion, it is very hard to treat with traditional methods. To effectively deal with this kind of emulsion with ultrafine oil droplets and highly concentrated surfactants, one kind of chemical method, demulsification is adopted. In this thesis, the synthesized 3rd generation polyamidoamine (PAMAM) was studied as a typical demulsifier. FTIR and MNR were applied to analyze the molecular structure of the demulsifier. The influences of temperature, settling time, and concentration of the demulsifier used on the efficiency of the demulsifier were investigated in detail. At the optimal operating condition (demulsification time: 120 min; demulsification temperature: 30 °C; demulsifier concentration: 1500 mg L-1), the oil removal rate of the demulsifier could reach more than 90%. The surface tension and interfacial tension at the diesel−water interface were also measured to give a basic understanding of the demulsification mechanism.

436869 Three Types of Filter Media Produced from Electrospun Cellulose Acetate (Cac)-Polystyrene (PS) Composite Membrane for Separating Oil-Water Mixtures

Abstract:

Three Types of Filter Media Produced from

Electrospun Cellulose Acetate (Cac)-Polystyrene (Ps)

Composite Membrane for Separating Oil-Water Mixtures

L. Baghernejad[1] Chem., E. Iski[2] Chem.

Department of Chemistry and Biochemistry

O. Shoham[3], Department of Petroleum Engineering

R. Mohan[4], Department of Mechanical Engineering

S. Odueyungbo[5], Chevron Energy Technology Company

The University of Tulsa

Tulsa, Oklahoma 74104

(918) 812-7956

lida-baghernejad@utulsa.edu

ABSTRACT

(261 Words)

Oil-water mixtures are encountered in many domestic, industrial and environmental activities and are often undesired and problematic. Particularly, oil-spills such as the sudden Mexican Gulf catastrophe have attracted tremendous interest towards the separation of oil-water mixtures. Membrane filtration is one of the most effective methods so far in separating oil-water mixtures. In this project, Cellulose Acetate (CAc) fibers were reinforced by Poly (styrene) (PS) fibers via co-electrospinning CAc and PS from one solution to produce a highly porous, nanofibrous and flexible composite membrane. Three types of filter media were developed from the electrospun CAc-PS membranes: 1) a hydrophobic-oleophilic filter medium (Filter #1), which separated oil from oil-water mixtures, 2) a hydrophilic-oleophobic filter medium (Filter #2), which separated water from oil-water mixtures and 3) a hydrophobic-oleophobic filter medium (Filter #3), which is expected to separate water and oil droplets from wet gas. These filter media were produced by heating and/or treating the electrospun CAc-PS membrane with two commercially available surfactants named PHOBOL-CP® and ULTRAPHIL® from the Huntsman-DuPont Textile Company. The filter media were characterized physically, mechanically and chemically. The filtration performances of the first two filter media were tested in a lab-scale batch filtration setup, while the performance of the third filter medium remains to be tested. These filter media are novel with respect to their material, structural topography and morphology, surface roughness and filtration performance. The uniqueness of these novel filter media can be attributed to their high reproducibility, scalability, low basis weight, flexibility, high separation efficiency (> 90%) and cost-effectiveness due to their abundant and cheap starting materials. In such terms, these filter media can compete with conventional filter media.

436152 Reducing Drilling Fluid Losses Using a Lab Scale Shale Shaker

Abstract:

The recovery of drilling fluid is essential in the oil and gas drilling under the environmental and economic point of view. These fluids are basically colloidal suspensions classified according to their base components in aqueous and non-aqueous fluids. The equipment typically utilized in solid-liquid separation units are divided into three groups sequentially arranged: shale shakers, hidrociclones (desander and desilter) and centrifugal decanter. Understanding the dynamic behavior of the shale shaker is very important to reduce the process load applied to the subsequent equipment. This paper presented a contribution on the experimental research on employment of g-force control loop dimensionless number applied to a vibrating screen prototype. The study was conducted based on the method traditionally employed in solids control units. The manipulated process variable is the rotation imposed on motor vibrators installed in the vibrating screen. The results obtained in the experiments indicate an appropriate and consistent operation of the proposed control loop. The dimensionless number g-force was maintained close to the set-point value, ensuring repeatability for the experiments.
The use of higher dimensionless values of g-force does not necessarily assured greater removing of residual liquid entrained with the solid material retained by the sieve screen. One possible reason for this behavior is the fact that the greater intensity of vibration provided by motor vibrators helped to increase the solid transport speed on the display screen, reducing the effect of elimination of retained fluid.

404743 Destabilization and Treatment of Produced Oil-Water Emulsions from EOR Application Using Polyacrylamides

Abstract:

Surfactant and polymer flooding technology can greatly enhance the oil recovery through the expansion of sweeping and displacing efficiency. The recovered oil from surfactant and polymer flooding emulsifies the residual chemical, which makes the separation of water from oil quite difficult, yet the impact of the enhanced oil recovery (EOR) chemicals on the produced water cycle is generally neglected in chemically-based EOR studies. This includes compatibility of EOR chemicals with the additives used to pre-treat the injected water or change reservoir wettability and result in producing oil/water emulsion after EOR breakthrough.

The largest waste produced in oil and gas industries is believed to be the produced water, as it contains different sort of organic and inorganic admixture. The discharging of produced water contaminates not only the water, both surface and underground, but also the soil. There are a number of treatment methods available for produced water. To separate water from oil in a much efficient manner and to reach the emission standard, a new class of water soluble polymer of polyacrylamides (PAMs) was used as destabilizing agents for water-oil emulsions, which have been stabilized by surfactant (Talowamine Acetate).

The impact of the surface charge form, the density of polyacrylamides in turbidity reduction , zeta potential, COD, viscosity of separated water were explored in this study. Different anionic polyacrylamide of different surface charge density and molecular weight were evaluated. Different anionic polyacrylamides were utilized, and at optimum dosage, anionic AN94 PAM at its optimum concentration was proved as the best way to reduce the residual turbidity compared with other PAMs mentioned in this research. The effects of electrolytes such as aluminum sulphates on produced water degree of flocculation in the existence of PAM were also investigated in terms of turbidity reduction and volume of separated water after jar test. The results showed that the volume of separated water increased more than 25% compared when only PAMs were used, and the turbidity of separated water improved significantly.

431325 Destabilization and Treatment of Produced Water-Oil Emulsions Using Anionic Polyacrylamide and Electrolyte of Aluminum Sulphate and Ferrous Sulphate

Abstract:

Surfactant and polymer flooding technology can greatly enhance the oil recovery through the expansion of sweeping and displacing efficiency. The recovered oil from surfactant and polymer flooding emulsifies the residual chemical, which makes the separation of water from oil quite difficult, yet the impact of the enhanced oil recovery (EOR) chemicals on the produced water cycle is generally neglected in chemically-based EOR studies. This includes compatibility of EOR chemicals with the additives used to pre-treat the injected water or change reservoir wettability and result in producing oil/water emulsion after EOR breakthrough.

The largest waste produced in oil and gas industries is believed to be the produced water, as it contains different sort of organic and inorganic admixture. There are a number of treatment methods available for produced water. To separate water from oil in a much efficient manner and to reach the emission standard, a new class of water soluble polymer of polyacrylamides (PAMs) was used as destabilizing agents for water-oil emulsions, which have been stabilized by surfactant.

The impact of the surface charge form, the density of polyacrylamides in turbidity reduction, COD, viscosity of volume separated water were explored in this study. The effects of electrolytes such as aluminum sulphates and ferrous sulphate on produced water degree of flocculation in the existence of anionic polyacrylamide were investigated in terms of turbidity reduction and volume of separated water after jar test. Different concentrations of both sulphates added into optimum concentration polyacrylamide selected from jar test were utilized, and at optimum dosage, anionic AN 934 PAM with aluminum sulphate at its optimum concentration was proved as the best way to reduce the residual turbidity compared with other additives mentioned in this research. The results showed that the volume of separated water increased more than 25% compared when only PAMs were used, and the turbidity and COD reduction of separated water improved significantly.

435174 Reliable Use of a Baghouse in High-SO3 Environment

Abstract:

Recent EPA regulations set very stringent particulate emission standards for coal-fired power plants, as well as limit mercury and acid gas emissions. This has led to an increased use of fabric filter baghouses, often in combination with sorbents for mercury and/or acid gas removal. EPRI and Southern Company have research collaboration around the use of a pilot-scale (20,000 ft3/min slip stream) baghouse on flue-gas from medium-to-high-sulfur fuel. This talk will review recent results around emission control, particulate filtration, pulse-jet cleaning, and baghouse O&M.

Solar Thermo Micro Chemical Pyrolysis of Natural Gas for Sustainable, Distributed Production of Hydrogen

H. Dai and R.S. Besser
Chemical Engineering and Materials Science
Stevens Institute of Technology, Hoboken, NJ

Natural Gas 2015 Scenario

Current Reserves : 2,266 trillion cubic feet (Tcf) in the United States, will last <u>87 years</u> (global oil, ≈25 years) (US EIA)

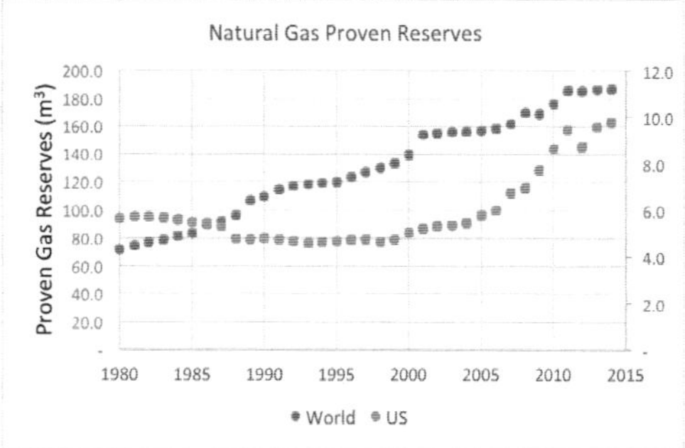

BP Energy Outlook 2015

Natural Gas Energy Conversion Approaches

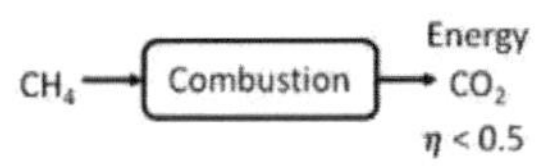

$CH_4 + O_2 \rightarrow CO_2 + H_2O + energy$

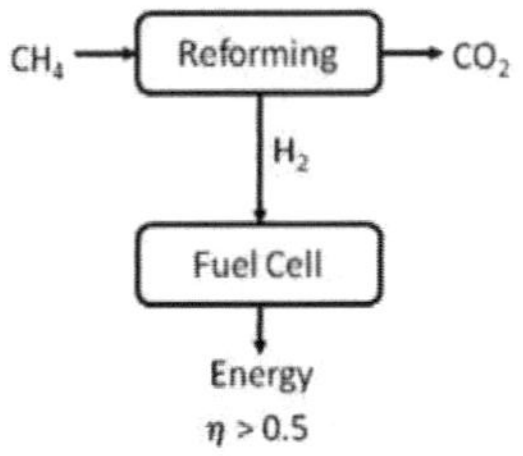

$CH_4 + 2H_2O + energy \rightarrow CO_2 + 4H_2$

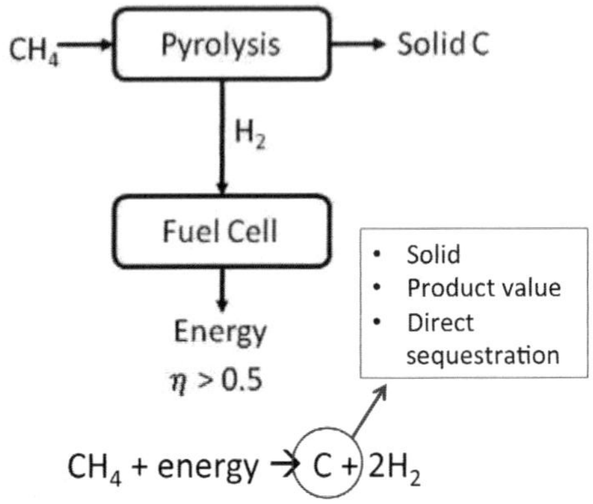

Centralized Concentrating Solar Processing

Brightsource Energy, Ivanpah, CA

Small Scale Approach: Solar Thermo Micro Chemical Process

Small footprint

Scalable by simple addition of units

Portability, mobile

≈Zero atmospheric carbon emission

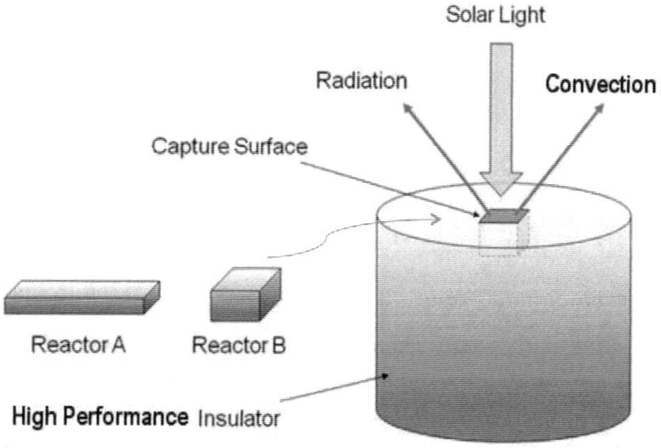

Microchemical Reactors - Process Intensification

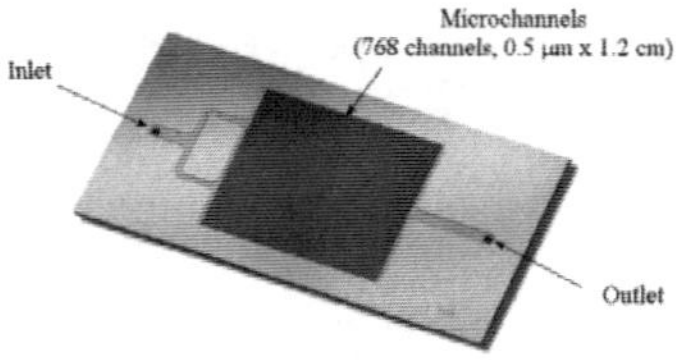

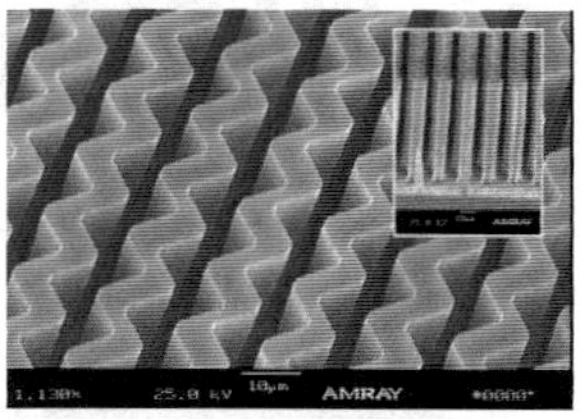

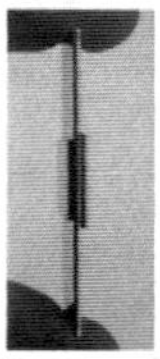

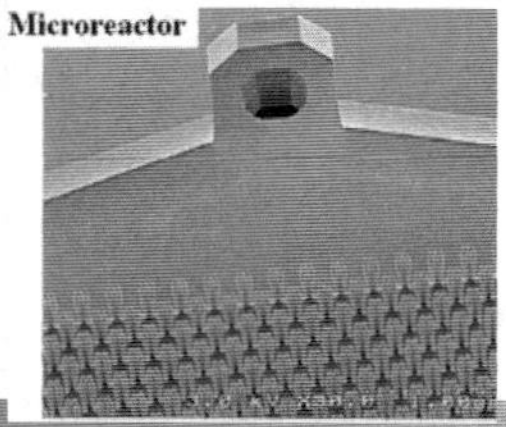

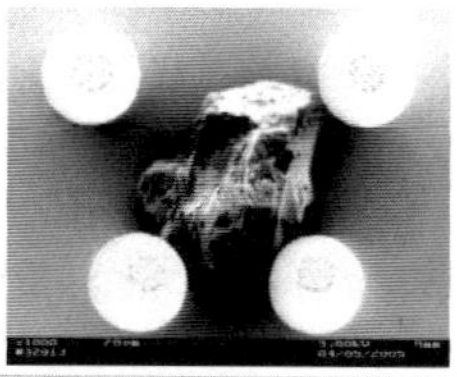

Material Considerations

- High reaction temperatures limit material options
  - Operating T of approximately <u>1000 C</u> (1300 K)
  - Ceramics (alumina, ceria, etc.)
  - Refractory metals (tungsten, tantalum, etc.)
  - Silicon
  - High T steels (e.g., inconel under low loads)
- Low emissivity surfaces help radiation loss
  - $\varepsilon \approx$ for ceramics, shiny surfaces 0.9-0.95
  - Lower emissivity metals, 0.65

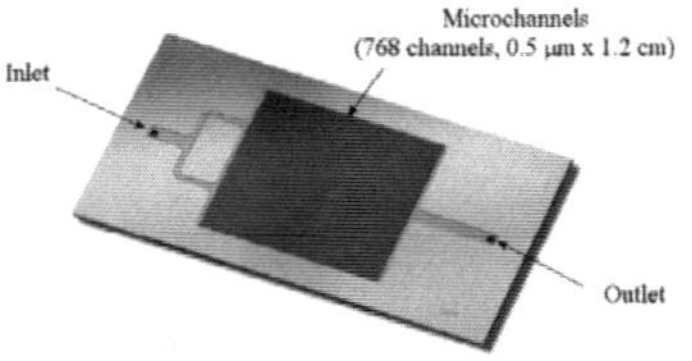

Natural Gas Pyrolysis - Catalysts

- Various active catalyst systems
  - Supported metal catalysts
    - Metals with activity Ni>Co>Fe
    - Supports SiO_2, graphite, TiO_2, ZrO_2, Al_2O_3, etc. (surface area effect vs. m-s interaction)
    - Cu, La_2O_3, MgO promoter
  - Carbon-based catalysts
    - Activated C
    - Graphite
    - Carbon black
- <u>Challenges</u>: deactivation, regeneration, recyclability, disposability, cost

Saint Gobain

Catalyst Regeneration/Reuse/Disposal Strategy

- Air: $C + O_2 \rightarrow CO_2$ $\Delta H_{1073} = -174.5$ kJ/mol
- Steam: $C + H_2O \rightarrow CO + H_2$ $\Delta H_{1073} = 135.9$ kJ/mol
- CO_2: $C + CO_2 \rightarrow 2CO$ $\Delta H_{1000} = 159$ kJ/mol

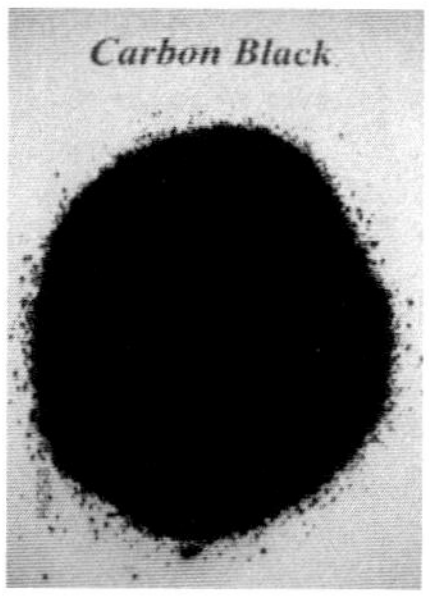

diytrade

- Fundamental approach:
  - Create no atmospheric carbon
  - Low cost sequestration (solid C vs. CO_2 capture)
  - Minimize additional energy demand
  - Low cost catalyst
  - Environmentally compatible sequestration

Steam Reforming with Carbon Capture

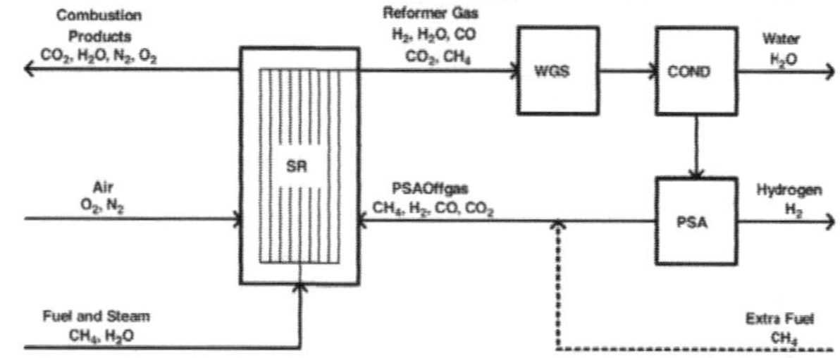

Ryd M., IJHE (2006)

Powermag

Model: Scheme

- Lens-diameter sets footprint
- Reactor
- Insulator
- Heat exchanger
- Fluidized catalyst particle feeding and recovery
- Solar tracking mechanism

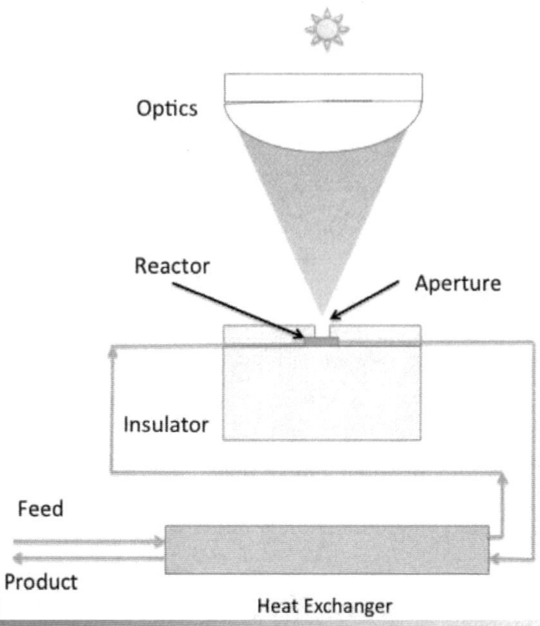

Solar Concentration: Plano Convex Lens

Commercially-available lens

Low-cost

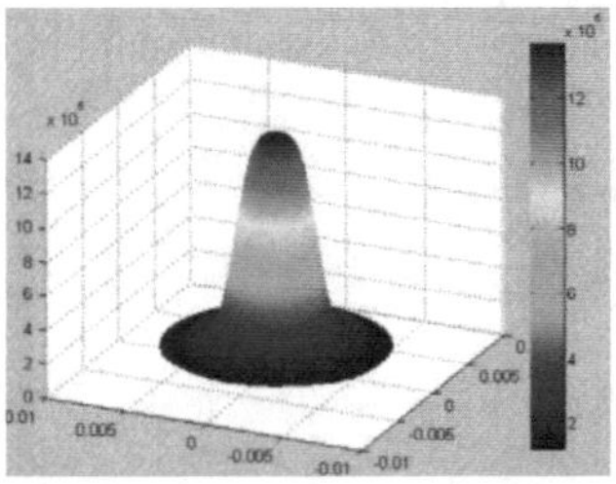

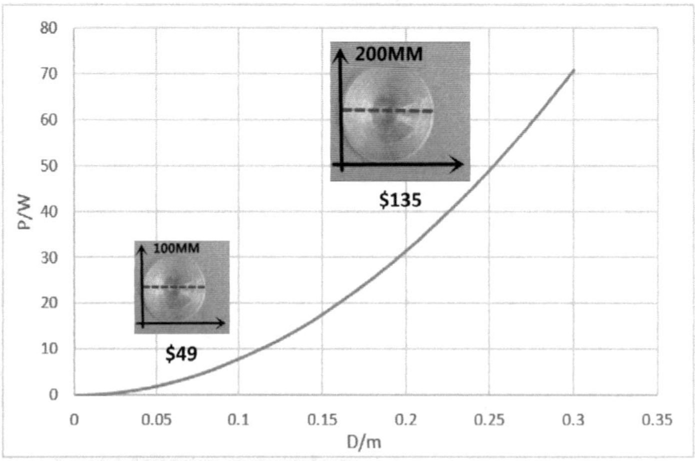

Preliminary results: solar heating experiment

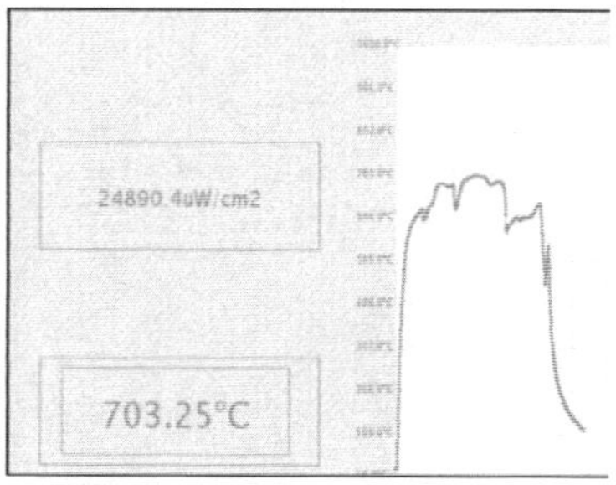

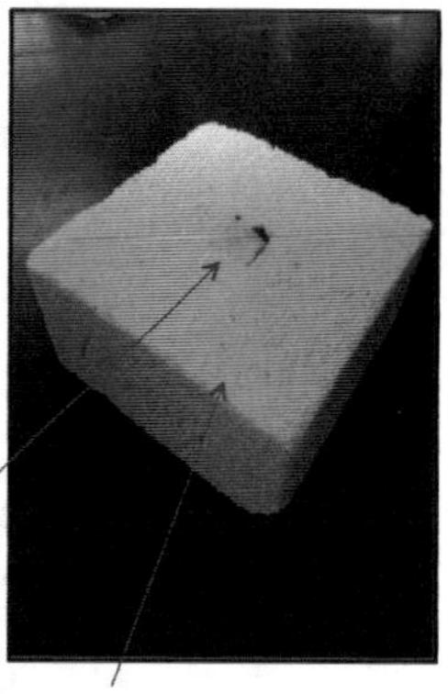

1 cm^2 ceramic substrate with thermocouple

High performance insulator

Zero-D Model Assumptions

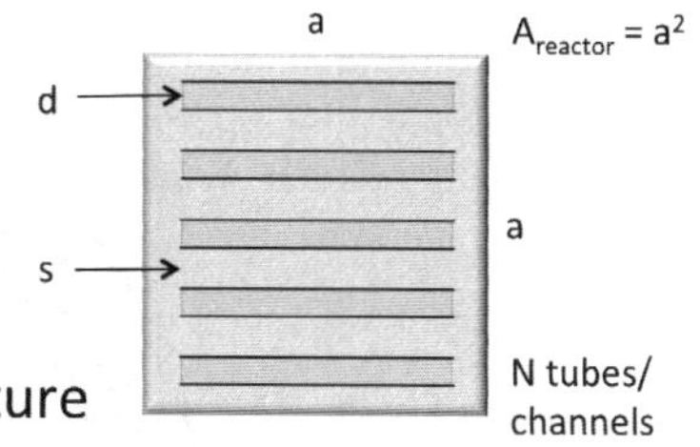

- Laminar Flow Reactor-Segregation Model
- Uniform reactor heat distribution and temperature
- Heat exchanger on outlet, product stream cooled to 400K, heat recuperated
- Heat effects of fluidized carbon catalyst ignored; heat effects of carbon product considered
- Full sun illumination (1000 W/m^2)
- Emissivity = 0.65 e.g., oxidized W
- Reactor adiabatic except at light receiving aperture

Model Equations

Overall Heat Balance

$$\dot{Q}_{solar} + \dot{n}_{CH_4} C_{P,CH_4} T_{in} - \dot{n}_{prod} \overline{C}_{prod} T_{out} - \varepsilon\sigma A(T_r^4 - T_\infty^4) - hA_c(T_r - T_\infty) - \dot{n}_{CH_4} X \Delta H_r = 0$$

$$k = k_0 \exp\left(\frac{-E_a}{RT}\right)$$

$$v = v_0(1+X)\frac{T}{T_0}$$

$$Da = \tau k$$

Reactor Conversion

$$\overline{X} = \frac{(4+Da)e^{0.5Da} + Da - 4}{(4+Da)e^{0.5Da} + Da}$$

Segregation Model-first order kinetics, Laminar Flow Reactor Residence Time Distribution

Catalyst

Carbon black: Vulcan Black Pearls BP-2000

Rate Law: ≈0.9 order – taken as first order

Activation Energy: E_a = 147 kJ/mol

Pre-exponential: k_o = 42.5 (mol/s) Pa^{-1} g_{cat}^{-1}

rate law: E.K. Lee, S.Y. Lee, G.Y. Han, B.K. Lee, T.-J. Lee, J.H. Jun, K.J. Yoon, Carbon, 2641-2648 (2004)

Model Results

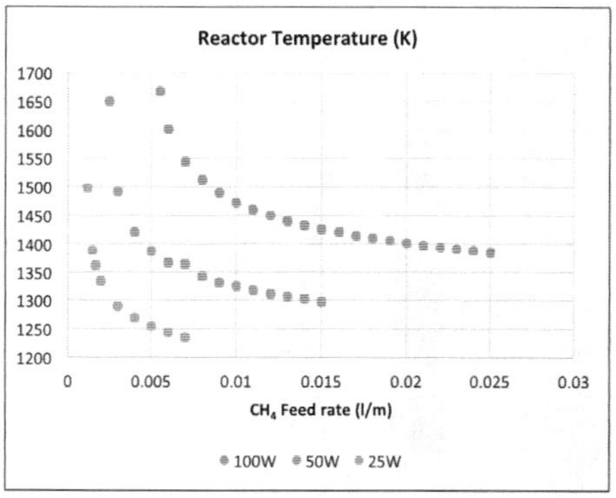

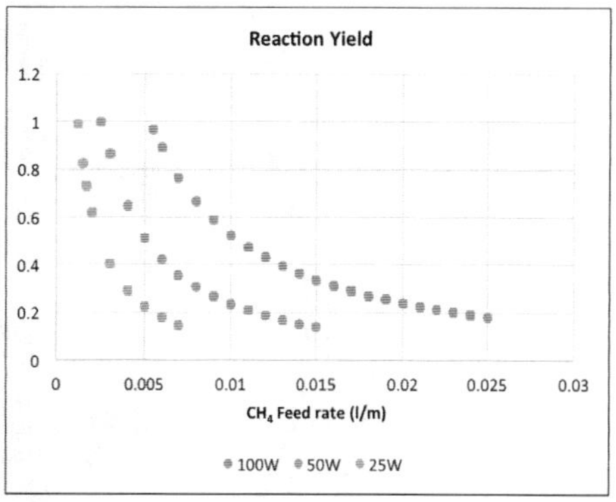

Higher reactor temperatures with higher solar flux

Higher yield with higher solar flux

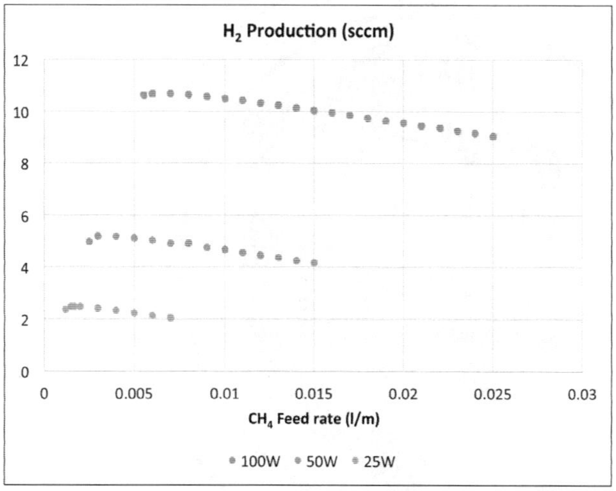

Highest production rates at the highest solar flux

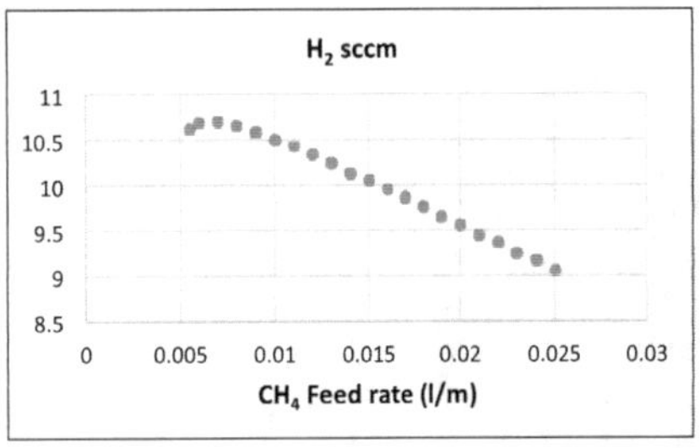

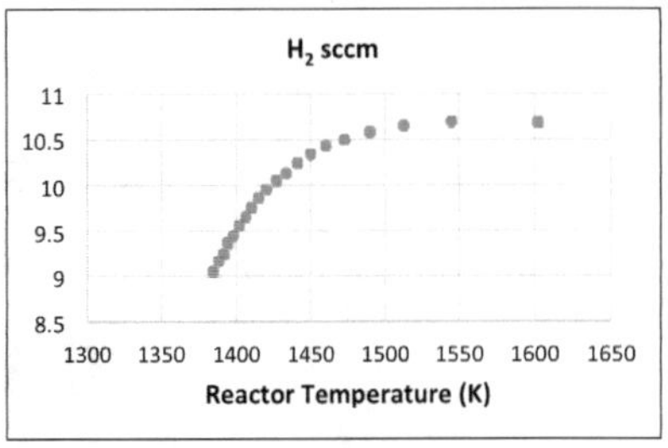

(100 W, 344 mm^3 reactor)

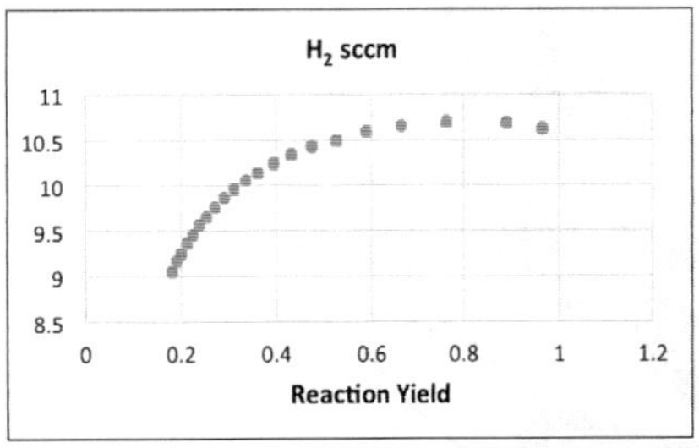

Highest H_2 production rate near:

- the lowest flow rate
- the highest yield
- the highest temperature

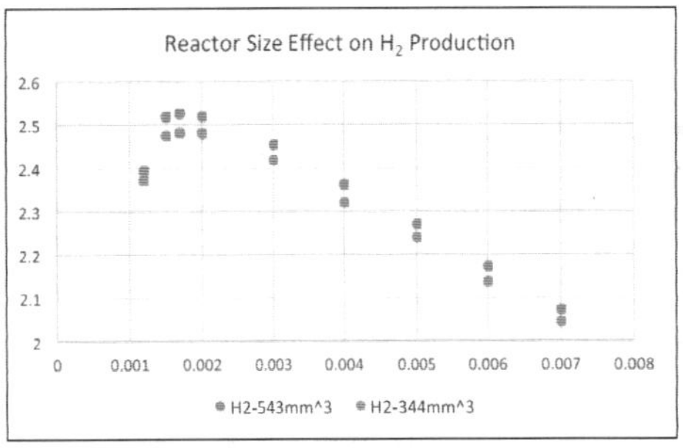

(25W solar)

Reactor size effect: 344 mm^3 vs 543 mm^3

Larger reactor size increases H_2 production slightly

Larger reactor size decreases reactor temperature

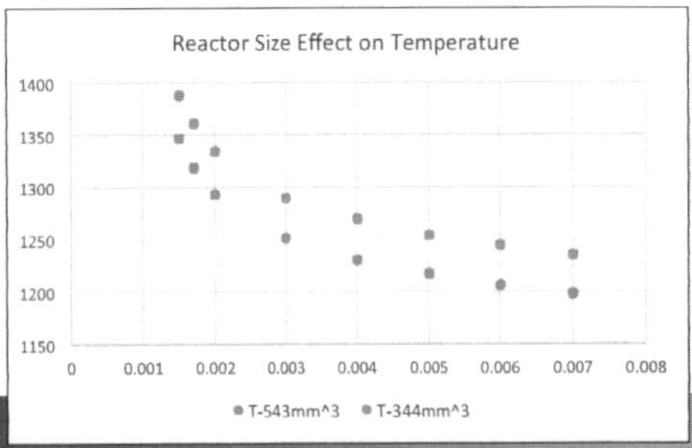

1300 K is max temperature target

Conclusions

Conversion of Sunlight to Chemical Energy (LHV) of Hydrogen

Incident Solar Power	Conversion Efficiency
100 W	1.93%
50 W	1.87%
25 W	1.82%

Achievement of reasonable operating conditions is possible

Achievement of useful hydrogen production is possible

Carbon dioxide-free, solid C sequestration approach is attractive

Small scale system concept is feasible enough to warrant further investigation

Acknowledgements

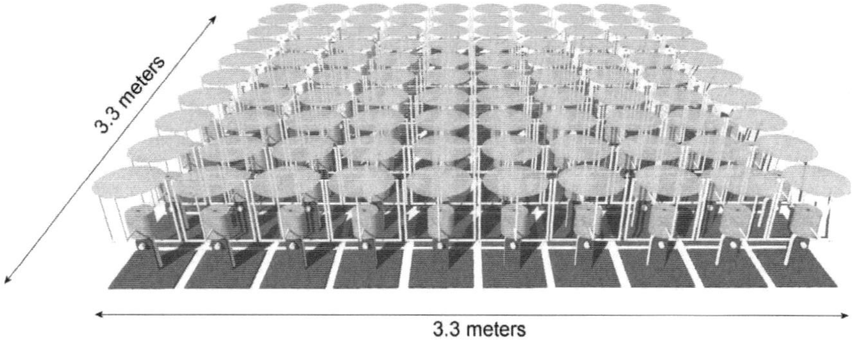

Siqi Hwang
Shafer School of Engineering and Science, Stevens Institute of Technology
Bob Glass, Lawrence Livermore National Laboratory

431659 Perovskite-Structured Redox Catalysts for Methane Partial Oxidation and Water Splitting in a Hybrid Solar-Redox Process

Abstract:

We report a perovskite promoted iron oxide as a highly effective redox catalyst in a hybrid solar-redox scheme for methane partial oxidation and water-splitting. In contrast to previously reported ferrite materials, which typically exhibit 20% or lower steam to hydrogen conversion, $La_{0.8}Sr_{0.2}FeO_{3-\delta}$ (LSF) promoted Fe_3O_4 is capable of converting more than 67% steam with high redox stability. Both experiments and a defect model indicate that the synergistic effect of reduced LSF and metallic iron phases is attributable to the exceptional steam conversion. To further enhance such a synergistic effect, a layered reverse-flow reactor concept is proposed. Using such a concept, over 77% steam to hydrogen conversion is achieved at 930 °C, which is 15% higher than the maximum conversion predicted by second law for unpromoted iron (oxides). When applied to the hybrid solar-redox scheme for liquid fuels and hydrogen co-generation, significant improvements in energy conversion efficiency can be achieved with reduced CO_2 emissions. Aside from LSF promoted iron oxide, effects of A-site and B-site material compositions on the redox performances of perovskite based redox catalysts are also investigated. A non-rare earth metal containing perovskite is found to be highly effective for the proposed hybrid solar-redox process. Over 95% syngas selectivity and 2:1 H_2:CO molar ratio with limited coke formation is found in the methane oxidation step. In addition, over 80% steam to hydrogen conversion is achieved in the water-splitting step. The current study demonstrates that structural and redox properties of perovskite-based redox materials can be optimized for the hybrid solar-redox process.

414536 Hydrogen Production Using Miec Membranes for Water Thermolysis with Partial Oxidation of Methane

Abstract:

In this presentation, we will discuss hydrogen production using oxygen permeable MIEC (mixed ionic-electronic conducting) membranes for water thermolysis on the feed side and nonreactive or reactive flow on the sweep side. Hydrogen production is driven by thermal energy and the chemical potential gradient across the membrane. Process intensification is facilitated by the addition of POM (partial oxidation of methane) on the sweep side for syngas production using the permeated oxygen.

In this work, a type of perovskite membranes made of lanthanum, calcium and iron LCF-91 ($La_{0.9}Ca_{0.1}FeO_{3-\delta}$) are used to prove the concept experimentally[1] and to develop a framework for reactor design based on the dependency of the hydrogen production rate on the design and operating conditions. The material was chosen because of its chemical stability in a reducing environment and in the presence of hydrocarbon fuels. Our results confirm that LCF-91 membrane has good oxygen permeability as well as good stability in both water and methane environment. In the presence of the membrane, hydrogen production from water thermolysis at 950 - 1030 ^{o}C is enhanced by several orders of magnitude compared to homogeneous thermolysis. Adding methane to the sweep side using a symmetric membrane hardly impacts hydrogen production rates because of the low reactivity of methane, even at 990 ^{o}C, confirming that the rate limiting process is oxygen consumption on the sweep side. With the addition of a porous layer of the same LCF-91 perovskite material on that side, the hydrogen production rate rises by two orders of magnitude. The syngas production from partial oxidation or reforming of methane on the sweep side is also studied. At high temperature, the production of soot by methane pyrolysis can be suppressed by adding CO_2 to promote the Boudouard reaction. Furthermore, a resistance-network model has been developed to model the hydrogen production rate in this membrane reactor at various conditions. Based on this model, the limiting steps for the hydrogen production process are identified and methods to enhance the production rate are proposed for the oxygen permeable membrane reactor. Other solutions to improve hydrogen production while preserving the membrane stability will also be discussed.

[1] X.Y. Wu, L. Chang, M. Uddi, P. Kirchen, A.F. Ghoniem, Toward enhanced hydrogen generation from water using oxygen permeating LCF membranes, PCCP, 17 (2015) 10093-10107.

423356 Kinetics and Reaction Steps of Autothermal Methanol Steam Reforming over CuO-ZnO-Al2O3 Catalyst

Abstract:

Kinetics and reaction steps of autothermal methanol steam reforming over CuO-ZnO-Al2O3 catalyst

Hyun Chan Lee, Jitae Lee, Dong Hyun Kim*

Department of Chemical Engineering, Kyungpook National University

Daegu, 702-701, Korea

1. Introduction

Methanol can be regarded as a hydrogen carrier as it can be easily reformed with steam (SRM) to produce hydrogen by

$$CH_3OH + H_2O \rightarrow CO_2 + 3H_2 \quad \Delta H = 49.5 \text{ kJ/mol} \qquad (1)$$

The reaction is endothermic and the reaction heat can be supplied by external heating or by oxidizing part of the methanol in the catalyst bed. The possible heat generation reactions are

$$CH_3OH + \frac{1}{2}O_2 \rightarrow CO_2 + 2H_2 \quad \Delta H = -192.5 \text{ kJ/mol} \qquad (2)$$

$$CH_3OH + \frac{3}{2}O_2 \rightarrow CO_2 + 2H_2O \quad \Delta H = -676.5 \text{ kJ/mol} \qquad (3)$$

The reaction (2) is partial oxidation (POX) and (3) is total oxidation (TOX) of methanol. Combination of either of the reactions, POX or TOX, with SRM can make an autothermal reaction (ATR) system.

$$CH_3OH + 0.795\ H_2O + 0.102\ O_2 \rightarrow CO_2 + 2.79\ H_2 \quad \Delta H = 0\ \text{kJ/mol} \quad (4)$$

As the heat is generated and consumed in the same catalyst bed, the ATR system can have a smaller, simpler reactor design, suitable for mobile hydrogen generation. Majority of studies on ATR assumed POX as the main oxidation reaction [1-5]. But a few studies have indicated TOX as the main oxidation reaction [6,7]. The objective of this study is to experimentally find the dominant oxidation reaction and its kinetics, as this information is critical in proper design of an ATR reactor.

2. Experimental

The reactor was made of 10 cm copper tube (3.175 mm OD, 1.55 mm ID). The catalyst in the reactor was a commercial $CuO\text{-}ZnO\text{-}Al_2O_3$ (Synetics 33-5) of a size of 0.3 – 0.43 mm (0.13 g). The feed rate was 150 ml/min (CH_3OH/H_2O= 25/37.5 ml/min, O_2 = 0 (SRM), 1.5 – 5.5 ml/min (ATR)), balance He). The measured temperature profile along the length of the reactor was flat due to the short length and the high thermal conductivity of the 1/8 inch Cu tube. The reaction products were mainly H_2 and CO_2 with traces of CO.

3. Results and discussions

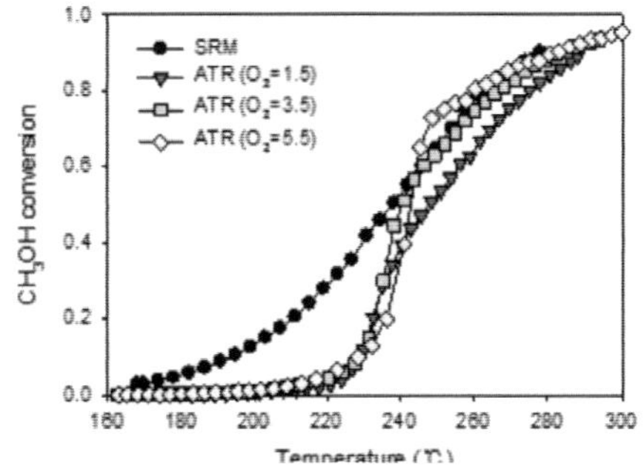

Fig. 1. shows the methanol conversion for various feed conditions. At temperatures below 240 C, SRM rate is seen to be much faster than ATR rate. This temperature region corresponded to incomplete O_2 conversion in the reactor as seen in Fig. 2. At above 240 C, the ATR rate became appreciable and comparable to the SRM rate. A power-law rate expression for the oxygen reaction rate was obtained by least square fitting of the experimental data of Fig. 2.

$$-r_{O2} = 1.1 \times 10^{12} \exp\left(-\frac{128\frac{kJ}{mol}}{RT}\right)\sqrt{P_{O2}}\ \text{mol/kg s} \qquad (5)$$

The oxygen rate is well described as a half-order reaction with an activation energy of 128 kJ/mol.

If oxygen is consumed by POX (Eq. (2)), 1 mole of oxygen consumes 2 moles of methanol. Or if oxygen is consumed by TOX (Eq. (3)), 1 mole of oxygen consumes 0.67 moles of methanol. We plotted the amount of oxygen and methanol consumption in Fig. 3. For each oxygen flow rate, the ratio of methanol/oxygen consumption followed the line of TOX to around 50 % oxygen conversion. At higher oxygen conversions, the ratio deviated from TOX and increased rapidly with increasing oxygen conversion. This can be explained by SRM occurring in parallel with TOX in the reactor. Under any conditions, the ratio of methanol/oxygen reaction ratio never followed the POX ratio, showing that POX as described by Eq. (2) does not occur directly over the catalyst.

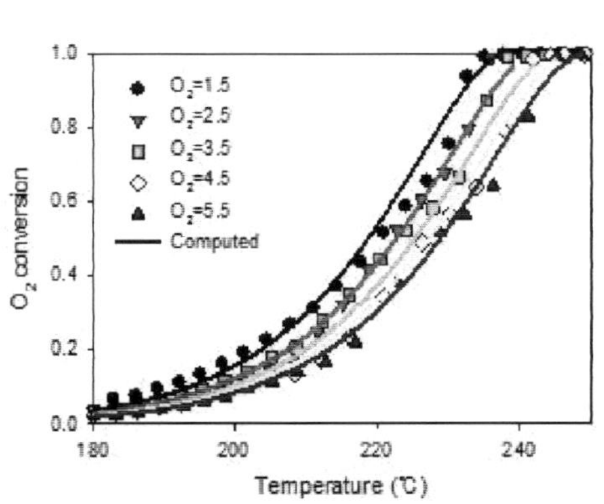

Figure 2. O_2 Conversion (O_2 rate (ml/min) shown)

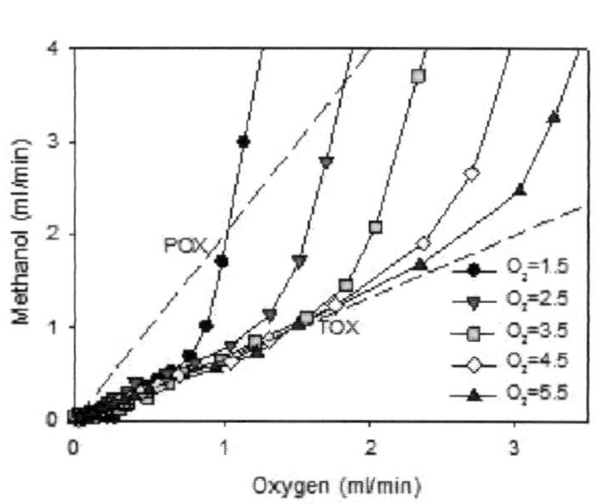

Figure 3. Methanol/Oxygen reaction ratio

4. Conclusions

In autothermal reforming (ATR) of methanol, the reaction between oxygen and methanol is experimentally found to be the total oxidation (TOX), Eq. (3), and hydrogen is formed only by the steam reforming of methanol (SRM), Eq. (1). No experimental evidence for the occurrence of the partial oxidation reaction (POX), Eq. (2), has been observed. It is thus concluded that ATR consists of TOX and SRM. The kinetics of TOX over the $CuO\text{-}ZnO\text{-}Al_2O_3$ is obtained as a half-order reaction with respect to the oxygen partial pressure with an activation energy of 128 kJ/mol.

References

1. S.Velu , K. Suzuki, M.P. Kapoor, F. Ohashi, T. Osaki, Applied Catal. A: General 213 (2001) 47.
2. J. B. Wang, Chia-Hao Li, Ta-Jen Huang, Catalysis Letters 103(2005).
3. M. Turco G. Bagnasco, U. Costantino, F. Marmottini, T. Montanari, G. Ramis, G. Busca, J. of Catal. 228 (2004) 56.
4. M. Turco, G. Bagnasco, C. Cammarano, P. Senese, U. Costantino, M. Sisani, Applied Catal. B: Env. 77 (2007) 46.
5. S. Patel, K.K. Pant, Applied Catal. A: General 356 (2009) 189.
6. J. Agrell, M. Boutonnet, J. L.G. Fierro, Appl. Catal. A: General, 253 (2003) 213.
7. M. Lyubovsky, S. Roychoudhury, Appl. Catal. B: Environmental, 54 (2004) 203.

431740 Layered-Double-Hydroxide-Derived Cu Catalysts for Methanol-Steam Reforming Reaction

Abstract:

Methanol-steam reforming (MSR) is one of the promising technologies for on-board production of hydrogen for portable fuel-cell systems, e.g. back-up power, due to its relatively low reaction temperature (200 – 300 °C), safe handling and low cost of the feed (methanol/water mixture). For the practical development of the methanol-steam reformers, this work deals with the design of highly efficient Cu-based reforming catalysts that have thermal stability and high activity around 200 °C in order to suppress CO formation by reverse water-gas-shift reaction. To design the Cu catalysts having the high dispersion with thermal stability, the layered double hydroxides (LDHs) comprising Cu, Zn, Ni and Al are synthesized via the ordinary co-precipitation method varying the composition of the components. The prepared LDHs are calcined at 400 °C to form mixed oxide materials and reduced at 250 °C. The catalytic performance tests of the prepared LDH-derived catalysts are carried out under following experimental conditions: water/methanol molar ratio = 1.1 and WHSV = 12 /h in the temperature range of 190 – 300 °C. The characterizations of the catalysts including XRD, TPR, chemisorption analyses, etc. are also performed to investigate their physicochemical properties corresponding to their catalytic performances. Among the prepared candidates, the LDH-derived Cu-Al catalyst shows the highest dispersion of the Cu particles (CuO crystallite size < 10 nm) as well as the best catalytic activity to the MSR reaction, suppressing CO formation.

430446 CO Preferential Oxidation and Methanation Catalysts and Their Performance for CO-Cleanup of Hydrogen-Rich Stream

Abstract:

In the last two decades, CO preferential oxidation (CO PROX) and CO preferential methanation (CO MET) are assumed as the most feasible reactions for CO removal from H_2-rich reformate for low temperature polymer electrolyte membrane fuel cells (LT PEMFC) feeding applications. Despite the existing advantages and disadvantages, both processes provide CO-cleanup to 10 ppm.

The present report summarizes the results obtained during a systematic study of catalysts, their performances and fundamental principles of the preferential CO oxidation and preferential CO methanation reactions. Catalytic performance of the prepared systems in a form of grains in the fixed-bed continuous flow reactor was studied. The most attention is concentrated on Cu/CeO_2 catalysts and their operation in a micro-channel reactor in preferential CO oxidation [1,2], and on Ni/CeO_2 catalysts supported over corrugated metal gauzes and their operation in a milli-channel reactor [3,4]. The role of nickel precursor was investigated and the positive effect of chlorine on the catalyst selectivity was demonstrated. The catalysts were characterized by BET, XRD, XPS, TEM, EDX, TPR and TPD techniques.

The effect of internal diffusion was estimated and the optimum grain size/coating thickness was found [8].

Based on the results obtained, the composition and structure of the catalysts' active components were determined, and reasonable suggestions on the reaction mechanisms of CO PROX and CO MET were made. The CO PROX and CO MET efficiencies were discussed comparatively.
The selected copper and nickel-ceria based catalyst provides efficient CO cleanup of H_2- rich gas streams suitable for feeding power generating units on the base of LT PEMFC.

The work was partially supported by MES (Russia), and RFBR Grant 14-03-00457-a.

References

[1] P.V. Snytnikov, M.M. Popova, et al., *Appl. Catal. A.* 350 (2008) 53.

[2] P.V. Snytnikov, D.I. Potemkin, et al., *Chem. Eng. J.* 160 (2010) 923.

[3] M.M. Zyryanova, P.V. Snytnikov, et al., *Chem. Eng. J.* 176-177 (2011) 106.

[4] M.M. Zyryanova, P.V. Snytnikov, et al., *Chem. Eng. J.* 238 (2014) 189.

[5] P.V. Snytnikov, V.A. Sobyanin, et al., *Appl. Catal. A: General* 239 (2003) 149.

[6] D.I. Potemkin, Yu.V. Shubin, et al., *Catal. Today* 235 (2014) 103.

[7] P.V. Snytnikov, S.D. Badmaev, et al., *Int. J. Hydrogen Energy* 37 (2012) 16388.

[8] D.I. Potemkin, P.V. Snytnikov, et al., *Kinet. Catal.* 52 (2011) 139.

425308 Coal Pyrolysis Gas Chemical Looping Combustion and Hydrogen Production in a Packed Bed Reactor: The Performance of Iron-Based Oxygen Carrier Using Al2O3 and Coal Fly Ash As Supports

Abstract:

Chemical looping combustion (CLC) is a promising approach to efficient use of energy and lost-cost capture of CO_2. In this study, a packed bad reactor was adopted to perform CLC tests on produced gas from coal pyrolysis. Al_2O_3 and coal fly ash were tested as the supports of an iron based oxygen carrier (OC) for both the combustion of coal pyrolysis gas and the production of hydrogen. The effects of different experimental parameters such as temperature, iron impregnate ratio, reactivity and duration of OC were investigated. Coal ash could be a cost-effective support for the OC used in this research.

433595 Comparative Study on Red Mud Gasification of Coal/Biomass Mixtures

Abstract:

Comparative Study on Red Mud Gasification of Coal/Biomass Mixtures

O. Hietsoi, C. G. Nyendu, F. Battaglia,* and F. A. Agblevor

USTAR Bioenergy Center, Department of Biological Engineering, Utah State University, Logan, UT 84322

*Department of Mechanical Engineering, Virginia Polytechnic Institute and State University, Blacksburg, VA 24061

Abstract

Coal is one of the major sources of fossil fuels used for power generation in the USA. Being one of the most polluting of all fuels, its annual use significantly contributes to environmental pollution and greenhouse gas emissions. Therefore, coal gasification is an alternative to mitigate pollutant emissions. Syngas produced from coal gasification can be used for electricity generation, chemical feedstocks, or for production of synthetic natural gas. Moreover, blending coal with biomass allows to control the hydrogen to carbon monoxide ratio, which is critical in converting syngas into transportation fuels such as gasoline and diesel by the Fischer-Tropsch process. Herein, in this work, mixtures of sub-bituminous coal and biomass feedstocks were gasified at atmospheric pressure in a bubbling fluidized bed reactor at 700—900 °C on sand and red mud. Three types of biomass were used, mainly corn stover, switchgrass, and poplar. It was shown that gasification of coal improved with increase in temperature and addition of biomass feedstock. De-fluidization of the reactor bed was observed during coal/corn stover gasification on sand at 900 °C due to formation of clinker-like material. Notably, gasification using the red mud improved gasification of the mixture compared to the baseline. The addition of biomass to the coal appeared to suppress the production of hydrogen but carbon monoxide yield increased. Additionally, the effect of dry coal/biomass gasification on red mud has been investigated.

430238 CO2 Utilization in the Production of Ethylene Oxide

Wednesday, November 11, 2015: 3:15 PM
250E (Salt Palace Convention Center)
Paul Mobley, Marty Lail and Jonathan Peters, Energy Technology Division, RTI International, Research Triangle Park, NC

A conventional hydrocarbon feedstock that is of particular interest for CO_2utilization is ethylene, which is used to make ethylene oxide. Ethylene oxide is an important feedstock for the chemical industry and is used to make many useful products such as polyurethanes, polyols, glycols, nitriles, alcoholamines, and ethers. Ethylene oxide is manufactured by several closely related industrial processes by major chemical makers worldwide (e.g., DOW Chemical, Japan Catalytic Company, Shell International Chemicals, Sumitomo Chemical, BASF, Scientific Design).

The chemical methodology by which ethylene oxide is currently produced is the partial oxidation of ethylene using oxygen (this process is called epoxidation). A silver-based catalyst is used, and either air or preferably oxygen is the oxidant. Ethylene epoxidation is performed at lower temperatures (220 to 280°C) compared with other selective oxidation processes and at high pressure (10 to 30 bar). [1]

The Captured CO_2 Catalyst for the Production of Ethylene oxide (C3-PEO), a technology being developed at RTI International, aims at producing ethylene oxide—a high-value chemical—while consuming CO_2—a greenhouse gas. This technology is based on novel catalysts utilizing the following key discoveries:

1) Abstraction of oxygen from CO_2using the reduced mixed-metal oxide catalysts

2) Transfer of the abstracted oxygen from $CO_2$to react with hydrocarbons to form desired products

3) Operation at temperatures which make these catalysts commercially practical to produce ethylene oxide

When considering the thermal stability of ethylene or ethylene oxide, both of these are highly unstable at high temperatures (i.e., greater than 300°C). Our proposed process is a two-step process detailed in Figure 1. In the first step, an oxygen atom is abstracted from CO_2 producing CO in the reducing zone of the reactor. In the second step, this highly reactive oxygen atom reacts with ethylene to produce ethylene oxide in the oxidizing zone of the reactor as shown in Figure 1 . One reason for this arrangement is the inherent reactivity between CO_2 and ethylene oxide, which could lead to the formation of undesirable ethylene carbonate in a co-feed system. Another reason for having separate reaction zones is that oxygen abstraction from CO_2typically requires higher temperature than ethylene oxidation, which is a relatively low temperature process.

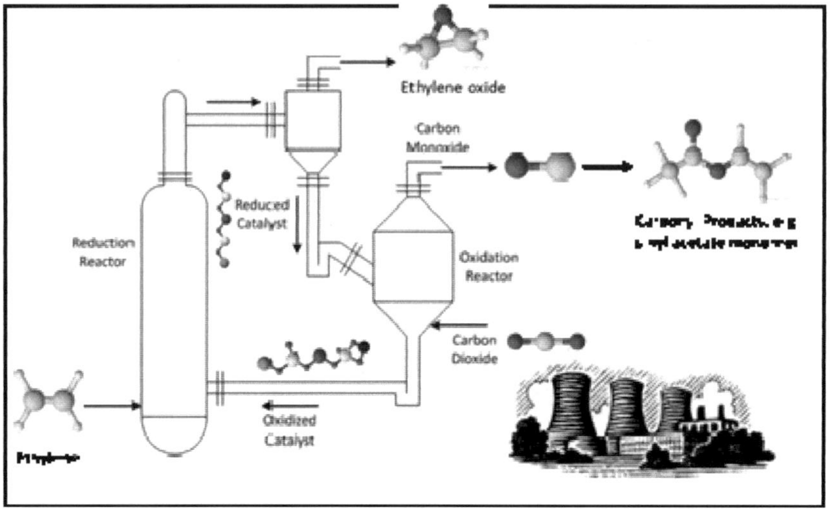

Figure 1. Conceptual schematic of the transport reactor process used in this process.

The proposed ethylene oxide process will have a concentrated CO byproduct stream, which could be used for the manufacture of many products such as methanol, dimethyl ether, acetic acid, acetic anhydride, vinyl acetate, styrene, terephthalic acid, formic acid, *n*-butanol, 2-methylpropanal, acrylic acids, neopentylacids, propanoic acid, dimethyl formamide, and Fischer-Tropsch hydrocarbons.[2] Therefore, the two marketable product streams from the proposed process are ethylene oxide and CO, which are both valuable intermediates for the petrochemical industry.

RTI has improved on its previous catalyst formulations for CO_2 utilization[3] and developed families of catalysts which can both remove oxygen from CO_2 and transfer the oxygen to ethylene to make ethylene oxide. The catalyst families are based on metal oxide phases which were found to be similar to iron in terms of reacting with CO_2, but are more selective than iron for ethylene epoxidation. Improvements on the production of ethylene oxide have been made by the use of promoters, probing the catalyst support to identify a correlation with support acidity, and observing the impact of surface area on dispersion. Ethylene oxide has been produced at small scale in the fixed-bed catalyst test reactor in both a continuous co-feed and a transport mode of operation.

(1) Chongterdtoonskul, A.; Schwank, J. W.; Chavadej, S. *Journal of Molecular Catalysis A: Chemical* **2013**, *372*.

(2) Kolb, K. E.; Kolb, D. *Journal of Chemical Education* **1983**, *60*, 57.

(3) Shen, J. P.; Mobley, P. D.; Douglas, L. M.; Peters, J. E.; Lail, M.; Norman, J. S.; Turk, B. *RSC Advances* **2014**, *4*, 45198.

433890 Dihydropteridine/Pteridine As a 2H+/2e- Redox Mediator for the Catalytic Reduction of CO2 to Methanol Via Hydride-Proton Transfer

Abstract:

The development of renewable energy technologies has been the focus of tremendous recent effort. One particularly desirable approach involves using solar energy to power the production of "solar fuels" by either splitting H_2O or reducing CO_2. These technologies would form the basis of either a hydrogen or, for example, methanol economy. In either case, the discovery of capable redox catalysts lies at the heart of both of these visions and remains as a grand challenge.

Dyer and coworkers have recently reported the use of a pteridine electrocatalyst, 6,7-dimethyl-4-hydroxy-2-mercaptopteridine (PTE), to catalyze the reduction of CO_2 to CH_3OH with a Faradaic efficiency of 10-23% and at low overpotentials; intermediate $2e^-$(formic acid) and $4e^-$ (formaldehyde) products are also observed. In this contribution, we use computational chemistry to discover that species **3** is the most stable tautomer of PTE and that it undergoes a concerted $2H^+/2e^-$ transfer to dominantly form **3a** (Scheme 1). We predict that PTE's ability to catalyze the reduction of CO_2 originates from a dearomatization-aromatization process of the **3**/**3a** redox couple, in which **3a** acts as a regenerable organo-hydride that reduces CO_2 to CH_3OH via three successive hydride and proton transfer (HTPT) steps (Scheme 1).

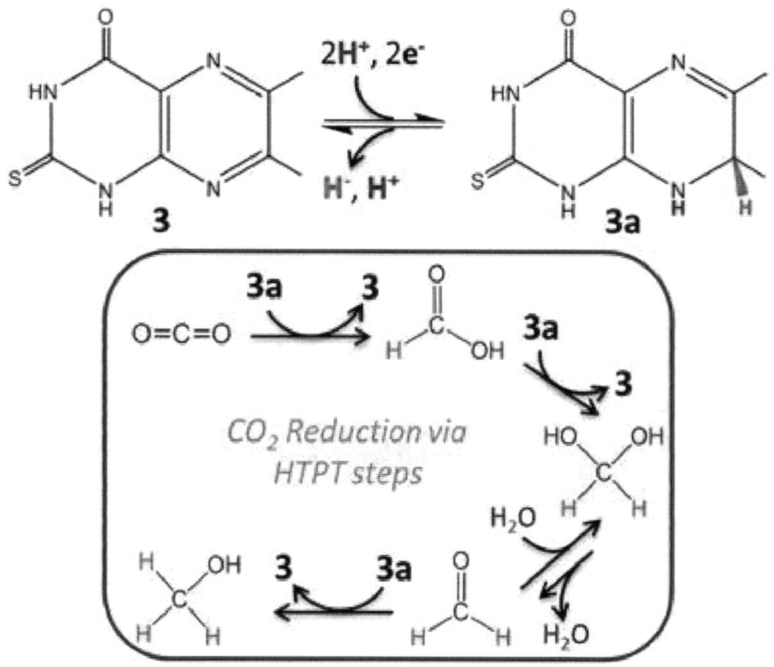

422915 Highly Efficient Catalytic Reactor for CO2 Conversion to Value-Added Chemicals

Abstract:

Carbon dioxide capture and utilization is becoming the key concern for investment in the energy and industrial sector due to ever increasing carbon emissions and growing concerns about climate change among public and government agencies around the world. While significant time and effort has been spent on finding new and innovative methods of capturing CO_2 from the atmosphere and industrial emissions, methods of utilizing this CO_2 remain to be fully developed. As Carbon Capture and Storage (CCS) technologies have advanced in recent years, the concept of CO_2 utilization has attracted more interest due to the potential of CO_2 as a useful chemical commodity to produce value-added chemicals, such as methane and methanol. The utilization of CO_2 to produce methane via the Sabatier reaction is one potential solution to this challenge. This process produces methane and water by reacting CO_2 with hydrogen. The methane generated can be used directly as a fuel or it can be converted into higher molecular weight hydrocarbons which can be transported or stored in liquid form.

Precision Combustion, Inc. (PCI), with support from NASA Small Business Innovation Research (SBIR) awards, has developed a novel efficient and compact Sabatier (i.e., CO_2 methanation) reactor based on its Microlith® catalytic technology to convert the captured CO_2 to produce methane. Our approach demonstrated the capability to achieve high CO_2 conversion and CH_4 selectivity (i.e., ≥90% of the thermodynamic equilibrium values) at high space velocities, low operating temperatures, and low H_2 requirements. The ability to achieve high CH_4 yield was made possible through the use of high-heat-transfer and high-surface-area Microlith catalytic substrates. PCI's novel approach permits reactor operation under exothermic conditions with efficient heat recuperation for high catalyst performance at high space velocities and extended catalyst lifetime. A key advantage is the ability of the reactor to efficiently sustain its temperature during the CO_2 methanation reaction via an efficient heat recuperator and heat exchanger, thus eliminating the need for an external heat source, resulting in minimal power usage during steady state operation.

Using this Sabatier reactor, PCI designed, developed, and demonstrated a stand-alone CO_2 methanation test system for demonstration and performance validation. The Sabatier reactor was integrated with the necessary balance-of-plant (BOP) components and control system, allowing an automated, single "push–button" start-up and shutdown. All of the BOP components were commercial off-the-shelf and the control system was developed in-house using National Instrument hardware and LabView software. The versatility of the test system prototype was demonstrated by operating it under H_2-rich (H_2/CO_2 of >4), stoichiometric (ratio of 4), and CO_2-rich conditions (ratio of <4) without affecting its performance and meeting the equilibrium-predicted CH_4 yield and throughput. In this presentation, the development of the Microlith-based CO_2 methanation reactor and the test system assembly for demonstration and performance validation will be discussed. Additionally, the performance results from testing the system for converting CO_2 to methane at various operating conditions and the results from the 1000-hr durability testing will be presented.

European Commission

Methanol and formic acid synthesis using captured CO_2 as raw material: techno-economic and environmental assessments

M. Pérez-Fortes

Energy Technology Policy Outlook (ETPO)
Joint Research Centre (JRC)

A. Boulamanti, G. Harrison,
J.C. Schöneberger*, E. Tzimas, S. Peteves

*CHEMCAD Stations

maria-del-mar.perez-fortes@ec.europa.eu

15AIChE
Annual Meeting, Salt Lake City, UT

European Commission

JRC's Mission and Role

... is to provide EU policies with independent, evidence-based scientific and technical support throughout the whole policy cycle.

Direct research:
JRC is the European Commission's in-house science service and the only DG executing direct research; providing science advice to EU policy.

Serving society, stimulating innovation, supporting legislation

13 November 2015 2

JRC Sites

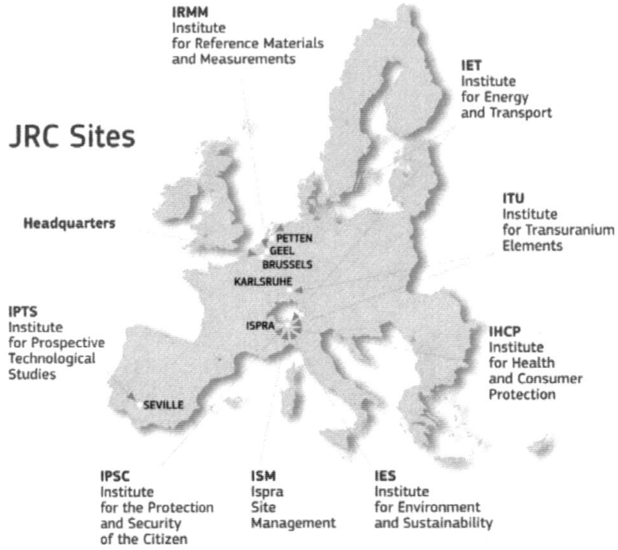

13 November 2015 3

Policy background

An Energy Union research priority - COM(2015)80:

"A forward-looking approach to CCS and CCU for the power and industrial sectors will be critical to reaching the 2050 climate objectives in a cost-effective way"

A SET-Plan Action priority – C(2015)6317:

"Step up R&I activities on the application of CCS and the commercial viability of CCU"

An Integrated Roadmap theme – JRC93056 / 2014:

"Enabling carbon capture, CO_2 utilisation and storage technologies ..." with advanced research and industrial research and demonstration actions to convert CO_2 to useful products.

13 November 2015 4

Policy issue

Is CCU profitable?

Potential to reduce CO_2 emissions?

Is CCU *sustainable*?

- Power industry - CCS
- Chemical industry
- Transport sector
- Renewable energy

13 November 2015 5

Introduction

- Two well-known products with mature markets
- Methanol economy, hydrogen economy → Prospect as fuels
- Motivating actions: manifold

Methanol (MeOH)

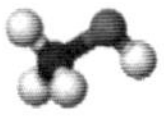

Production scale:	400 kt/yr
World production (2012):	62 Mt/yr
Product price:	350 €/t
CCU-Technology Level:	Dem scale

Formic Acid (FA)

Production scale:	10 kt/yr
World production (2013):	0.65 Mt/yr
Product price:	650 €/t
CCU-Technology Level:	Lab scale

13 November 2015 http://commons.wikimedia.org/ 6

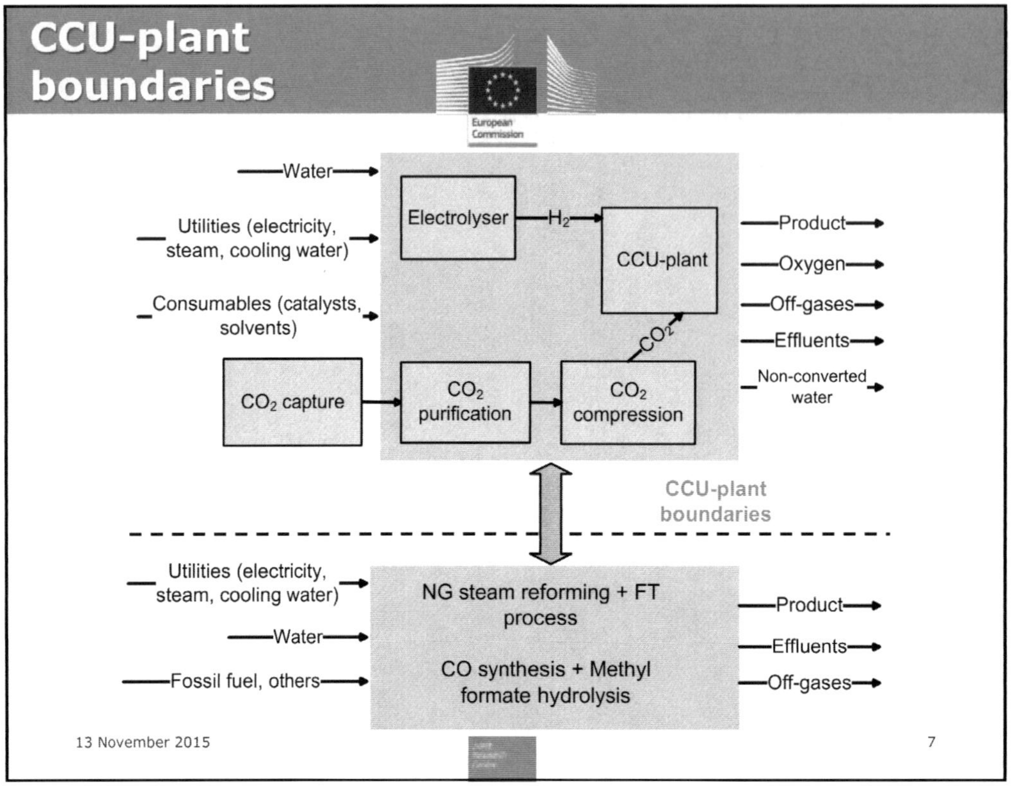
CCU-plant boundaries
European Commission
Water
Utilities (electricity, steam, cooling water)
Consumables (catalysts, solvents)
Electrolyser
H_2
CCU-plant
CO_2
CO_2 capture
CO_2 purification
CO_2 compression
Product
Oxygen
Off-gases
Effluents
Non-converted water
CCU-plant boundaries
Utilities (electricity, steam, cooling water)
Water
Fossil fuel, others
NG steam reforming + FT process
CO synthesis + Methyl formate hydrolysis
Product
Effluents
Off-gases
13 November 2015
7

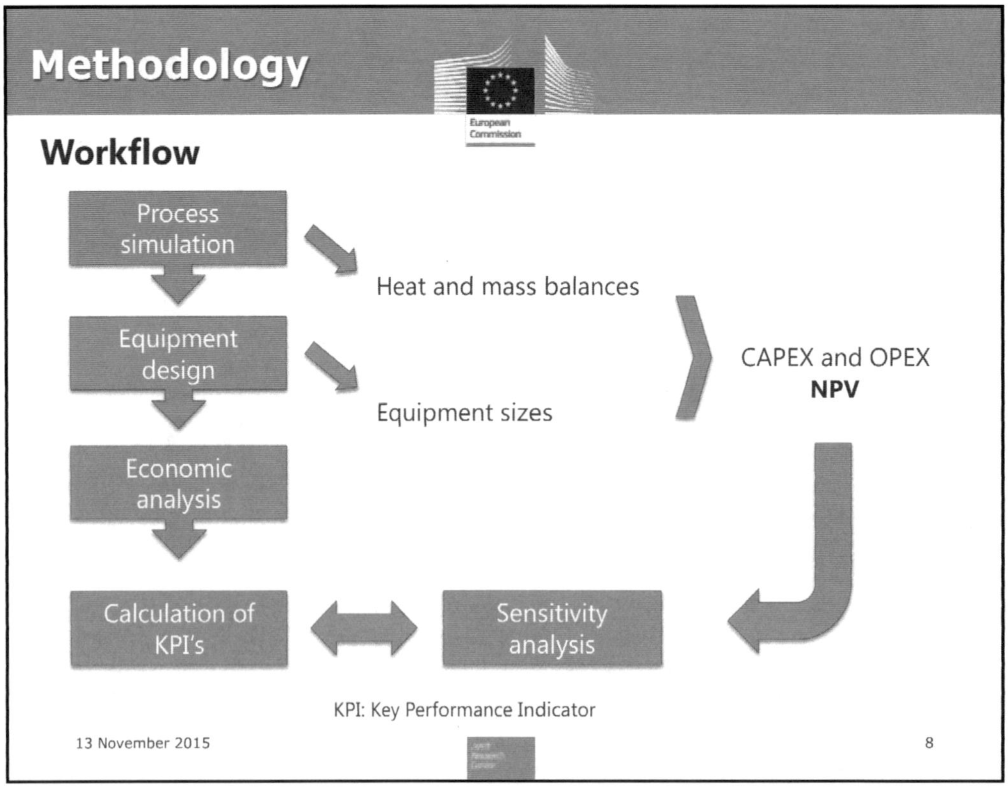
Methodology
European Commission
Workflow
Process simulation
Equipment design
Economic analysis
Calculation of KPI's
Heat and mass balances
Equipment sizes
CAPEX and OPEX
NPV
Sensitivity analysis
KPI: Key Performance Indicator
13 November 2015
8

MeOH synthesis from captured CO_2

- **Plant production:** 440 kt_{MeOH}/yr
- Reaction at **76 bar**
- Distillation at **atmospheric pressure**
- NRTL

$$CO_2 + 3H_2 \leftrightarrow CH_3OH + H_2O$$
$$CO_2 + H_2 \leftrightarrow CO + H_2O$$

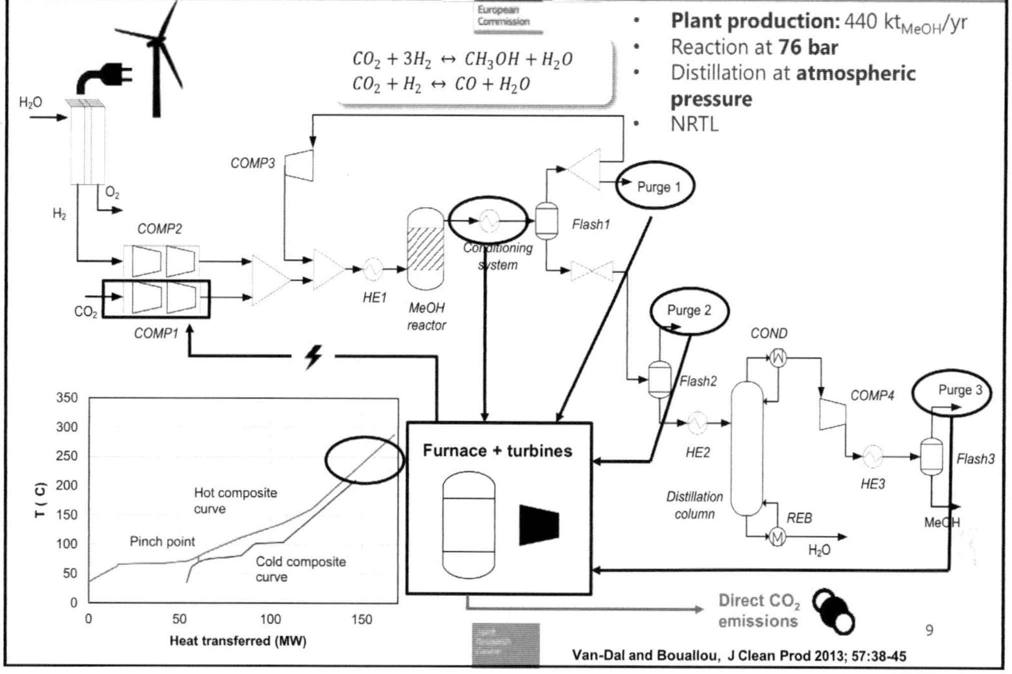

Van-Dal and Bouallou, J Clean Prod 2013; 57:38-45

9

MeOH CCU-plant results

CO_2 converted (%)	94
H_2 converted (%)	100
Electricity consumption (MWh/tMeOH)	12
Electricity consumption (w/o electrolyser)	0.2
CO_2 used (tCO_2/tMeOH) (renewable electrolyser)	1.3
CO_2 used (tCO_2/tMeOH) (EU mix electrolyser)	-4.7

CO_2 not-produced = 0.6 t/tMeOH

CO_2 emissions reduction = 77%

NPV < 0

Installed cost

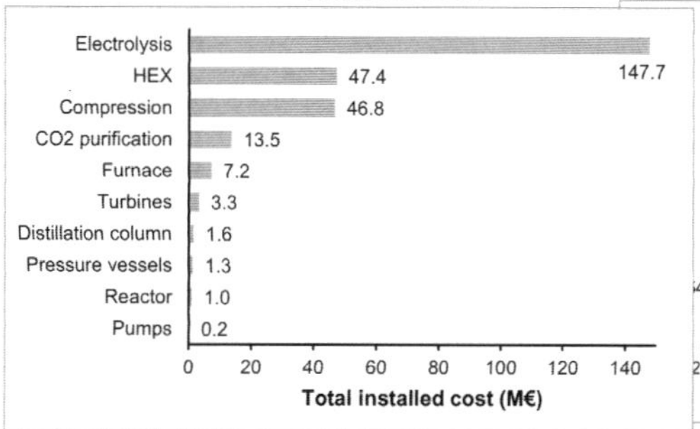

Cost of production

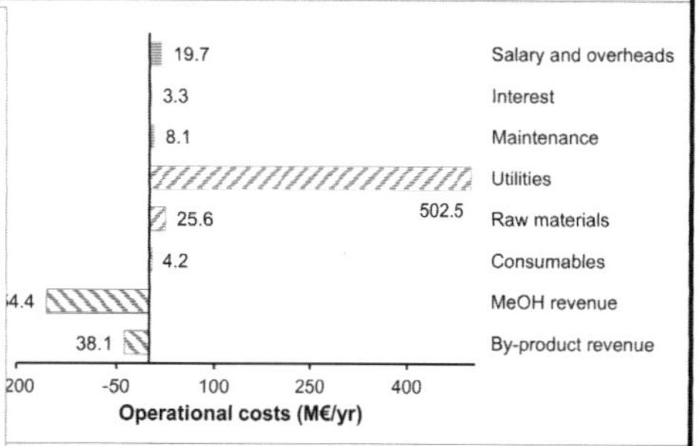

13 November 2015

10

FA synthesis from captured CO_2

European Commission

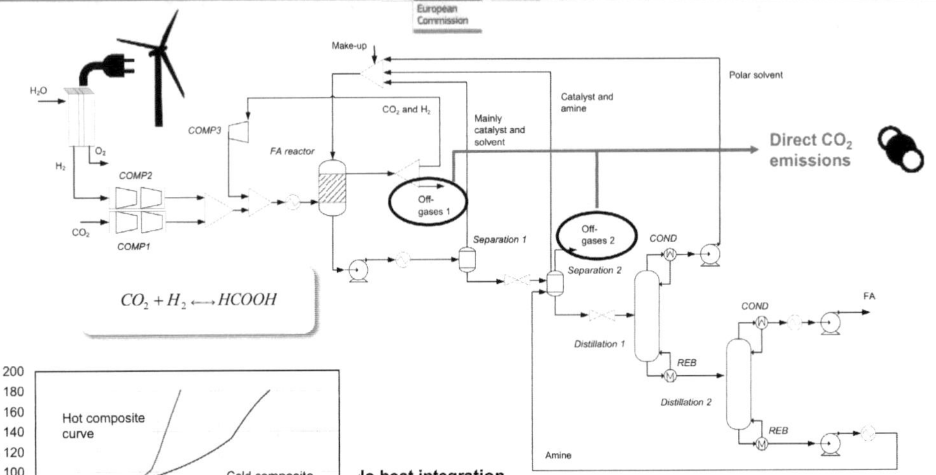

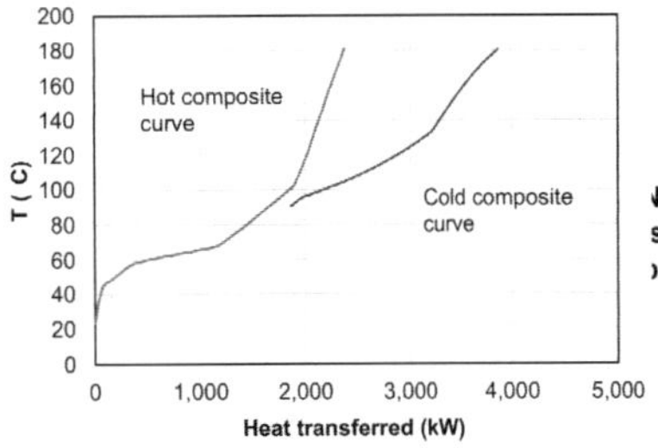

No heat integration is feasible in this process

- **Plant production:** 11.4 kt_{FA}/yr
- Reaction at **105 bar**
- Four separation steps (the final, at 0.1 bar)
- Predictive Soave-Redlich-Kwong (PSRK), UNIFAC

Schaub et al., United States patent number 8791297 B2 (2014)

FA CCU-plant results

European Commission

CO_2 converted (%)	94
H_2 converted (%)	99
Electricity consumption (MWh/tFA)	2.8
Electricity consumption (w/o electrolyser)	0.3
CO_2 used (tCO_2/tFA) (renewable electrolyser and steam)	0.7
CO_2 used (tCO_2/tFA) (EU mix electrolyser, steam emissions)	-1.3

CO_2 not-produced = 1.98 t/tFA

CO_2 emissions reduction = 91%

NPV < 0

Installed cost

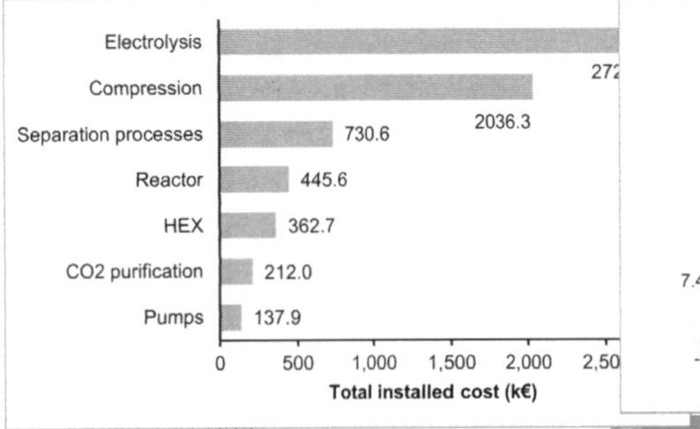

Cost of production

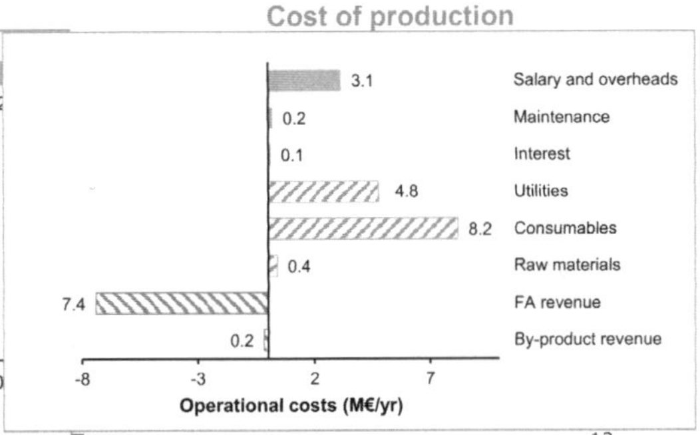

12

Main economic assumptions

MeOH	350 €/t_{MeOH}
FA	650 €/t_{FA}
CO_2 captured	38.4 €/tCO_2
Electricity	95.09 €/MWh
HP steam	25.12 €/tH_2O
Oxygen	54.2 €/tO_2
Copper-based catalyst (MeOH)	95.2 €/kg_{cat}
Ruthenium-based catalyst (FA)	210,000 €/kg_{cat}
Phosphino-based catalyst (FA)	84,900 €/kg_{cat}
72 MW Electrolyser (MeOH)	638 €/kW H_2
0.6 MW Electrolyser (FA)	1980 €/kW H_2

Economic assessment

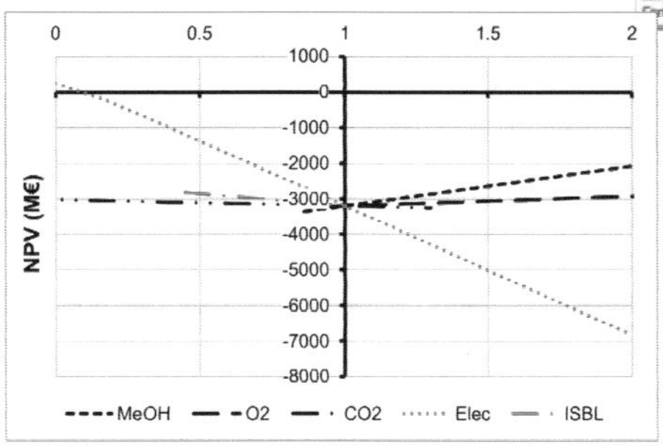

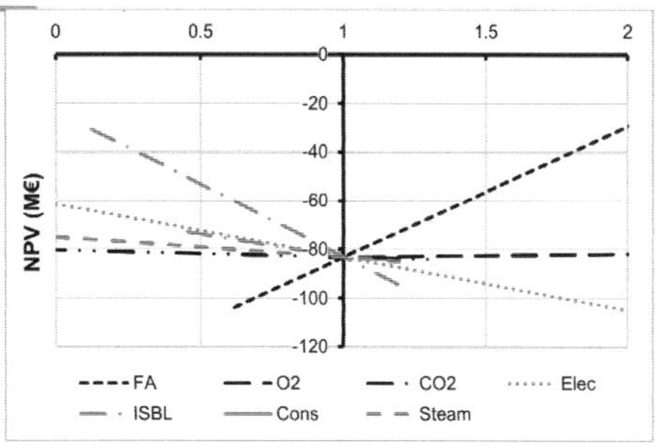

Δy/Δx
- ✓ **Price of the product**
- ✓ **Electricity**
- ✓ **Consumables**

	CCU (€/t)	Market (€/t)
MeOH production		
MeOH price	1,378	350
CO_2 price	-665	38.4
FA production		
FA	1,656	650
CO_2	-1,100	38.4

For an NPV=0:

- An increase of MeOH and FA prices **up to 2.5 to 4 times**
- Or **very high** CO_2 prices

Economic analysis: Sensitivity analysis

European Commission

Methanol

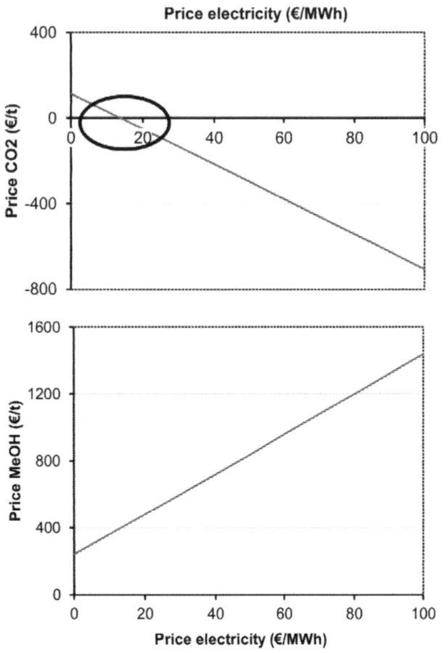

Formic acid

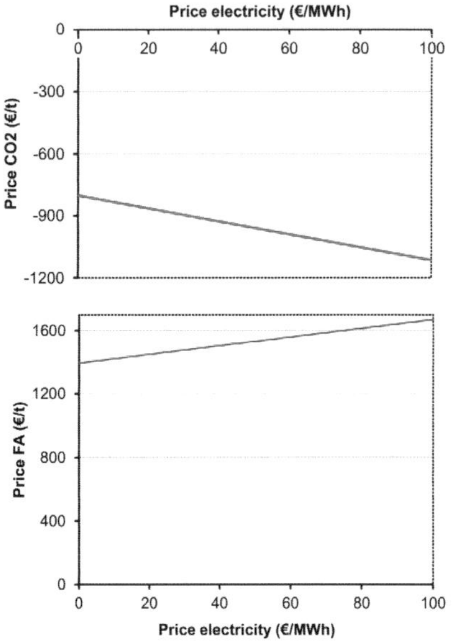

Market analysis – Year 2030

European Commission

➢ **MeOH and derivatives**;
MTBE, ethanol, DME, biodiesel

➢ Predictions based on the EC Roadmap, 2013; use of the JRC – Power Train Transition Market Agent Model (PTTMAM)

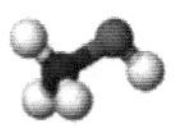

MeOH

- 2018: Current uses
- Shipping sector
- Transport – Private and light commercial vehicles
- Stationary applications – residential and industrial sectors

➢ **FA is considered as hydrogen supplier**

➢ Ad-hoc transport scenario with the PTTMAM

FA

- 2018: Current uses
- Growing demand MeOH
- Transport – Private and light commercial vehicles
- Growing demand H_2
- Stationary applications – residential and industrial sectors

Council Regulation (EU) number 630/2012 amending Regulation 503 (EC) No 692/2008, Directive 2009/30/EC

16

Market analysis – Year 2030

According to the *predicted* market:

How much CO_2 is needed?

MeOH — **9.3 coal power plants** of 750 MW

FA — **1.6 coal power plants** of 750 MW

The coal power plant emits 5 $MtCO_2/yr$

13 November 2015 17

Market analysis – Year 2030

According to the *predicted* market:

How much CO_2 emissions are not-produced?

MeOH — **16.5 Mt/yr**

FA — **15.8 Mt/yr**

The coal power plant emits 5 $MtCO_2/yr$

13 November 2015 18

Conclusions

Is CCU *sustainable*?

Large scale re-use of captured CO_2 for methanol and formic acid, under the hypotheses of this study:

- The CCU-plant is efficient in terms of CO_2 and H_2 conversion.
- Nowadays, CCU is not competitive.
- Competitiveness comes with favourable conditions: low cost electricity and consumables, simpler configurations, high integration.
- R&D is crucial
- At plant level, CO_2 emissions are decreased when compared to conventional plants, only if renewable electricity is used to power the electrolyser, and steam contributes with zero emissions.
- Plant scale should match available renewable energy sources.
- CO_2 re-use offers a relatively small, but a net contribution to the CO_2 emission reductions.

19

Future work

Is CCU competitive?

Potential to reduce CO_2 emissions?

Is CCU *sustainable*?

- Power industry - CCS
- Chemical industry
- Transport sector
- Renewable energy

JRC work

Phase 1: Methanol and formic acid as fuels

Phase 2: Urea, polymers, mineralisation

Phase 3: Methane - system integration

13 November 2015

20

Horizon Prize for CO_2 Reuse

Further information:
DG Climate Action:
http://ec.europa.eu/clima/

European Commission

Aim:

- Accelerate innovation in CO_2 utilisation technologies
- Contribute to climate change mitigation
- Increase transparency about technology readiness, barriers, costs, environmental performance and innovation needs

The winner:

The most significant and measurable improvements in the technology for CO_2 utilisation

- while overcoming technical, commercial and/or financial barriers
- and demonstrating future commercialisation potential and scalability

Prize: 1,500,000 €

Timeline:

- Detailed rules of contest published: 4Q 2015
- Opening of registration: 3Q 2016
- Deadline submission of applications: 2Q 2019
- Award of the prize: 4Q 2019

21

European Commission

Thank you for your attention!

http://setis.ec.europa.eu/

maria-del-mar.perez-fortes@ec.europa.eu

Many thanks to:

TIMES model team, specially Pablo Ruiz and Alessandra Sgobbi, Michel Honselaar, Andreas Uihlein, Jette Krause, Iratxe González, Bergur Sigfusson, Panagiota Dilara, Christian Thiel, Marta Santamaria, Kostas Kavvadias, Antonio Espuña, Aliki Georgakaki

Responsibility for the information and views set out in this work lies entirely with the authors and do not necessarily represent the view of the European Commission.

15AIChE
Annual Meeting, Salt Lake City, UT

Icons from: http://thenounproject.com/

(602e) Modeling of Particle Size Distribution in Supercritical Antisolvent Recrystallization Process

Kumar, R.,
Mahalingam, H.,

Originally presented on: 11/11/2015 16:39:00 - 17:00:00

ABSTRACT

In the pharmaceutical industry, there are conventional techniques such as milling, spray drying, liquid antisolvent process to micronize the drug particles in order to increase the bioavailability by increasing the surface area. But the particles produced from these techniques have certain disadvantages such as a broad particle size distribution, thermal degradation, presence of toxic solvent on drugs etc. Therefore, there is a need to develop effective particles technologies which could overcome the disadvantages of traditional micronization techniques. Supercritical fluid-based micronization techniques are an attractive alternative to the traditional micronization techniques capable of producing pharmaceutical compounds with a narrow particle size distribution and without any thermal and chemical degradation as well as no residual solvent content.

A fluid is described as a supercritical fluid when its temperature and pressure is above its critical temperature and pressure. Carbon dioxide (CO_2) is the most widely used fluid in supercritical fluid-based micronization techniques as its critical temperature and pressure are mild. In this work, a supercritical antisolvent process in which carbon dioxide is acting as an antisolvent has been modeled. The drug molecule considered in this work is rifampicin. This drug is first dissolved in dimethyl sulfoxide and this solution is then atomized into droplets through a nozzle into a vessel which is filled with supercritical CO_2. These droplets, which are a binary mixture of drug molecules and solvent, become a ternary mixture in the vessel as supercritical CO_2 diffuses into the droplet. Rifampicin is insoluble in supercritical CO_2 and there is a high miscibility between supercritical CO_2 and the solvent which leads to a high and fast diffusion of supercritical CO_2 into the droplet. Due to this antisolvent effect of supercritical CO_2, the solubility of rifampicin in the ternary mixture decreases sharply leading to the start of nucleation and growth. In the mathematical model presented in this work, the ternary system of rifampicin-dimethyl sulfoxide- carbon dioxide has been considered. Supersaturation is calculated as the ratio of actual mole fraction of rifampicin to the equilibrium mole fraction of rifampicin at any time interval. When the supersaturation becomes greater than one, the ternary mixture becomes unstable and the amount of rifampicin excess to equilibrium solubility starts depleting by nucleation and growth. Supersaturation is a driving force for nucleation and growth. The supersaturation profile has been predicted in the model. The model equations have been solved by MATLAB 7.1. These equations have been solved for a very small time interval (10^{-3}s). In this model, the average particle size and particle size distribution has been predicted. The effects of process parameters such as flow rate, nozzle diameter and solution concentration has been studied on the average particle size and particle size distribution. From the results obtained from our simulation, it is found that the results show the same effects of process parameters on the particle size and particle size distribution as observed by many researchers experimentally.

429247 The Effects of Variation in CO2 Stream Composition and Flow Rate on Enhanced Oil Recovery and Geologic Storage

Abstract:

Electricity generation and industrial processing typically produce large quantities of carbon dioxide (CO_2). These CO_2 streams may vary, sometimes substantially, in terms of both composition and mass flow rate. The impact of varying composition and flow rate on injection infrastructure and geologic formations during enhanced oil recovery and/or storage activities is not well known.

Literature searches were conducted by researchers at the University of North Dakota Energy & Environmental Research Center (EERC) to identify some of the effects of CO_2 stream variability on the operability of CO_2 injection infrastructure (e.g., in-field distribution pipeline system, wells) as well as on the geologic formation in which enhanced oil recovery or geologic storage is occurring. Fatigue and corrosion are considered to be the most likely effects of intermittent flow on wellbore integrity. Injectivity of a reservoir can be diminished or project infrastructure damaged as a result of pressure and temperature changes related to intermittent CO_2 supplies. Salt precipitation and hydrate formation, which can be formed as a result of these types of pressure and temperature changes, can significantly impact injectivity in a geologic formation.

The impacts of CO_2 stream impurities on enhanced oil recovery or geologic storage are reservoir-specific and depend on both the mineralogical composition and the type and amount of the impurities. Noncondensable impurities in a CO_2 stream reduce the density of the gas, leading to a drop in total storage capacity for a reservoir. Fractures and pore spaces can be blocked by mineral precipitation. The formation of acids because of the presence of water and SO_2 or H_2S can reduce the pH of the geologic formation water. This can form a zone in which mineral dissolution of carbonate and silicate minerals takes place, altering the porosity of the formation.

This paper provides a broad overview of the potential challenges to enhanced oil recovery and CO_2 utilization that could arise from variability in CO_2 streams. Opportunities for research that could reduce the cost and improve the value of the carbon capture, utilization, and storage chain will be identified.

408451 Low Temperature Hydrogenation of Amine Captured CO2

Abstract:

Low-temperature Hydrogenation of Amine Captured CO_2

*Ji Su, Mi Lu, Lisha Yang, Hongfei Lin**

Chemical & Materials Engineering, University of Nevada, Reno, NV 89557, USA

HongfeiL@unr.edu

Climate change resulting from the emission of CO_2 has become a widespread concern in the recent years. Though various CO_2 capture technologies have been proposed, chemical absorption and adsorption are currently believed to be the most suitable ones for post-combustion power plants.[1] Carbon Capture and Storage (Sequestration) (CCS) is a promising technology that has the potential to address the 40% of emission emanating from large-point sources such as power plants.[2–7] The utilization of CO_2 as a raw material in the synthesis of chemicals and liquid energy carriers offers a direct way to mitigate the increasing CO_2 build-up[8,9] and catalytic reduction of CO_2 to fuels or value added chemicals is one of the greatest challenges facing today's society.[8–12] Recently, increasing efforts have been devoted to the transformation of CO_2 into value added substances such as formic acid or formate, which can be conveniently obtained by hydrogenation of CO_2 and is highly attractive as an important chemical used as preservative and antibacterial agent as well as a clean source for H_2 generation.[13–21]

Herein, effective low temperature processes for hydrogenating amine captured CO_2 to formate over supported transition metal nano-catalysts has been developed in this study. Five representative amine monoethanolamine (MEA), diethanolamine (DEA), Triethanolamine (TEA), 2-amino-2-methyl-1-propanol (AMP) and Piperazine (PZ) were used to capture CO_2 and then the solution were conducted to hydrogenation reaction over different metal on carbon catalysts. The effect of amine structure and co-solvent was investigated. At the optimized reaction conditions, a high yield of formate, >99%, was achieved by the reduction of AMP captured CO_2 over the carbon supported Pd nanoparticles catalyst. The stability of the Pd on carbon catalyst was tested and the possible reaction mechanism was proposed. The effects of amine structure, solubility, and functional group were profound on the conversion of amine-CO_2 adducts. The yield of formates is highly dependent on the bicarbonate and alkyl carbonate ion intermediates in an solution of amine captured CO_2.

References

(1) Yu, C.-H. *Aerosol Air Qual. Res.* **2012**, 745–769.

(2) House, K. Z.; Harvey, C. F.; Aziz, M. J.; Schrag, D. P. *Energy Environ. Sci.* **2009**, *2*, 193–205.

(3) Stone, E. J.; Lowe, J. a.; Shine, K. P. *Energy Environ. Sci.* **2009**, *2*, 81–91.

(4) MacDowell, N.; Florin, N.; Buchard, A.; Hallett, J.; Galindo, A.; Jackson, G.; Adjiman, C. S.; Williams, C. K.; Shah, N.; Fennell, P. *Energy Environ. Sci.* **2010**, *3*, 1645–1669.

(5) Orr, Jr., F. M. *Energy Environ. Sci.* **2009**, *2*, 449–458.

(6) Goeppert, A.; Czaun, M.; Surya Prakash, G. K.; Olah, G. a. *Energy Environ. Sci.* **2012**, *5*, 7833–7853.

(7) Boot-Handford, M. E.; Abanades, J. C.; Anthony, E. J.; Blunt, M. J.; Brandani, S.; Mac Dowell, N.; Fernández, J. R.; Ferrari, M.-C.; Gross, R.; Hallett, J. P.; Haszeldine, R. S.; Heptonstall, P.; Lyngfelt, A.; Makuch, Z.; Mangano, E.; Porter, R. T. J.; Pourkashanian, M.; Rochelle, G. T.; Shah, N.; Yao, J. G.; Fennell, P. S. *Energy Environ. Sci.* **2014**, *7*, 130.

(8) Markewitz, P.; Kuckshinrichs, W.; Leitner, W.; Linssen, J.; Zapp, P.; Bongartz, R.; Schreiber, A.; Müller, T. E. *Energy Environ. Sci.* **2012**, *5*, 7281–7305.

(9) Mikkelsen, M.; Jørgensen, M.; Krebs, F. C. *Energy Environ. Sci.* **2010**, *3*, 43.

(10) Centi, G.; Quadrelli, E. A.; Perathoner, S. *Energy Environ. Sci.* **2013**, *6*, 1711–1731.

(11) Dorner, R. W.; Hardy, D. R.; Williams, F. W.; Willauer, H. D. *Energy Environ. Sci.* **2010**, *3*, 884.

(12) Kondratenko, E. V.; Mul, G.; Baltrusaitis, J.; Larrazábal, G. O.; Pérez-Ramírez, J. *Energy Environ. Sci.* **2013**, *6*, 3112.

(13) Boddien, A.; Gärtner, F.; Federsel, C.; Sponholz, P.; Mellmann, D.; Jackstell, R.; Junge, H.; Beller, M. *Angew. Chem. Int. Ed. Engl.* **2011**, *50*, 6411–6414.

(14) Cai, Y.-Y.; Li, X.-H.; Zhang, Y.-N.; Wei, X.; Wang, K.-X.; Chen, J.-S. *Angew. Chemie Int. Ed.* **2013**, *52*, 11822–11825.

(15) Grasemann, M.; Laurenczy, G. *Energy Environ. Sci.* **2012**, *5*, 8171.

(16) Hull, J. F.; Himeda, Y.; Wang, W.-H.; Hashiguchi, B.; Periana, R.; Szalda, D. J.; Muckerman, J. T.; Fujita, E. *Nat. Chem.* **2012**, *4*, 383–388.

(17) Joó, F. *ChemSusChem* **2008**, *1*, 805–808.

(18) Loges, B.; Boddien, A.; Junge, H.; Beller, M. *Angew. Chem. Int. Ed. Engl.* **2008**, *47*, 3962–3965.

(19) Ojeda, M.; Iglesia, E. *Angew. Chem. Int. Ed. Engl.* **2009**, *48*, 4800–4803.

(20) Papp, G.; Csorba, J.; Laurenczy, G.; Joó, F. *Angew. Chem. Int. Ed. Engl.* **2011**, *50*, 10433–10435.

(21) Wang, Z.-L.; Yan, J.-M.; Ping, Y.; Wang, H.-L.; Zheng, W.-T.; Jiang, Q. *Angew. Chem. Int. Ed. Engl.* **2013**, *52*, 4406–4409.

(22) Mani, F.; Peruzzini, M.; Stoppioni, P. *Green Chem.* **2006**, *8*, 995–1000.

(23) Monrnz, B.; Pnrr-rp, A. S.; Plpnncurh, W.; Nrns, R. J. *Am. Mineral.* **1989**, *74*, 1152–1158.

(24) Wen, N.; Brooker, M. H. *J. Phys. Chem.* **1995**, *99*, 359–368.

(25) Pelkie, J. E.; Concannon, P. J.; Manley, D. B.; Poling, B. E. *Ind. Eng. Chem. Res.* **1992**, *39*, 2209–2215.

435365 Natural Gas to Chemicals

Abstract:

This research encompasses production of value added chemicals including benzene, ethylene , naphthalene etc from natural gas via nonoxidative conversion process. This process is a thermodynamically limited reaction. If co product hydrogen is Insitu removed, the equilibrium can be shifted to the right.

The presentation will focus on investigations of a reactive separation method and pertinent findings. Over system model and approximate process economics will be discussed.

419710 Nature of the Active Mo Phase of Mo/ZSM-5 Catalysts during Non-Oxidative Conversion of Methane Reaction

Abstract:

Supported MoO_3/ZSM-5 is the most promising catalyst for dehydro-aromatization of methane (DHA) due to its near equilibrium conversion. Despite extensive studies of this catalyst system in the past decades, the nature of the initial surface MoO_x structure and the catalytic active Mo phases in the supported Mo/ZSM-5 catalyst have been under debate in the literature. To resolve the molecule details of Mo/ZSM-5 catalysts, a systematic *in situ* molecular spectroscopic study under fully dehydrated conditions and *operando* spectroscopic study during methane DHA were investigated over a series of supported MoO_3/ZSM-5 catalysts with varying Mo loading and Si/Al ratio of the ZSM-5 support.

The anchoring sites, electronic and molecular structure of the initial MoOx phase of dehydrated Mo/ZSM-5 were characterized under *in situ* condition by DRIFTs, UV-vis, Raman and X-ray absorption spectroscopy. The *in situ* FTIR results suggested that surface MoOx species prefer to anchor at Brønsted acid sites ($Al\text{-}(OH)^{+}\text{-}Si$) but also at external Si-OH and extra framework Al-OH sites at high Mo loading or low framework Al content of ZSM-5. Combined *in situ* UV-Vis DRS, *in situ* Raman and *in situ* XAS results, the surface MoOx species are essentially isolated and fully oxidized species. The absence of any d-d transition in the *in situ* UV-Vis spectra, the absence of crystalline MoO_3 or Mo-O-Mo vibration in the *in situ* Raman spectra, and the sharp pre-edge characteristic of tetrahedral coordination and the absence of Mo-Mo absorption in the *in situ* EXAFS spectra confirmed that surface MoOx species are predominate isolated dioxo $(O=)_2MoO_2$ species with tetrahedral coordination. Only a small amount of polymeric (MoOx)n species are present on catalysts with high Si content (Si/Al=140) as evident by the presence of weak Mo-Mo absorption in the *in situ*EXAFS spectra.

Operando Raman-MS investigated the relative reactivity of four distinct MoOx species on supported MoO_3/ZSM-5 catalysts during methane DHA reaction. The reducibility of the four distinct surface MoOx species decreases in the order of dioxo $(O=)_2MoO_2$ on double Al-Al sites >dioxo $(O=)_2MoO_2$ on single Al sites >>dioxo $(O=)_2MoO_2$ on double Si-Si sites~ mono-oxo $O=MoO_4$ species on extra-framework Al sites. The *operando* Raman-MS observed that the initial reactivity of methane DHA does not correlate to different MoOx structures but increases with decreasing Si/Al ratio of the ZSM-5 support. The absence of correlation between reactivity and surface MoOx structure might be due to the dynamic structural change of surface MoOx species during methane DHA reaction. The correlation between reactivity and Si/Al ratio suggests that presence of Brønsted acidity is necessary for the catalytic oligomerization of intermediate C_2H_xspecies to benzene which is the reason for zeolites being applied to oil cracking.

The *in situ* EXAFS/XANES during methane DHA reaction condition provides the molecular information of catalytic active Mo phase of supported Mo/ZSM-5 catalysts. *In situ* EXAFS/XANES demonstrated that the active site is the poorly ordered molybdenum oxycarbide (MoOxCy) nanoparticles with predominate Mo^{2+} oxidation state with small domain size (~4 Mo atom) rather than crystalline Mo_2C phase.

The present study provides a better understanding about the isolated surface MoO_x molecular structures of the supported Mo/ZSM-5 catalyst and their transformation to active phases during the methane dehydroaromatization reaction.

(673c) Oxidative Coupling of Methane on Na2WO4-Mn/SiO2: Impact of Reactor Type

Aseem, A., University of Houston
Harold, M. P., University of Houston
Originally presented on: 11/12/2015 9:20:00 - 9:45:00

Oxidative Coupling of Methane on Na_2WO_4-Mn/SiO_2: Impact of Reactor Type

Aseem, Michael P. Harold

University of Houston, Department of Chemical and Biomolecular Engineering, Houston, TX 77204

The oxidative coupling of methane (OCM) to higher molecular weight hydrocarbons has great appeal as an alternative route to chemicals and polymers. The reaction system is challenging because of large differences in the reactivity of the reactant methane and desired products (e.g. Ethylene, ethane etc.) at the elevated temperatures (> 700°C) needed to activate methane. Moreover, the reaction system is highly exothermic and is noted for significant parametric sensitivity. The goal is to achieve a C_{2+} yield in excess of 30% for the process to be potentially viable.

In spite of the large number of catalysts and reactor configurations have been studied over the past few decades, a catalyst and reactor combination has yet to be found out that gives sufficiently high C_{2+} yield. The most studied catalysts to date for the OCM reaction system have been alkali-promoted alkaline earth metal oxides, transition metal oxides and rare earth metal oxides. Na_2WO_4-Mn/SiO_2 catalyst has attracted attention because of its comparatively favorable catalytic performance [1]. To date there have been no systematic studies that have compared the performance of different rector types for the Na_2WO_4-Mn/SiO_2 catalyst.

In this study a systematic study is underway to evaluate different reactor types using the Na_2WO_4-Mn/SiO_2 catalyst. We have evaluated the fixed bed reactor and monolith reactor systems and obtained C_{2+} hydrocarbon yields between ~20% for the former to ~6% for later. The data show that as the reaction system is oxygen limited beyond a particular temperature, with the methane conversion limited to ~32%. Thus distributed oxygen feed might be helpful to enable further reaction with the goal of achieving higher selectivity for desired products. Previous studies have shown favorable results through distributed oxygen addition [2].

In our current work we are using a modified porous asymmetric γ-alumina tubular membrane to contain the catalyst and provide distributed oxygen to a continuous stream of methane. The permeability of membrane is modified to achieve an optimum flux of oxygen through membrane [3]. Membranes are impregnated by silica sol and characterized by permeability measurements with nitrogen gas. The fresh membrane has a permeability of $28cm^3/cm^2$minbar. After impregnation with silica the permeability reduces 4-fold to $6cm^3/cm^2$minbar. The permeability reduction requires a relatively high imposed transmembrane pressure gradient which facilitates homogeneous distribution of oxygen along the reactor.

Figure 1 shows that on increasing the pressure gradient across the membrane nitrogen flow to the tube side increases which correspondingly decreases the back diffusion of methane to shell side. By distributing the feed the aim is to maintain a low but uniform local concentration of oxygen along the catalyst bed length in order to avoid complete oxidation of desired products which in turn will increase the selectivity of C_{2+} hydrocarbons. Experiments are underway for a range of residence times and temperature. The performance of this fixed bed membrane reactor will be compared with conventional fixed bed reactor at same conversion under identical conditions.

Figure 1. Dependence of permeation flux of nitrogen to tube side and methane to shell side on transmembrane pressure gradient.

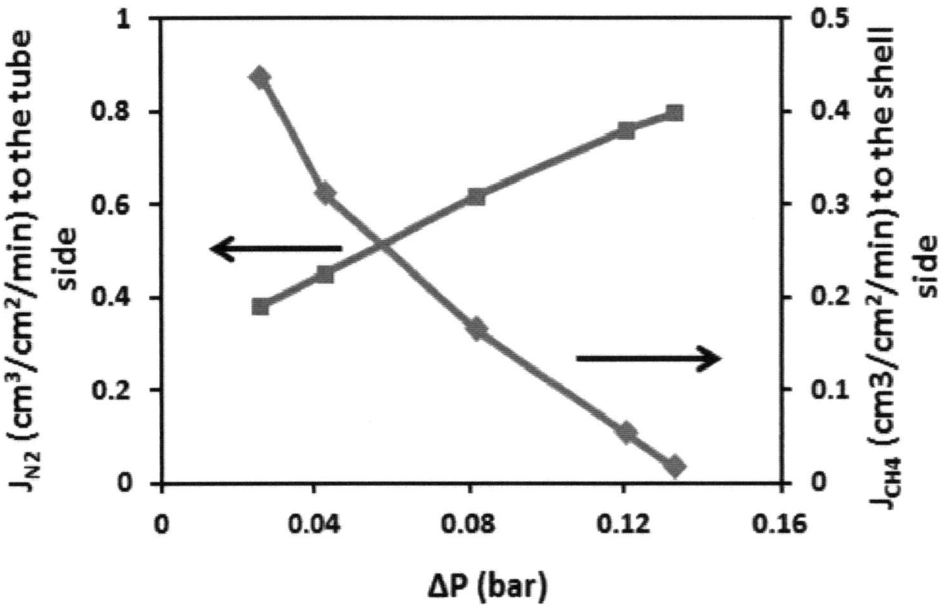

References

[1] Arndt, S., Otremba, T., Simon, U., Yildiz, M., Schubert, H., Schomäcker, R, *Mn-Na_2WO_4/SiO_2 as catalyst for oxidative coupling of methane. What is really known?* (2012) Applied Catalysis A: General, 425-426, pp. 53-61

[2] Lu Y., Dixon A.G., Moser W.R., Ma Y.H. ,*Oxidative coupling of methane in a modified γ-alumina membrane reactor* (2000) Chem. Eng. Sci., 55 (21), pp. 4901?4912

[3] Papavassiliou, V., Lee, C, Nestlerode, J., Harold, M. P., *Pneumatically Controlled Transport and Reaction in Inorganic Membranes.* (1997) Ind. Eng. Chem. Res., 36, pp. 4954-4954

435571 Ceramic Membrane Reactor for Hydrogen and Chemicals Production from Methane

Abstract:

The utilization of methane has become more attractive due to the decline in reserves of crude oil and the abundance of natural gas. The non-oxidative dehydroaromatization of methane (MDA) has great potential to directly convert methane into higher hydrocarbons and hydrogen compared to the energy intensive 2-step steam reforming/Fishcher-Tropsch process. However, intrinsic kinetics and thermodynamics of the reaction have been problematic in trying to commercialize the process. This talk considers the technique of shifting the reaction equilibrium in order to increase the methane conversion by continuously removing hydrogen from the product stream via ceramic membrane reactor. We will discuss the emerging technologies of the synthesis of the metal/silica catalysts and the fabrication of the thin hydrogen permeable membrane (SrCe0.7Zr0.2Eu0.1O3-δ) on tubular support. The integrated reaction/separation system can be optimized by manipulating the composition of the catalysts, and composition and thickness of the SrCe0.7Zr0.2Eu0.1O3-δ tubular membrane. This work offers a unique opportunity to develop a novel membrane reactor for the very active MDA process.

432547 Examination of Oxidative Coupling of Methane By Traditional Catalysis and Chemical Looping with Manganese-Based Oxides

Abstract:

The need to reduce international reliance on oil and the increased access to natural gas reserves due to improved technological developments have simultaneously elevated recent interest in utilizing natural gas and its main component, methane, to invaluable products. Ethylene, formed from the steam cracking of petroleum, and ethane are critical for the petrochemical industry. Oxidative coupling of methane (OCM) has exhibited potential in producing these higher hydrocarbons since its conception in the early 1980s. Traditional OCM schemes use gaseous oxidants from energy-intensive air separation units to catalytically convert methane to C_2 products known as the co-feed approach. An alternative approach is the chemical looping technology that poses an interesting technique to convert hydrocarbons to flexible products utilizing an intermediate, metal-based oxygen carrier that provides oxygen and catalytic activity towards higher hydrocarbons.

This study compares the traditional co-feed approach to the chemical looping approach for selectively controlling the OCM reactions. For both approaches, reducible manganese-based oxide catalysts or oxygen carriers were synthesized and compared. Co-feed and chemical looping experiments were conducted in a high-pressure, fixed bed reactor with pressures between atmospheric and 5 atm and at varied gas-hourly space velocities. The oxidation and reduction reactions are also investigated in a high-pressure thermogravimetric analyzer (TGA). The effects of pressure on the oxygen carriers were characterized by scanning electron microscopy (SEM) and nitrogen physisorption (BET). The experimental results are used in process simulations in order to model the theoretical commercial OCM co-feed system and the commercial OCM chemical looping system. Previous work demonstrated that pressure and steam have an effect on methane conversion and C_2+ selectivity with the OCM co-feed and chemical looping processes. Initial simulations demonstrate that with proper heat integration, the OCM chemical looping process exhibit promising reactor and process operations for an alternative direct approach to utilize methane. The processes are further proposed and examined for various final products such as ethylene and liquid fuels.

412347 Catalytic Effects of CeO2 and/or La2O3 on Fe2O3-Al2O3 Based Chemical-Looping Dry (CO2) Reforming of CH4

Abstract:

Chemical-looping dry (CO_2) reforming (CLDR) technology has been proposed which utilizes CO_2 as oxidant instead of air or steam to replenish reduced oxygen carrier (OC) with oxygen. The challenge of CLDR lies in the low oxidation degree of OC and slow oxidation rate due to the weaker oxidation ability of CO_2 compared to air and steam. Moreover, it is also critical to select appropriate OC materials that favor partial oxidation of CH_4 (POM) for syngas production. In this regard, the aim of this study is to use CeO_2 and/or La_2O_3 for catalyzing Fe_2O_3-Al_2O_3 based CLDR. The OCs were prepared via ultrasonication and freeze-drying assisted co-precipitation (UFDAC) method. The reaction performance of all the composite OCs was carried out in a fixed-bed reactor under atmospheric pressure condition by periodically switching CH_4 and CO_2. The influencing factors, including temperature, time-on-stream (TOS) and number of cycles were investigated. The characteristics of OCs were checked with Brunauer-Emmett-Teller (BET) analysis, X-ray diffraction (XRD), X-ray photoelectron spectroscopy (XPS) and transmission electron microscopy (TEM). The reducibility, basicity and oxidation ability of the composite materials were elucidated with temperature-programmed reduction by H_2 (H_2-TPR) and CH_4 (CH_4-TPR), temperature-programmed desorption of CO_2 (CO_2-TPD) and temperature-programmed oxidation of CO_2 (CO_2-TPO), respectively. Experimental results suggest that the simultaneous presence of CeO_2 and La_2O_3 can not only enhance the selectivity of Fe_2O_3-Al_2O_3 toward POM and its oxygen releasing rate for fast reaction kinetics, but also improve the reactivity of the reduced OC toward CO_2 splitting reaction (CSR).

Rare Earth Occurrences Proximal to the Cretaceous/Tertiary* Boundary in the Raton Basin, South-central Colorado**

Dr. H.T.Andersen[1], Dr. Rex Bryan[2], Mr. Thomas Gray[2], & Dr. Dave Richers[2]

*Presented at the AiChe Conference, November 8-13, 2015, Salt Lake City, Utah

[1] Digitus International, Golden, CO.
[2] Tetra Tech, Inc., Golden, CO.

Abstract

The Raton Coal Basin of Colorado and New Mexico is one of the premier coal basins where the Cretaceous/Paleogene (K-Pg) or the Cretaceous/Tertiary (K-T) boundary can be viewed and samples of coals, tonsteins, and shaly materials proximal to this boundary can be readily obtained. Near Trinidad, CO, extensive coal deposits exist that straddle this boundary. Further, the presence of near-by igneous features such as the Sangre de Cristo Mountains and associated dikes and sills provide a possible source of anomalous geochemical features found in the coals. Reconnaissance geochemical investigations utilizing a hand-held X-Ray Fluorescence Spectrometer (XRF) indicate that appreciable amounts of titanium, zircon, yttrium (Ti, Zr, Y), and light rare-earth elements (REE) are present within the coals and coal partings near the boundary of the Cretaceous Vermejo Formation and the overlying Tertiary Raton Formation. In some instances, semi-quantitative determinations indicate elevated quantities of Y and light REE (hundreds of parts per million) are present in select raw coal samples in the area. Earlier studies of the REE content of coal samples in the basin were conducted by the USGS and presented in their CoalQual database. The results of the XRF spectroscopy corroborates the findings reported in that database.

Altered kaolinitic pyroclastic ash is present in the basin that is interpreted to be in-part sourced from the Cretaceous/Tertiary impact. This thin claystone or tonstein is present in at least 20 locations. This thin layer is generally overlain with a thin layer of coal reported to be 4 to 16 cm thick, (Izett, 1987). Interestingly, 2 samples of this material did show detectable and elevated concentrations of iridium, titanium, and other exotic heavy metals, the REE were not detected suggesting possible other sources of the REE in the coals. Further, aside from the purported impact ash fall materials from localized volcanic episodes within the basin, intrusive igneous activity is also reported in the general area. Geochemical studies of dikes, stocks, and other intrusive material in the Spanish Peaks and the Sangre de Cristo Mountains show similarly elevated REE content. There is an apparent proximal relationship between elevated Y and REE in coal samples to these tonstein features. Additional study of REE occurrences associated with tonstein deposits and kaolinized ash within the coal as well as in the coal material itself is warranted. If the REE occurrence is found to be widespread and stratigraphically predictable, then strategies for mining the coal and segregating coal ash and other coal waste products might be developed to exploit and extract REE resources when processing the coal.

Introduction

The Raton Basin is a Laramide structural basin (Baltz, 1985). It is an asymmetrical, elongated sedimentary basin formed covering approximately 4,000 mi^2 in southwestern Colorado and northeastern New Mexico. The oldest sedimentary units in the basin are marine beds of Devonian and Mississippian age. Units of interest in this paper are the late Cretaceous and early Tertiary (Paleogene) sediments of terrestrial origin. Baltz, 1985 reports these units reached an aggregate thickness of 12,000 feet in the northern portion of the basin. Geographically the basin extends from Ute Park, New Mexico to Huerfano Park, Colorado and is bound on the western side by the Sangre de Cristo Mountains, on the north by the Wet Mountains and the Apishipa Arch, and on the southeast by the Sierra Grande Uplift (Keighin, 1995). Figure 1 (from Flores and Bader, 1999) shows the generalized Cretaceous/Tertiary boundary (K-T or K/T) geology and areal extent of the basin. Thick coal deposits are present in portions of the basin. Flores and Bader, 1999 report that lower coal units in the Raton Formation in the Raton Basin are interspersed over a 100 to 250 foot thick lower unit comprised of coal, carbonaceous shale, mudstone, siltstone, and sandstone. An upper tertiary coal-bearing unit is also present and is comprised of between 600 to 1100 ft. thick unit of sandstone, siltstone, mudstone, coal and carbonaceous shale. Coal units in these sequences can range up to 10 ft. or more in thickness with total coal thicknesses ranging between 10 and 140 feet (EPA, 2004). The underlying Vermejo Formation (upper Cretaceous) contains numerous coal bearing seams range between a few inches to up to 14 feet with total coal thicknesses ranging between 5 to 35 feet (EPA, 2004). Note also that the Spanish Peaks igneous complex lies within the northwestern portion of the Basin. This complex is composed of intermediate to silicic Cenozoic volcanics

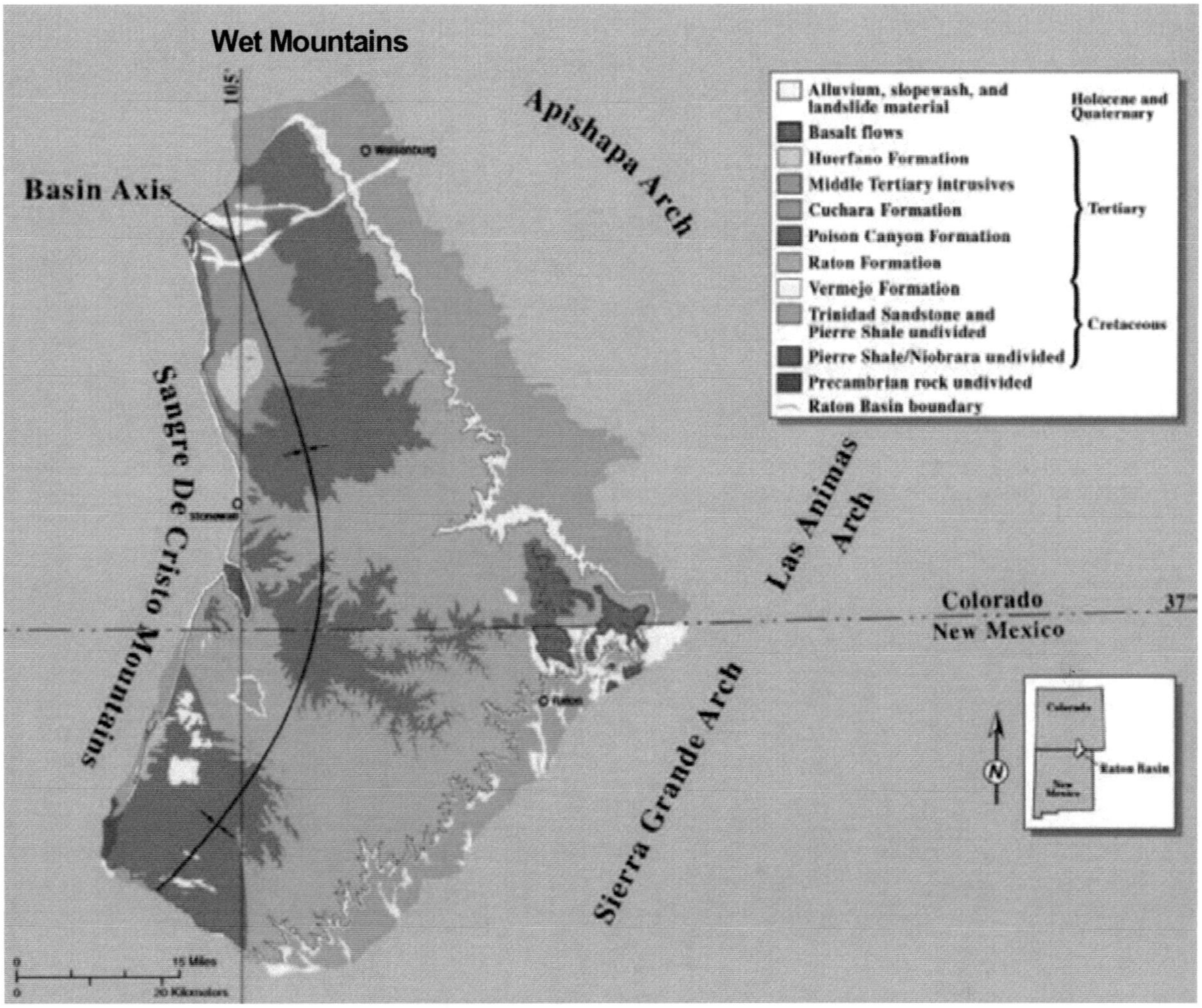

Figure 1 Generalized Geologic Map of the Raton Basin, Colorado and New Mexico (After Flores and Bader, 1999)

Rare Earth Associations in the Raton Basin

Rare earth elements (REE) are a group of 17 elements with similar chemical properties made up of the 15 lanthanides and the elements yttrium and scandium. Figure 2 from Rare Element Resources (2014) shows where they fall within the periodic chart of elements. Because of their particular fractionation behavior, most rare earth deposits appear to occur in rocks that are under-saturated with respect to silica and are believed to be products of differentiation of partial melts derived from undepleted metosomatically enriched mantle material (Sheard, 2010). As such nepheline syenite lamprophyre, and other alkaline rocks have been known to favor REE enrichment. Both the Spanish Peaks and Wet Mountains of Colorado exhibit elevated REE content. Petrographically, the Spanish Peak rocks are a continuous serial gradation from granite to gabbro, with syenodiorite being the most common rock type (Johnson, 1968) and alkaline rocks high in both Th and REE are present in the Wet Mountains (Armbrustmacher, 1984).

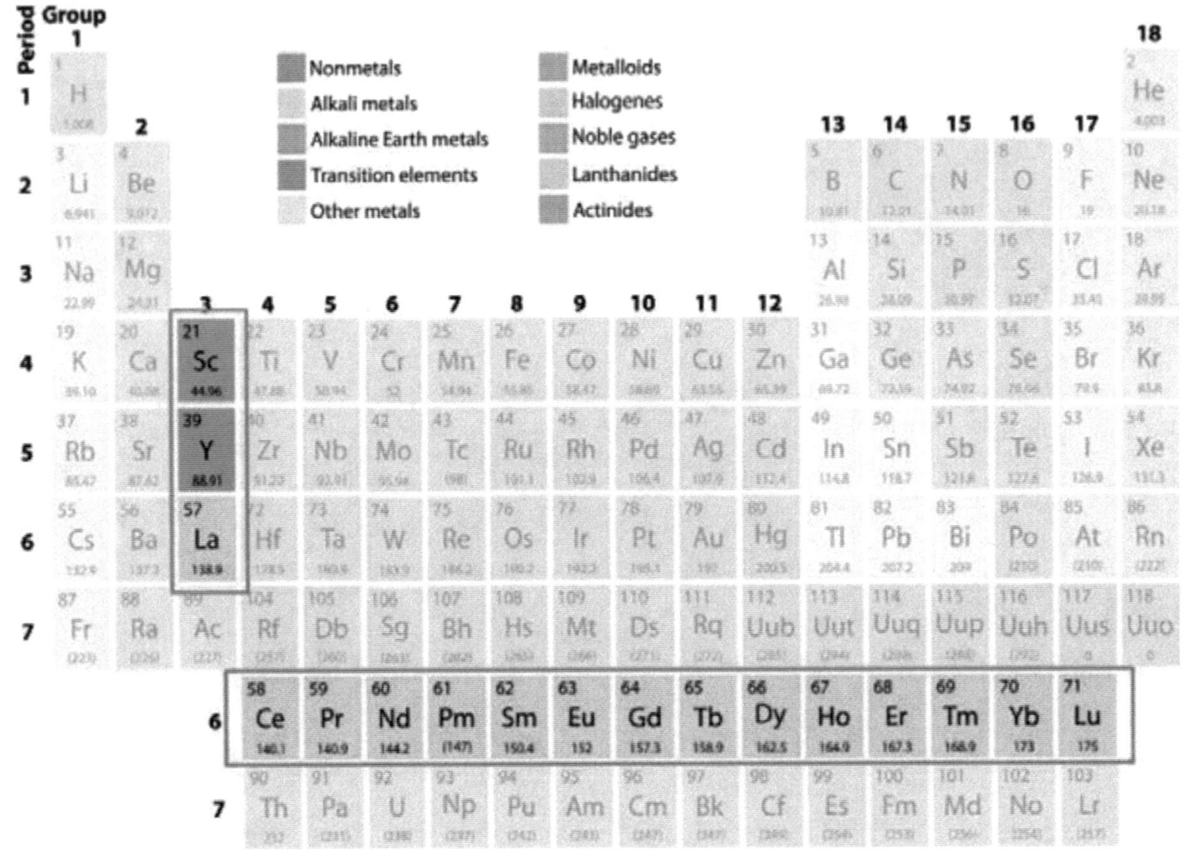

Figure 2 General Periodic Chart showing Rare Earth Elements (REE) (Rare Element Resources, 2014)

When considering possible mechanisms on how REE material can be introduced into coal and coal bearing basins, recent studies by Andersen et al., (2015) suggest three plausible means.

These include

1) introduction of REE via widespread volcanic ash falls (or tuff),
2) introduction of REE from igneous intrusions (diapirs, dikes, sill, and pipes), and
3) introduction of REE with the fluvial-detrital transport of resistate heavy black sands.

Figure 3 schematically shows the expected influence and distribution of REE in a basin using these three mechanisms (Andersen, et al, 2015). Based on the known geology history and character of the Raton basin, the influence of kaolinitic-rich volcanic ash/tuff deposits or tonsteins layers (for example the K-T layer tonstein) plus local igneous activity during the middle Tertiary resulted in the formation of igneous dikes and sills associated with the development of the Spanish Peaks are likely sources for the observed REE content in both the coals and coaly shales. Many of these igneous features "coked" coal in place in portions of the basin. If they contained REE then their influence on the observed REE in sediments may be substantial.

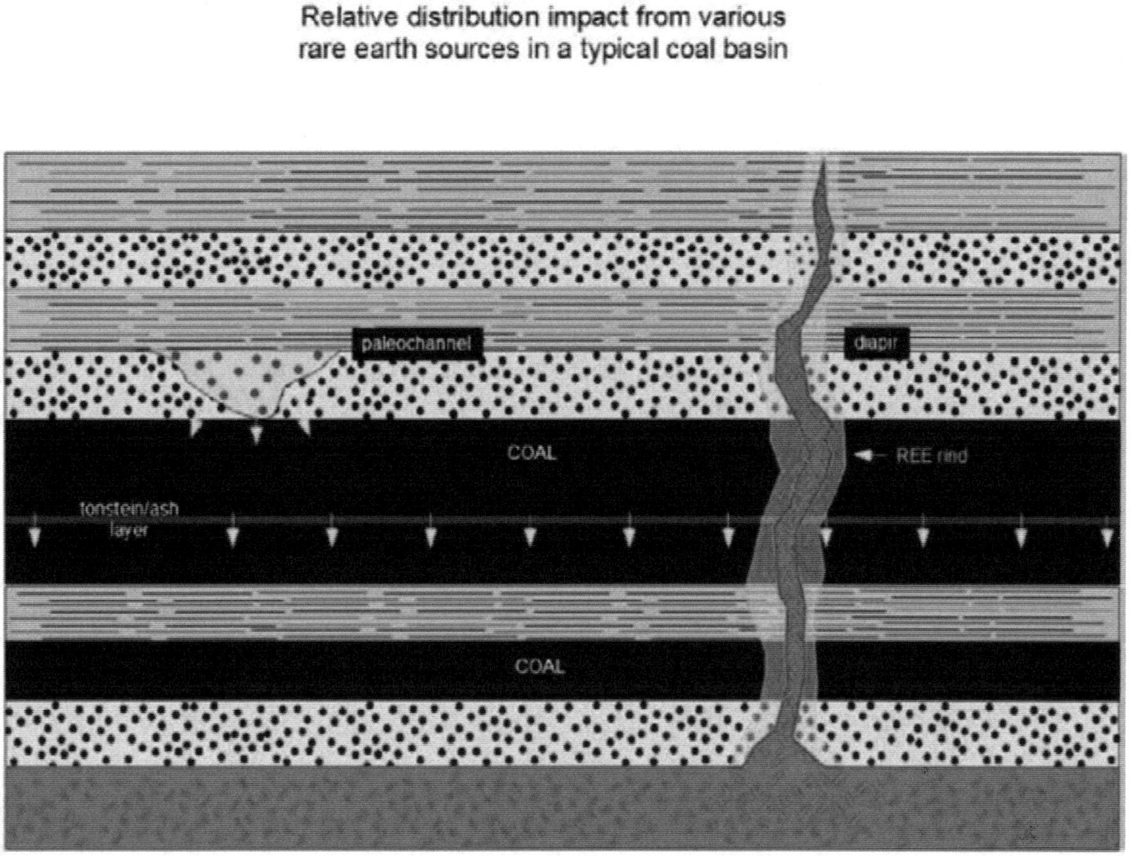

Figure 3 Mechanisms for introducing REE into Coal Basins (Andersen, et al, 2015)

When considering the Raton Basin, certainly ash falls associated with either the K-T impact, as well as earlier and later volcanic activity could be involved as could direct involvement of intruding dikes and sills. While not discounting fluvial influence as a possibility, this mechanism was not investigated in this report. In the Colorado portion of the Raton Basin, elevated occurrences of light REE in coals and coaly shales are found close to K-T boundary. This boundary is defined by a thin claystone (tonstein) layer, (Izett, 1990). In a recent field trip (Taylor, 2015) by the authors to this area, 2 samples of the actual K-T boundary clay were sampled by XRF. Figure 4 shows one site visited located in the Trinidad Lake State Park. The boundary is depicted by the thin white clay layer just beneath the massive sandstone in this figure. Interestingly enough, while samples from both sites exhibited extremely elevated levels of iridium and tungsten, as well as traces of other heavy metals, no rare earth elements were seen in the analysis. This area however, the basin is punctuated by extensive igneous activity associated with the formation of the Spanish Peaks and the associated radial lamprophyre dikes. In fact, coal

Figure 4 Cretaceous-Tertiary Boundary Clay at Trinidad Lake State Park

has actually been coked naturally in some areas of the basin suggesting that both volcanic ash as well as igneous intrusion could be responsible for the elevated REE content in the basin coals and coaly shales. Published REE data (Knight, 2011) on the rocks of the Spanish Peaks as well as new data obtained by the authors utilizing a portable X-Ray Fluorescence Spectrometer (XRF) on dike material and coal/coaly shale along the K-T boundary of the area. Knight's data appears in Table 1. These data were obtained utilizing instrumental neutron activation analysis (INAA) along with literature derived values.

Table 1. Rare Earth Elemental Data of Select Spanish Peak Igneous Samples

Sample	La	Ce	Nd	Sm	Eu	Gd	Tb	Dy	Ho	Er	Tm	Yb	Lu
bj15	74.31	138.2	62.4	10.43	2.82	8.2	5.25	2.46	1.99	0.288			
bj19	36.5	69.4	32.9	6.4	2.04	5.75	4.62	2.26	1.89	0.277			
bj2	64.55	127.6	64.2	11.23	2.93	8.25	5.42	2.51	1.98				
bj23	33.61	56.32	19.47	3.09	0.737	2.45	1.7	0.9	0.79	0.117			
bj25	58.08	110.6	44.4	7.41	1.985	5.56	3.99	1.933	1.692	0.258			
bj29	36.4	70.82	34.5	6.43	1.86	5.61	4.28	2.4	2.16	0.271			
bj37	55.2	101.8	44.7	7.77	2.08	6.47	4.65	2.3	1.84	0.272			
bj5	97.3	195.1	92	14.18	3.6	9.5	4.85	1.94	1.41	0.192			
bj64	34.03	59.28	19.15	3.3	0.56	2.76	3.2	2.17	2.72	0.475			
bj65	68.7	122	46.9	7.61	2.14	5.73	4.23	1.97	1.7	0.26			
bj8	95.46	199.2	96.3	14.69	3.72	8.86	4.61	1.844	1.48	0.196			
bj8-18	91.2	154.5	50.1	6.95	1.98	4.32	3.33	1.74	1.65	0.254			
bj8-40	68.1	132.4	60.9	10	2.82	7.4	4.88	2.16	1.77	0.226			
bj8-w	22.17	42.59	19.64	4.1	0.8	3.66	3.97	2.63	2.89	0.467			
rj108	61.6	108	6.27	1.93	6.1	0.91	1	1.96	0.36				
rj113	71.4	113	6.01	1.64	7.4	0.804	1.1	2.22	0.44				
rj137	55.7	108	8.48	2.97	9.4	1.26	1.4	2.54	0.45				
rj16	0.59	3.33	0.9	0.348	2.1	0.358	0.48	1.03	0.18				
rj168	85	185	13.5	3.54	12	1.5	2.3	2.16	0.31				
rj173	84.6	159	11.3	3.21	11	1.36	2.3	2.36	0.33				
rj178	29.7	74.4	7.13	2.28	8.4	1.05	0.92	2.2	0.36				
rj5	1.39	4.45	0.78	0.299	1.5	0.204	0.23	0.43	0.08				
rj78	58	68	3.24	0.754	4	0.435	0	1.35	0.35				
rj86	80.3	130	6.77	1.88	8.3	0.957	1.3	3.21	0.58				
rj97	200	403	23.5	6.5	19	1.94	3.6	2.66	0.38				
rjESP	64.6	76.6	5.04	0.707	6.3	0.821	0	4.72	1.02				
rjMtMest	4.06	6.23	0.96	0.349	0	0.212	0	0.91	0.21				
rjSWPeak	18.8	40.5	3.08	0.911	2.4	0.361	0.35	0.94	0.15				
sp101	55.9	111	47	8.44	2.16	6.82	0.927	1.2	0.408	2.59	0.363		
sp118	17.8	34.4	15	2.6	0.653	1.82	0.245	0.312	0.118	0.688	0.096		
sp121	39.9	89.2	42.1	8.21	1.91	6.23	0.857	0.886	0.292	1.74	0.252		
sp123	33.5	69.9	31.8	5.61	1.48	3.72	0.548	0	0.269	1.54	0.236		
sp124	3.7	8.19	4.31	1.31	0.424	1.57	0.223	0.27	0.079	0.455	0.053		
sp131	94.6	206	98	14.2	3.56	8.95	0.969	0.779	0	1.4	0.185		
sp134	57.7	110	40.5	6.46	1.46	5.68	0.76	1.08	0.469	2.91	0.431		
sp18	30.4	65.6	32.3	6.26	1.74	5.05	0.658	0.777	0.293	1.73	0.252		
sp20	42.2	88.1	41.6	7.87	2.06	6.3	0.847	1.02	0.42	2.65	0.393		
sp21a	39.7	92.7	46.8	9.61	2.36	8.22	1	1.03	0.313	1.67	0.226		
sp21b	70.3	142	63.9	10.7	2.77	7.21	0.926	0.921	0	1.69	0.263		
sp23	40.8	81	34.6	5.82	1.58	4.37	0.574	0.753	0.299	1.81	0.264		
sp33	51.2	97.9	39.1	7.18	2.03	5.64	0.755	0.881	0.324	1.8	0.253		
sp34b	70	150	71.5	11.3	3.07	9.11	1.1	1.29	0	2.61	0.382		
sp34ts	48.9	107	51.2	9.48	2.4	6.62	0.884	0.988	0.344	2.01	0.297		
sp37	52.5	116	60	11.2	2.88	7.66	0.958	0.976	0.302	1.77	0.246		
sp41	44.5	94.9	43.6	8.47	2.22	6.86	0.878	0.924	0	1.86	0.269		
sp42	62.1	132	57.1	9.67	2.23	6.6	0.803	0.892	0.292	1.66	0.242		
sp46	49.3	101	42.9	8.01	2.11	6.33	0.776	0.731	0	1.44	0.206		
sp5	69.8	160	76.5	12	3.09	7.7	0.857	0.788	0	1.44	0.183		
sp51	59.1	119	53.6	9.4	2.5	7.91	0.965	1.08	0	1.9	0.259		
sp52	66.2	142	65.3	11.1	2.83	8.13	0.939	0.91	0.303	1.8	0.262		
sp57	41.9	87.9	38.3	6.82	1.8	5.41	0.721	0.93	0	2.57	0.383		
sp58	53.1	113	51.7	8.7	2.28	6.09	0.752	0.691	0	1.26	0.181		
sp58b	53.1	111	52.1	8.86	2.3	6.42	0.748	0.688	0	1.25	0.169		
sp65	52.5	114	52.4	9.25	2.54	7.21	0.867	0.866	0.303	1.72	0.244		
sp70	107	262	129	18.9	4.39	11.3	1.05	0.786	0	1.15	0.154		
sp70b	109	271	132	19.5	4.51	11.2	1.06	0.807	0	1.21	0.151		
sp95	60.2	119	51.2	8.69	2.29	7.3	0.932	1.13	0	2.5	0.357		

In the above table, samples beginning with "bj" are from Jahn (1973), and samples beginning with "rj" are from Johnson (1968).

Surface outcrop samples of coal and shaly-coal material from the Colorado portion of the Raton basin were analyzed for their light REE content using a Niton XL3T Gold++ portable X-ray fluorescence (XRF) spectrometer. Table 2 shows the results of several of these samples as well as some from the Florence, Colorado, area. What is interesting about the Raton samples is they were collected near the Cretaceous/Tertiary boundary and are situated next to the Spanish Peaks volcanics and associated lamprophyre dikes. The Florence samples, collected northwest of the Raton Basin are adjacent to the Wet Mountains, which are reported to contain appreciable thorium and REE (Long et al., 2010) in syenitic and carbonatite dikes.

Coal samples were collected by USGS in the Raton Basin as part of their CoalQual Database effort (USGS, 2014). That study sent coal samples for laboratory analysis using inductively coupled plasma/mass spectrometry (ICP/MS) to determine their REE contents. Comparison of XRF results for samples collected by Tetra Tech in the Raton Basin with the reported USGS values shows that the Raton Basin consistently has elevated REE contents. Figure 5 shows the Colorado portion of the Raton Basin with both light REE values from the USGS CoalQual database coal samples (colored squares) and the XRF spectrometer-derived values of surface coal, coaly shale, and igneous dike samples (colored diamonds). As reported, four USGS coal samples had light REE contents greater than 100 ppm, and two samples had light REE contents greater than 200 ppm. The results for several other samples of coal and coaly shale obtained using the field-portable XRF unit found light REE contents in excess of 1,000 ppm. Samples obtained from a Tertiary lamprophyre dike just southwest of the Town of La Veta showed similarly high light REE contents as the coal samples. This implies that, in addition to introducing REE in the basin via tonstein deposits within the coal, igneous injection of dikes and sills could also be a possible sourcing mechanism for introducing REE. In some portions of the basin, coal has actually been converted to coke due to igneous activity. Further study should include obtaining samples of this naturally occurring coke to ascertain its REE content. Pertinent REE data and other elements derived from XRF analysis within the Raton basin in south-central Colorado appears in Table 2.

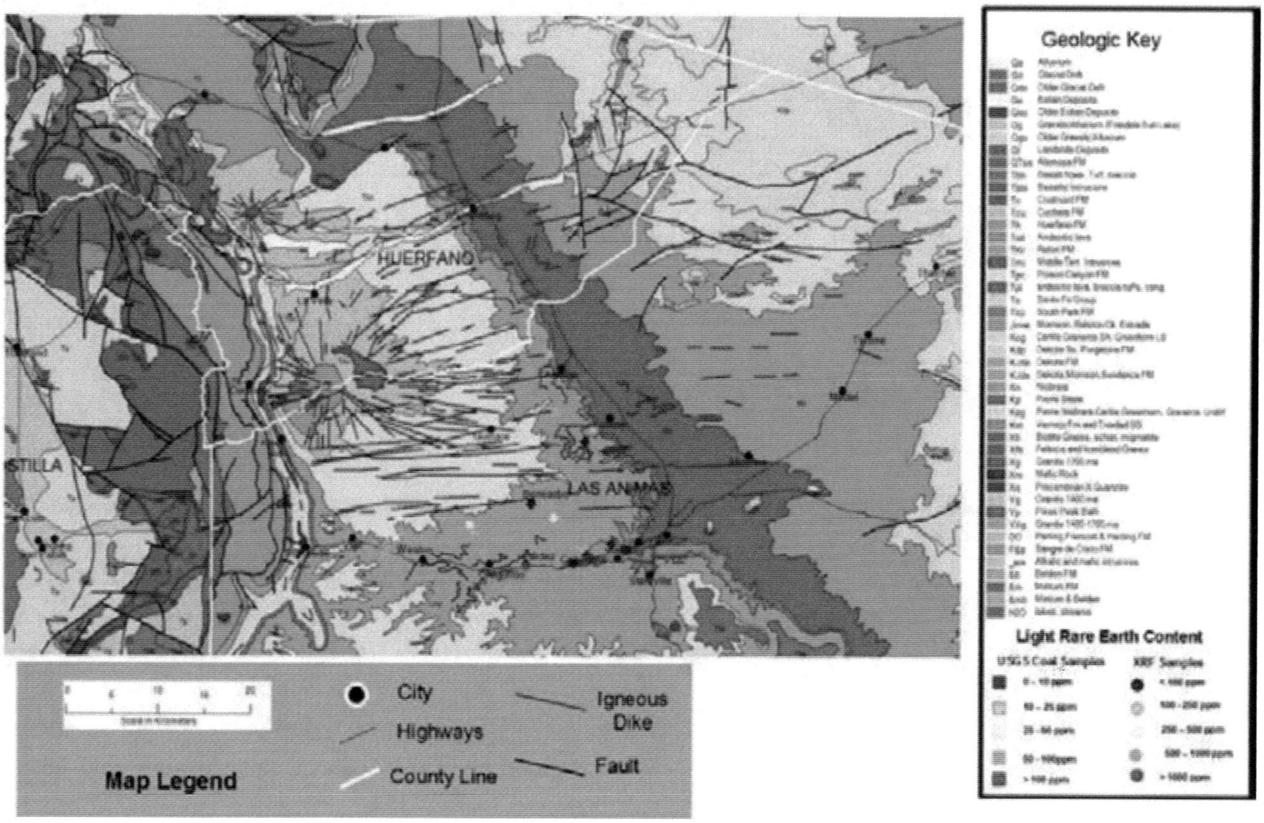

Figure 4 Rare Earth Element Outcrop Samples in Raton Basin

Table 2. Rare Earth Elemental analysis of Raton Basin samples by XRF Spectroscopy

SAMPLE	Ti	V	Th	U	Y	Sc	La	Ce	Pr	Nd
cosotr002	762.42	64.17		15.35	317.17					
tr001	541.23	60.44			218.14					
tr001a	196.44				185.81					
tr001ta	400.11	21.27	10.78		137.67					
tr001zn			8.9		117.16					
tr001b	332.34				116.32					
cosotr002					67.15					
cosogj001	1118.37	123.35		13.75	47.21	103.68				
cosofp0001	917.41	67.08	21.79		25.16	13.03				321.42
kr007	3131.35		18.81		24.86		237.69	242.94	404.73	599.37
kt003	1738.04				24.04					
kt004	5632.97	198.96	9.92		22.98		119.04	135	176.96	304.69
nmsofw0001	1745.39	113.88	13.63		22.28	86.22	74.24	102.28	119.33	286.9
fl011	670.36			7.58	20.25					
cosofr0001	1793.21	89.22	9.62		18.59	125.77				
kt001	4269.5	39.87			16.88					
cosofr0002	1055.23	75.71		79.75	15.75		98.68	126.25	180.57	325.9
fl011	1729.24	95.93	4.93		15.03	51.23				
cosogj0001	493.88	25.24			14.85	21.45				
kt006	1654.15				14.71					
tr003	480.38			16.05	13.92	243.96				
nmsofr0003	1525.3	76.5	13.7		13.76	173.32				
fl011	3905.29	171.28	10.39	8.07	13.47	126.87				117.67
kt005	4295.69	129.15	6.85		11.06		182.35	197.7	300.84	513.28
fl011	80.83	22.36			10.21	6.9				
lu008			10.57		9.98					
lu006	2366.67	148.12			9.13		227.93	252.71	434.37	702.07
lu008	1543.5	66.6	10.01		8.52					
tr004a	290.25	14.22			7.78	5.85				
cocdfr0001	2189.01	99.36		5.82	7.26	167.27				
kt002	3818.34	86.46			7.17					
kt008	2449.95		10.87	5.12	5.93					
l dike	1573.09		21.03	9.81	5.13		159.79	158.63	348.55	549.88
l dike2	988.02	21.22	16.79	10.68	5.08		192	199.65	351.15	535.1
tr005	3226.66	121.95		11.04	4.8	40.6				

Conclusions

Coal, shaly coals and tonstein samples collected in the greater Raton basin of Colorado exhibited elevated rare earth content. Several of these samples were collected proximal to the K-T layer inferring that ash derived from this impact which resulted in the formation of a thin tonstein layer could be a plausible source. Volcanic and other igneous activity in the basin could also be responsible and post-dates the K-T boundary, suggesting that hydrothermal fluids and/or direct contact with igneous dikes and sills could introduce REE into the samples. Certainly, samples of Cenozoic lamprophyre dike material collected at the northern edge of the basin exhibited REE contents that were as high as or higher than those observed in the coal and coaly shale samples.

If a systematic survey over the entire basin can be conducted, then the association of REE-rich layers with coal might provide a means to formulate a strategy to exploit these elements as a by-product of coal production. Future work should include careful scrutiny of exploration cores in the basin if available to identify samples of tonstein, the K-T clay, and fresh coal above and below these layers to determine if the source of REE in the coals are from volcanic ash, intrusive, or the K-T impact.

Bibliography

Armbrustmacher, T.J., 1984, Alkaline Rock Complexes in the Wet Mountains Area, Custer and Fremont Counties, Colorado, USGS Prof. Paper 1269, 38 p.

Andersen, H.T, Bryan, R, Gray, T, and Richers, D, 2015, Rare Earth Occurrences and their Associations with Tonsteins, Diapirs, and Igneous Activity, presented at the 2015 Pittsburgh Coal Conference, October 6, 2015, Pittsburgh, PA., 13 p.

Baltz, E. H., 1985,Stratigraphy and history of the Raton Basin and notes on San Luis Basin, Colorado-New Mexico, AAPG Bulletin, Volume 49,No. 11, pp. 2041-2075.

Flores, R.M., and Bader, L.R., 1999, A Summary of Tertiary Coal Resources of the Raton Basin, Colorado and New Mexico, USGS Professional Paper 1625-A, pp. SR-1,SR-28.

EPA, 2004, Attachment 9, "The Raton Basin" in Evaluation of Impacts to Underground Sources of Drinking Water by Hydraulic Fracturing of Coalbed Methane Reservoirs. EPA 816-R-04-003, 19 p.

Izett, G.A, 1987, The Cretaceous-Tertiary (K-T) boundary interval, Raton Basin, Colorado and New Mexico, and its content of shock-metamorphosed minerals: Implications concerning the K- T boundary impact-extinction theory, U.S.G.S Open-File Report 87606.

Johnson, R.B., 1968, Geology of the Igneous Rocks of the Spanish Peaks Region, Colorado, USGS Professional Paper 594-G, 58 p.

Keighin, C.W., 1995, Raton Basin-Sierra Grande Uplift Province (041), USGS, 126p.http://certmapper.cr.usgs.gov/data/noga95/prov41/text/prov41.pdf

Rare Element Resources, 2014, http://www.rareelementresources. com/rare-earth-elements#.VeXBqpRUyMx

Sheard, E.R., 2010, Behavior of zirconium, niobium, yttrium, and the rare earth elements in the Thor Lake rare-metal deposit, Northwest Territories, Canada, Masters Thesis, Department of Earth and Planetary Sciences, McGill University, Montreal, QC, Canada, 151 p.

Taylor, L. H, 2015m Field guide for the Cretaceous-Paleogene Boundary (Formerly the Cretaceous – Tertiary Boundary), Ludlow and Trinidad, Colorado. A guide prepared for the KT Field Trip, September 26, 2015, Denver Museum of Nature and Science, 48p.

437626 Recovery of RARE Earth Minerals and Elements from Coal and Coal Byproducts

Abstract:

Recent studies have shown that rare earth and other critical elements exist in coal and coal byproducts at concentrations as high as 2000 ppm on an ash-basis. The rare earth elements have been found in mineral form as monazite, xenotime and bastnaesite and also as an element substitute in clay. An investigation has been conducted to evaluate the potential of concentrating the rare earth (RE) minerals using physical concentration technologies such as density-based separators, froth flotation and magnetic separators. Distribution of the RE minerals as a function of the physical properties varied significantly between coal sources. However, the results indicate the potential to achieve concentration to a level dependent on the source using physical concentration techniques

437131 Uncertainties and Optimum Detection Modes for Rare Earth Analysis in Coal and Coal Ash Using Inductively Coupled Plasma Mass Spectrometry (ICP-MS)

Abstract:

Inductively Coupled Plasma Mass Spectrometry (ICP-MS) is a very powerful technique that can detect multiple elements at very low concentrations in a variety of matrixes. The advantages of low detection limits and multi-element analyses make ICP-MS an indispensable tool for rare earth elements (REEs) analysis in soils, sands, clays, ashes, coals and other solid and liquid matrixes. However, low concentrations of REEs and complicated matrixes of coals and coal ashes can induce a significant amount of error in the analysis if the digestion of the sample is incomplete and detection modes are not optimized. In this study standard reference materials of ash and coal were analyzed for REEs and other trace elements using ICP-MS with a lithium metaborate fusion digestion method. ICP-MS methods tested for optimum REE recovery included standard operating mode as well as advanced cell technologies including kinetic energy discrimination (KED) and dynamic reaction cell (DRC) mode. Each mode had advantages for certain rare earth elements indicating that using certain modes for specific REE may provide the most accurate method for analysis. The use of different modes for specific REEs is a relatively easy way to enhance the accuracy of REE measurements in coal and coal ash and may be a good technique for other complicated matrixes as well.

The instrument used in this study is a Perkin Elmer Nexion 300D ICPMS. The Nexion is designed with an internal cell that allows the user to apply gases which either chemically or physically react with a sample to minimize interferences with the analyte(s) of interest. In this study, KED mode uses ultra-high purity helium gas to physically collide with ions in a sample and minimize the concentration of interferents caused by high total dissolved solids. The result is fewer total ions reaching the detector, but the effect of removal is much greater for large diameter ions such as Na+ and Cl-. Alternatively, DRC mode using ammonia removes molecular ions, such as 35Cl16O-, that may overlap with metals of interest. For each complicated sample matrix, the exact conditions which optimize interference removal and maximize the signal for analytes of interest need to be experimentally determined.

437136 Analysis of Twenty Coal Ashes for Rare Earth Element Content Using Inductively Coupled Plasma Mass Spectrometry (ICP-M)

Abstract:

Rare earths elements (REE) are widely employed in magnets, electronics, phosphors, metal alloys, glasses, batteries, computer hard drives, automobile emission and petroleum refinery catalysts, and sorbents, with adequate substitutions often unavailable. Unfortunately there have been concerns with the supply and environmental issues associated with the mining and production of REE from ores in the U.S. and around the world. Therefore, already mined and processed secondary sources of REE are of interest. Coal ash is seen as a possible source for REEs since the U.S. produces over one hundred million tons of ash from coal-fired power plants annually, and some coals and coal ashes have been shown to have relatively high concentrations of REEs. Additionally, it is also important to determine the concentration of critical or more valuable rare earth elements, since many of the elements are in good supply.

In this study twenty coal ash samples were analyzed for total rare earth content, individual REE concentrations, as well as percent critical REE. These data were compared to data reported on the NETL EDX website, and uncertainties in the data are discussed. All of the ashes studied were reported to have at least 500ppm total REE content based on the EDX website data on a whole sample basis (i.e. no further ashing was conducted). Other trace elements were also analyzed to see any association with the REE as well as identify any other valuable elements that may be economically co-extracted. Results of the REE amount and uncertainties in coal fly ash and bottom ash are discussed.

437135 Characterization of Rare Earth Elements in Canadian Oil Sand Process Streams

Abstract:

The concentration of rare earth elements in an Canadian oil sand ore and six oil sand waste streams were examined using Inductively Coupled Plasma Mass Spectrometry (ICP-MS). The results indicated that the rare earth elements (REEs) are largely concentrated in the TSRU tailings, similar to typical froth floatation tailings, with a total rare earth concentration of 1380 ppm (1380 μg/g). This is a 13.5 fold increase in concentration compared to the oil sand ore itself, and an 8 fold increase compared to average Clarke value of sedimentary rocks. Not surprisingly the process water used for extracting the oil from oil sands and the water fraction associated with the different waste streams had very low values of REEs that were near or below the detection limits of the instrument. The highest total concentration of REEs in the water fraction of the samples tested was from the mature fine tailings with a total rare earth concentration of ~ 7 ppb (7 μg/kg). These results give insights into the possibility of recovering rare earth elements from REE concentrated waste streams generated from oil sand processing.

437132 Deep Eutectic Solvents and Ionic Liquids for Extraction of Rare Earth Elements from Coal Ash

Abstract:

Deep eutectic solvents (DESs) are a relatively new class of ionic liquids (IL) that differ from traditional ionic liquids by mixing a hydrogen bond donor with a salt. One of the first deep eutectic solvents studied in depth was a 2:1 mixture of urea and choline chloride. An interesting property of this DES is its ability to dissolve metal oxides. The low volatility, inexpensive starting materials and ability to dissolve metal oxides could make deep eutectic solvents an interesting solvent for recovery of rare earth elements (REEs). Additionally, ionic liquids alone also have many interesting solvent properties and could provide a novel separation method for the recovery of REEs from many solid fossil fuel sources specifically coal ash. However, some ionic liquids are potentially very expensive, and would need to be cycled many times for an economical extraction process.

This study reports preliminary data on the total amount of rare earth elements extracted from a sample of coal ash using a DES (choline chloride and urea) and an ionic liquid. Specifically, an ionic liquid with a reactive carboxylic group was synthesized and studied. The percentage of each REE extracted is reported and measured using Inductively Coupled Plasma Mass Spectrometry (ICP-MS). The viscosity and percentage of water, which are known to affect the kinetics of dissolution, will also be reported. Other parameters that are also important to consider are contact time, reactor configuration, extraction temperature, ratio of solvent to coal ash, stirring speed (mixing), pH, and the recovery of REEs from the solvent (regeneration of solvent), will also be discussed.

409996 U.S. Doe Carbon Storage R&D Program: Advancing Carbon Storage Technologies Towards Commercialization

Abstract:

2015 AIChE Annual Meeting **Salt Lake City, Utah, U.S.A.**
November 8-13, 2015

U.S. DOE Carbon Storage R&D Program: Advancing Carbon Storage Technologies Towards Commercialization

Kanwal Mahajan[1], Traci Rodosta[1], Derek Vikara[2]

[1] *National Energy Technology Laboratory, 3610 Collins Ferry Road, Morgantown, WV 26507-0880, U.S.A*

[2]*KeyLogic Systems, Inc. National Energy Technology Laboratory, 626 Cochrans Mill Road, Pittsburgh, PA 15236-0940, U.S.A*

Abstract: The U.S. Department of Energy (DOE) continues to be a world leader in advancing carbon capture and storage (CCS) technologies towards commercialization. The Carbon Storage Program implemented by the U.S. Department of Energy's (DOE) Office of Fossil Energy and managed by the National Energy Technology Laboratory (NETL) is helping to develop technologies that safely and permanently store carbon dioxide (CO_2) without adversely impacting natural resources or hindering economic growth. Specifically, the Carbon Storage program is developing and advancing CCS technologies both onshore and offshore that will significantly improve the effectiveness of the technology, reduce the cost of implementation, and be ready for widespread commercial deployment. Since 1997, DOE's Carbon Storage program has significantly expanded the CCS knowledge base in selected technology areas through a diverse portfolio of applied research projects and field demonstrations. The Storage program is comprised of three primary technology components: (1) Core Storage R&D, (2) Storage Infrastructure, and (3) Strategic Program Support. These areas work together to address significant technical challenges in order to meet program goals that support the scale-up and widespread deployment of CCS.

Core Storage R&D involves applied laboratory- and pilot-scale research focused on developing new technologies and systems for geologic storage, and encompasses three technology areas: (1) Geologic Storage Technologies and Simulation and Risk Assessment; (2) Monitoring, Verification, Accounting (MVA), and Assessment; and (3) Carbon Use and Reuse.

The Storage Infrastructure technology component is comprised of three technology pathways: (1) the Regional Carbon Sequestration Partnerships (RCSP) Initiative; (2) Characterization field projects (onshore and offshore); and (3) Fit-for-Purpose projects. All three technology pathways address the opportunities and challenges associated with the range of geologic formations potentially available for large-scale CCS development across the United States. The RCSPs include more than 400 distinct organizations, spanning 43 states and four Canadian provinces. These partnerships form the core of a nationwide network developing the framework needed to validate geologic carbon storage technologies. The RCSPs are actively involved in disseminating research information and public outreach, both domestically and internationally, by integrating research initiatives and lessons learned into Best Practices Manuals that highlight technical and nontechnical lessons learned from these field projects. The RCSPs also play an integral role in providing regional storage data to the National Carbon Sequestration Database and Geographic Information System (NATCARB) that provide an interactive visual representation of CCS potential in the United States. Additionally, the DOE Carbon Storage Atlas provides updates on CO_2 storage resources and showcases research being conducted through the RCSP Initiative, with emphasis on large- and small-scale field activities.

The Strategic Program Support component promotes an integrated approach to ensure that CCS technologies are cost-effective and commercially available. The program relies on research conducted by NETL as part of the national laboratory network to complement the program approach to reducing CO_2 emissions. NETL research focuses on fundamental and applied fossil energy R&D, performed by Government engineers and scientists, and offers a venue for participation in collaborative research and provides an evaluation of new technology concepts, products, and materials.

CCS and other clean coal technologies can play a critical role in mitigating CO_2 emissions while supporting energy security throughout the world. DOE's Carbon Storage program has positioned the United States on a path toward ensuring that the enabling technologies will be available to address the demands of new regulations and impact CCS projects in the 2020–2030 timeframe. Continued U.S. leadership in technology development and future deployment is important to the cultivation of economic rewards and new business opportunities, both domestically and abroad.

Neeraj Gupta
Lydia Cumming
Mark Kelley
Jacqueline Gerst

Midwest Regional Carbon Sequestration Partnership: Importance of Field Projects and Regional Mapping to Demonstrate Geologic Storage Potential

Battelle, Columbus, Ohio,
gupta@battelle.org, 1-614-424-3820

DOE/NETL Cooperative Agreement # DE-FC26-0NT42589

AIChE Annual Meeting, Salt Lake City, November 12, 2015; session T4G, Paper #425989

Battelle
The Business of Innovation

MRCSP Field Tests and Regional Characterization
Presentation Outline

About the MRCSP

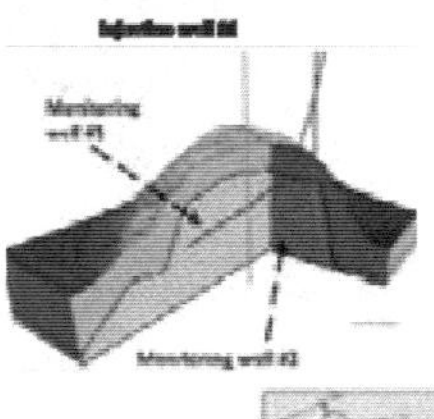

Large Volume CO_2 Injection Test

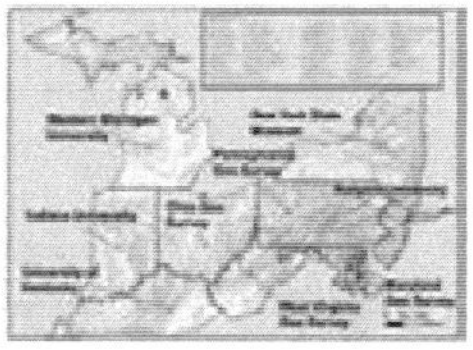

Regional Characterization

2

Battelle
The Business of Innovation

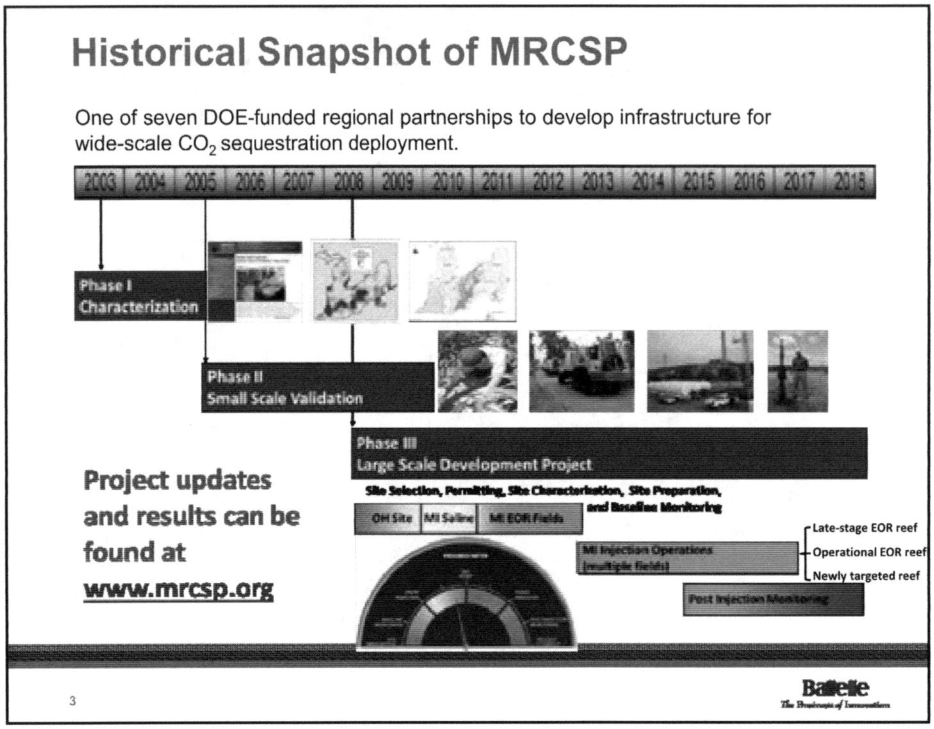
Historical Snapshot of MRCSP
One of seven DOE-funded regional partnerships to develop infrastructure for wide-scale CO_2 sequestration deployment.
2003 2004 2005 2006 2007 2008 2009 2010 2011 2012 2013 2014 2015 2016 2017 2018
Phase I
Characterization
Phase II
Small Scale Validation
Phase III
Large Scale Development Project
Site Selection, Permitting, Site Characterization, Site Preparation, and Baseline Monitoring
OH Site
MI Saline
MI EOR Fields
MI Injection Operations (multiple fields)
Late-stage EOR reef
Operational EOR reef
Newly targeted reef
Post Injection Monitoring
Project updates and results can be found at www.mrcsp.org
3
Battelle

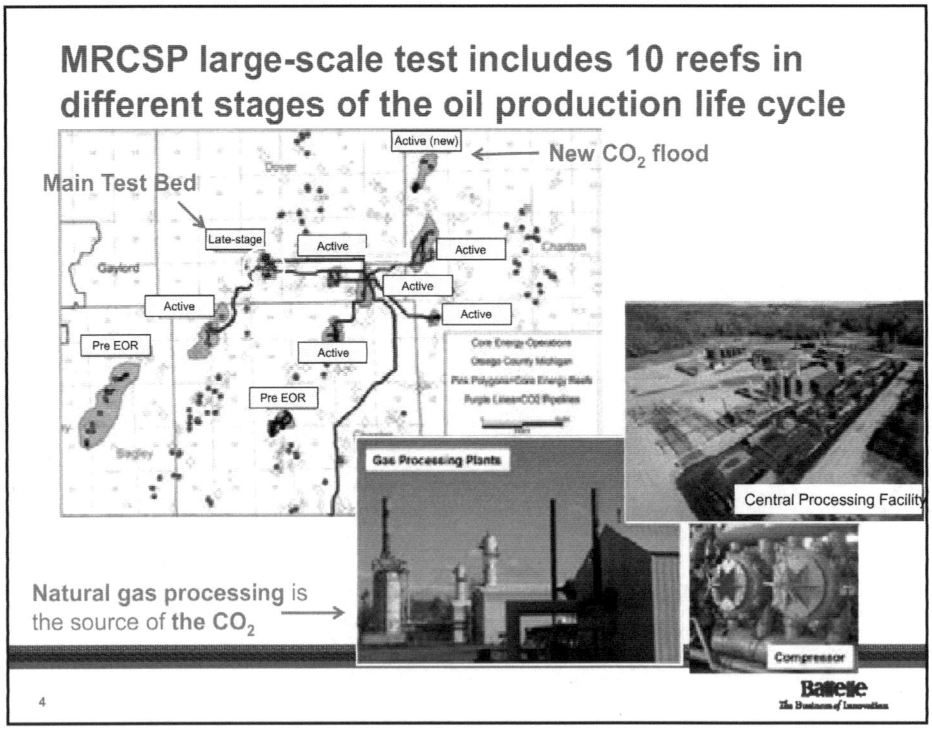
MRCSP large-scale test includes 10 reefs in different stages of the oil production life cycle
Active (new)
New CO_2 flood
Main Test Bed
Late-stage
Active
Active
Active
Active
Active
Pre EOR
Active
Pre EOR
Gas Processing Plants
Central Processing Facility
Compressor
Natural gas processing is the source of the CO_2
4
Battelle

Late-Stage Reef is the Main Test Bed for MRCSP Large-Scale Injection

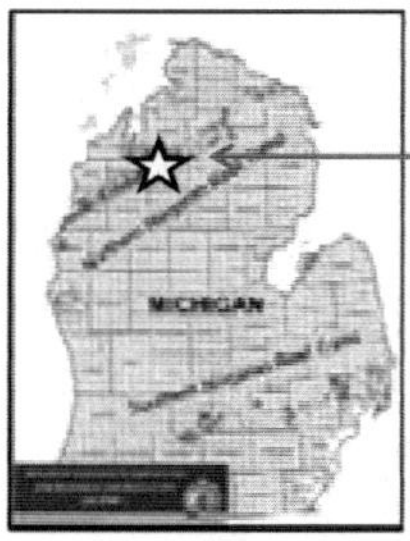

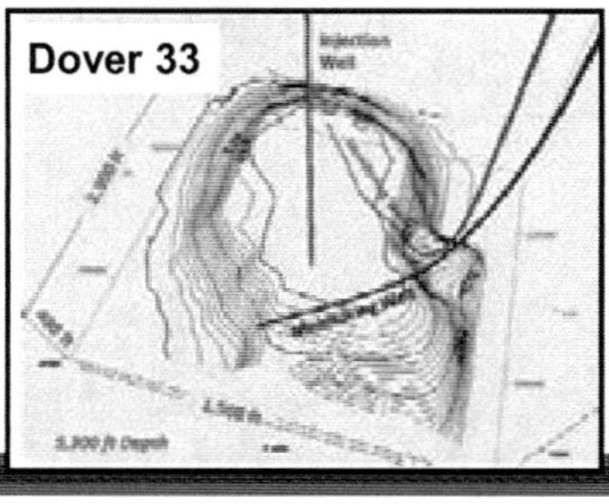

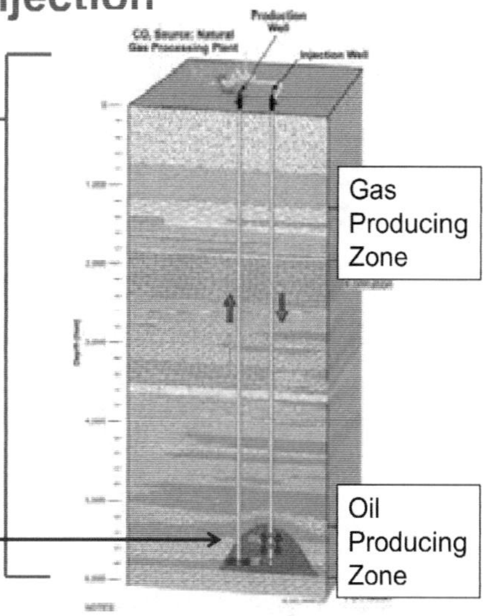

MRCSP is collecting data on four key elements relevant to CO_2-EOR and storage projects

- Injectivity – Measure of the rate at which CO_2 can be injected (factor of porosity, permeability, and prevailing pressure)
- Capacity – Calculated from geological data, or from production and reservoir pressures
- Performance – Variety of MVA tools to monitor injection operations and verify storage.
- Safety – Understanding and mitigating risks

Verifying these four elements are important to demonstrate permanent storage and increase public acceptability

Monitoring and Accounting for CO_2 in a Closed Loop CO_2 EOR Cycle

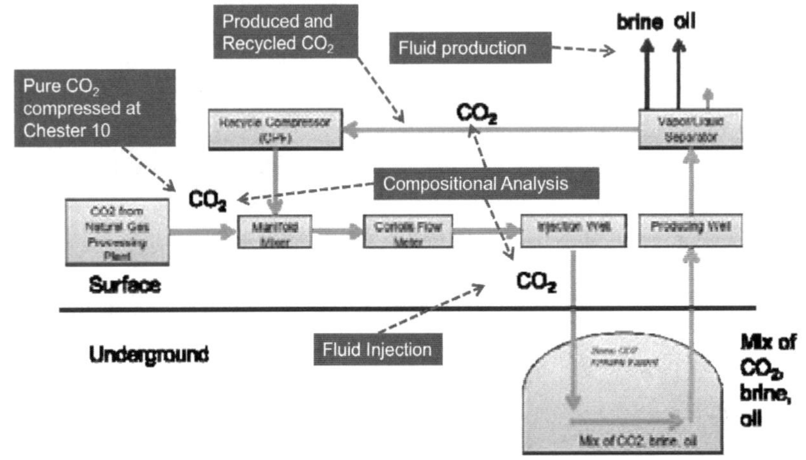

All produced CO_2 is recycled back into system.

Depleted oil and gas field

7

Historical and Recent CO_2 Injection Data

Late Stage Reef EOR Operations

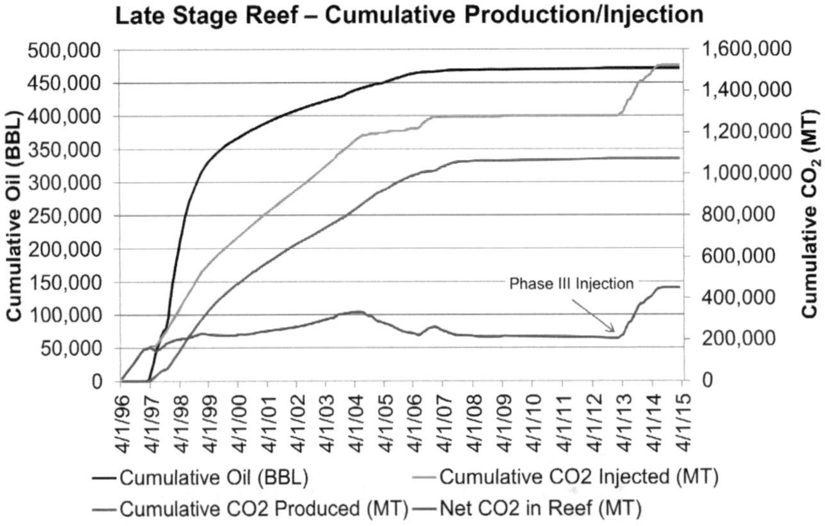

8

Net CO_2 **Stored** in Reef Complex Over Time

~1.5M metric tonnes retained since 1996

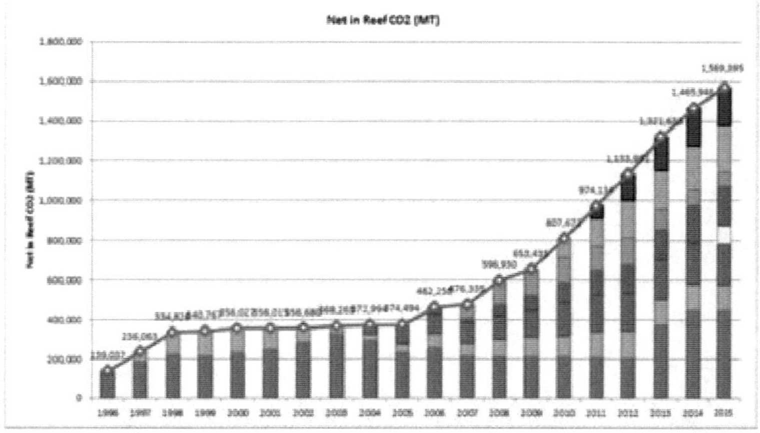

Accurate historical injection and production records is crucial for CO_2 accounting and credits

Battelle
The Business of Innovation

Monitoring Status for Late Stage Reef

A portfolio of technologies is being tested

Activity	Before Injection	Early Injection	Mid Injection	Late Injection	After Injection
CO_2 flow		X	X	X	
Pressure and temperature		X	X	X	X
Wireline logging	X		X		X
Borehole gravity	X				X
Fluid sampling	X		X		X
Vertical seismic profile	X				X
Microseismic	X			Under planning	
Satellite radar	X	X	X	X	X

Lessons learned will be applied to design the MVA plan for the newly targeted field

Battelle
The Business of Innovation

InSAR monitoring results consistent with no meaningful surface displacement

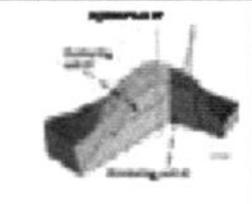

Vegetation and snow are challenging for radar, but there were a reasonable number of natural reflectors

Artificial reflectors augmented the data for injection monitoring

Time series displacement data show no correlation to injection

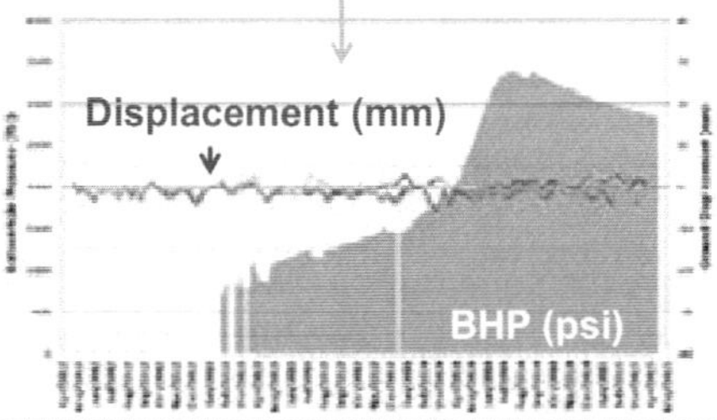

Pressure Monitoring in Late-Stage Reef

Slow, long-term pressure decline 9 months after injection

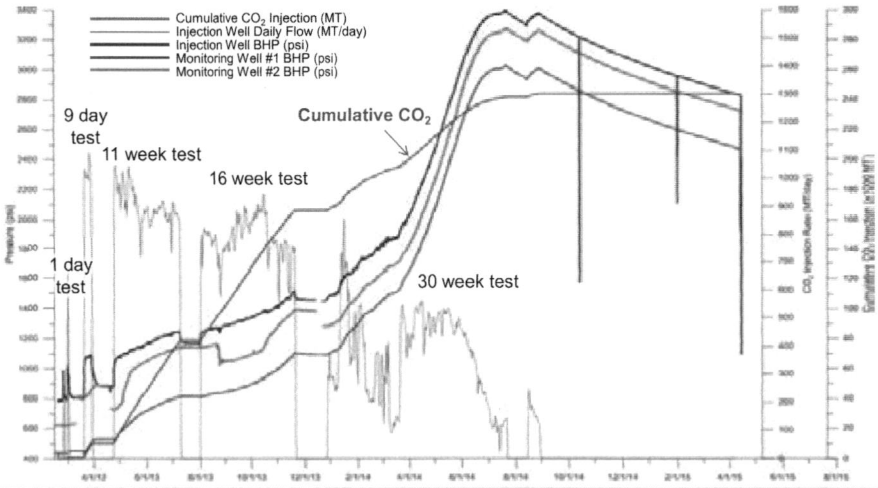

Modeling Objectives for CCUS Projects

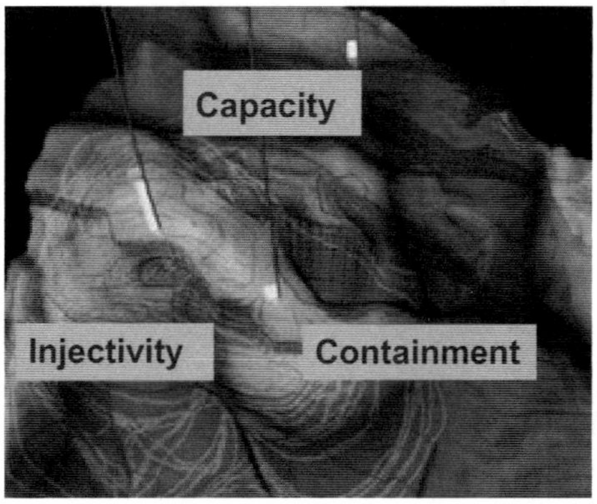

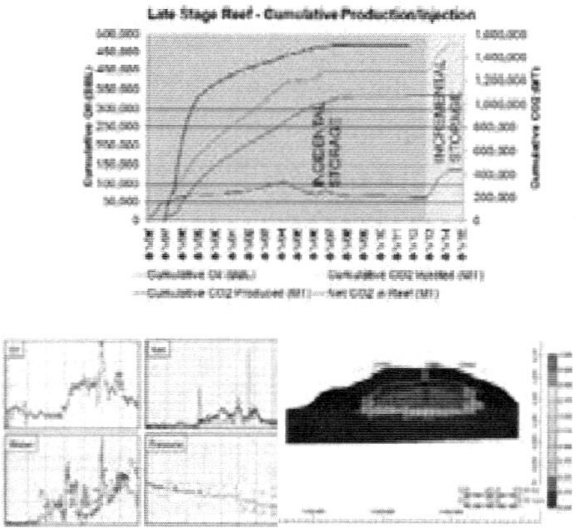

Battelle
The Business of Innovation

Alternative Numerical Model Conceptualization

Equivalent Homogeneous Task 3 Reef Models

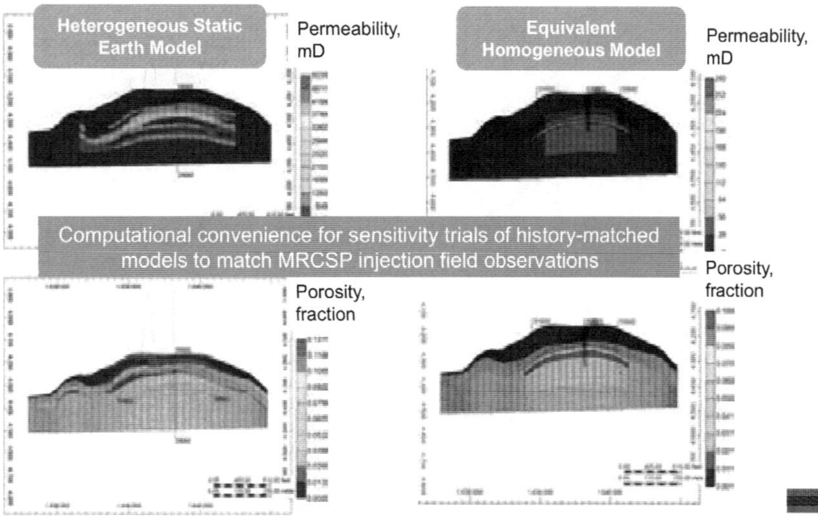

Battelle
The Business of Innovation

Simplified Compositional Model History-Match

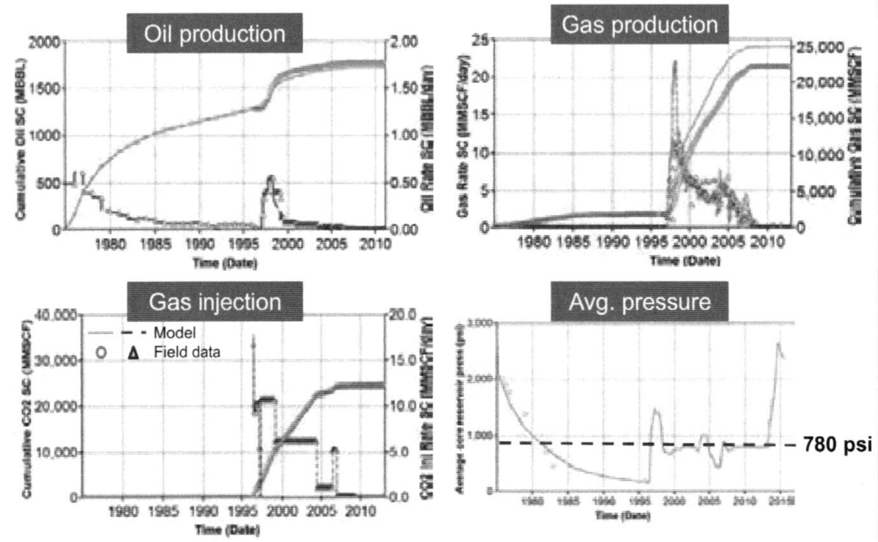

15

Modeling Pressure Response

Equivalent Homogeneous Compositional Reservoir Model
MRCSP Injection Response Validation

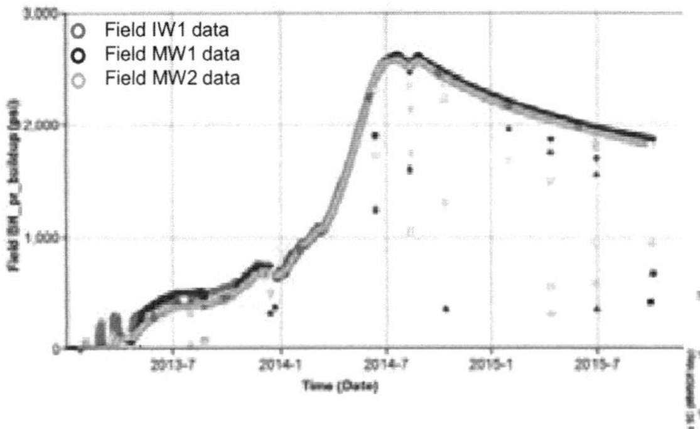

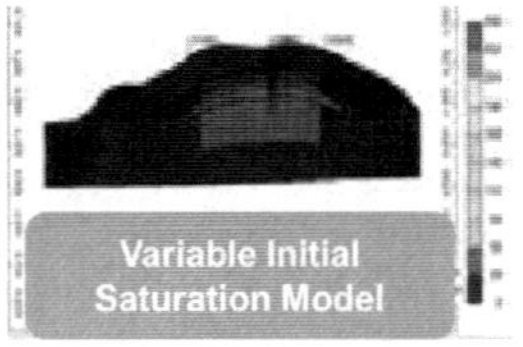

Injection Schedule

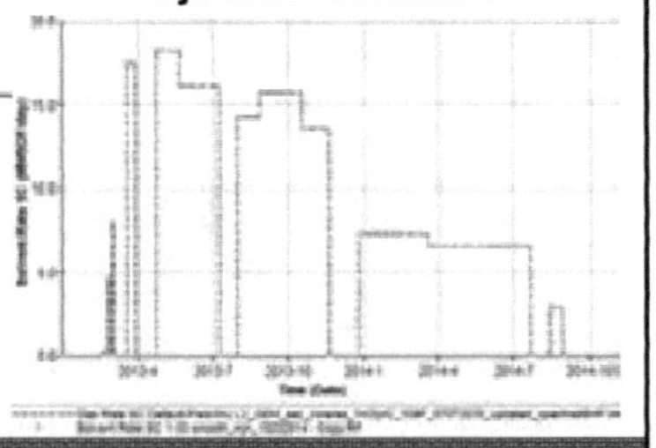

Reasonable match except near the end of injection

16

Scenarios to explain MRCSP Injection Phase Pressure Buildup

- Tighter Rock
  - Lower rock compressibility
  - Tighter i.e. less permeable flanks
- Smaller pore volume for HC fluids-in-place and CO_2
  - Lower model pore volume
  - Higher water saturation outside the core reservoir in the flanks
  - Amount of CO_2 present in the system

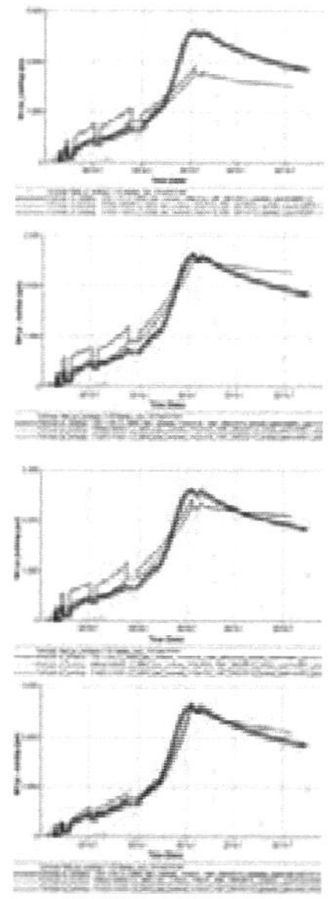

17

Battelle
The Business of Innovation

Simulated CO_2 Distribution within the Reef

Migration into reef flanks over time

At beginning of MRCSP Injection | At end of MRCSP Injection @10/2014 | Post-MRCSP Injection @06/2015

J-K

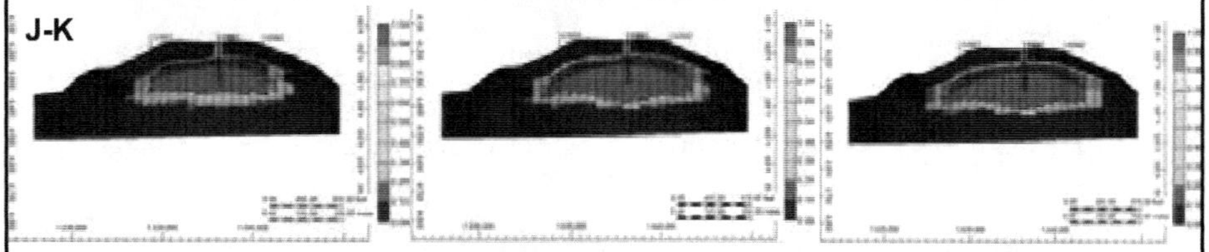

Difference Maps

For MRCSP Injection Period until 10/2014 | Fall-off post MRCSP Injection Period until 06/2015

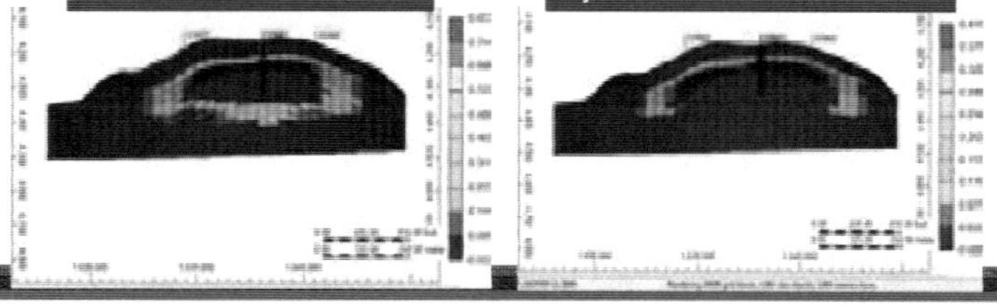

18

Battelle
The Business of Innovation

Geochemical Studies: Stable Isotope as Tracers

Late-Stage reef shows impact of CO_2 relative to new reef

- Collected gas and brine samples from a new EOR reef, which has not received CO_2.
- General brine chemistry is similar between new EOR reef and late-stage reef with very high TDS.
- Isotopically, the brines are different. Differences in the $\delta^{13}C$ for dissolved carbonate suggest the brine chemistry is altered by the injection of CO_2.
- Note: the $\delta^{13}C$ value has been corrected for fractionation resulting from dissolution and dissociation.

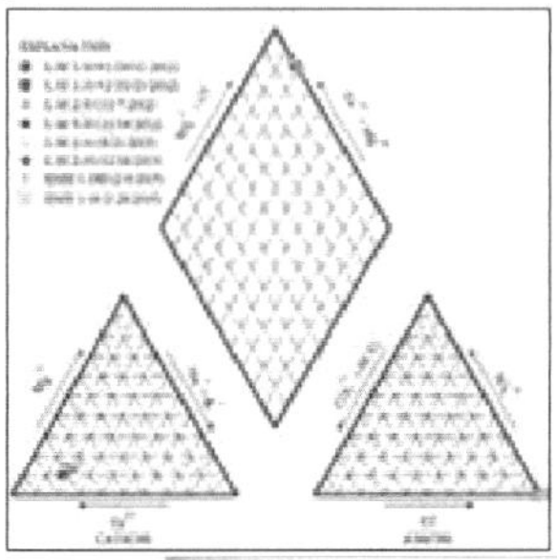

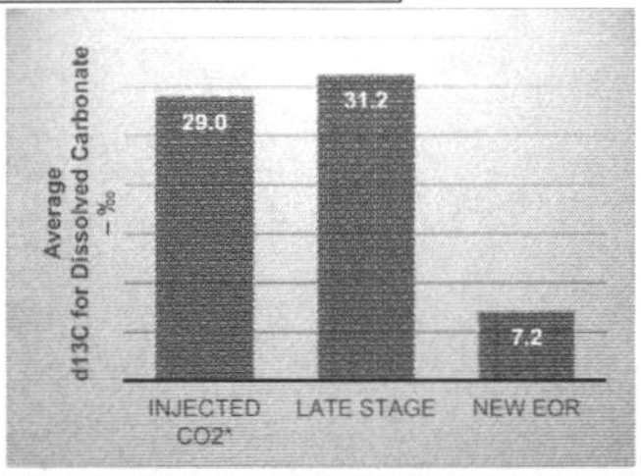

19

Battelle
The Business of Innovation

Regional Characterization – A Geologic Storage Potential Mapping Collaboration

State geological surveys are:

- Compiling/interpreting data
- Developing common terminology for formations across state boundaries
- Creating maps and other tools

MRCSP's geology team

20

Battelle
The Business of Innovation

Selected Regional Geology Team Activities

Cambro-Ordovician Storage Potential
Led by Indiana

East Coast Offshore and Onshore Storage Targets
Led by Rutgers

Silurian Pinnacle Reef Reservoirs
Led by W. Michigan University

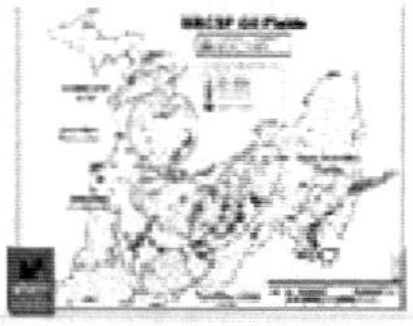

CCUS Opportunities in Appalachian Basin
Led by Pennsylvania

Storage and Enhanced Gas Recovery for Organic Shale
Led by Kentucky

Reservoirs for CO_2-EOR, EGR, and other Commercial Uses
Led by West Virginia

21

Battelle
The Business of Innovation

Mid-Atlantic U.S. Offshore Carbon Storage Resource Assessment Project

New Project (FOA 1246)

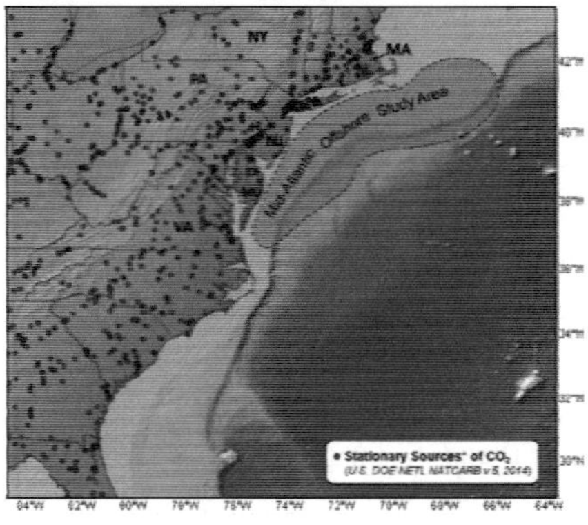

Project Team:

Battelle; geological surveys of MD, DE, and PA; USGS; Rutgers; Harvard; and Columbia

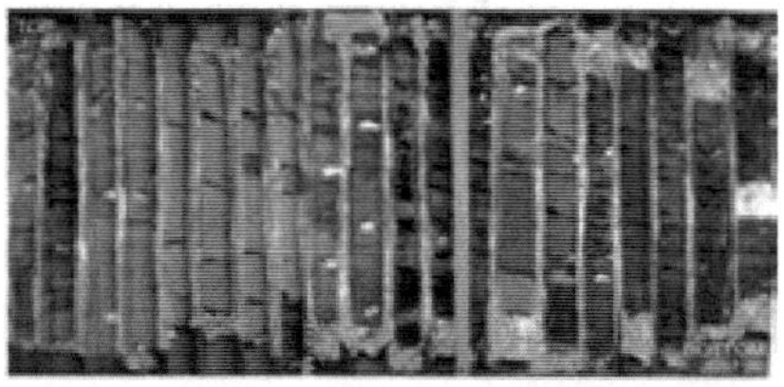

Example of existing core material, COST G2 well

22

Battelle
The Business of Innovation

Assessment of CO_2 Storage/Utilization in Ohio and Adjacent Areas

- An estimated 200 million tonnes of CO_2 are emitted a year within 75 miles of the Ohio River from point sources
- While significant oil and gas exploration has occurred in the region, there are many uncertainties
  - Much of the production is from historical wells
  - Little exploration has occurred in the Ordovician-Cambrian Section
  - Little is known about the regional extent of carbonate storage capacity
- Projects co-funded by Ohio Coal Development Office and DOE Over 10 years; Jointly with Ohio Geological Survey

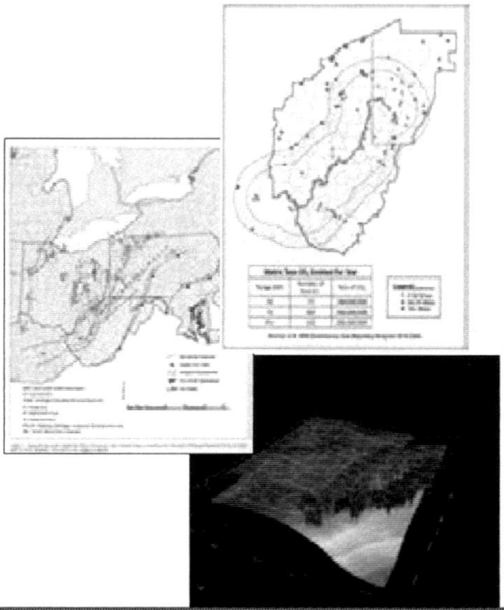

Battelle
The Business of Innovation

CO_2 Sequestration in Saline Reservoirs in the Upper Ohio River Valley

- Determine extent of potential reservoirs and caprocks
- Characterize and map petrophysical and geomechanical properties
- Continue gathering new data through piggyback opportunities

MRCSP Piggyback Wells

Battelle
The Business of Innovation

Characterizing Storage Zones

Learning from Brine Injection Wells

Vug Probability in Lower Copper Ridge and Rome Dolomites

- Local view of vug probability across three closely spaced wells
- There is better probability of vugs and potential connectivity between Wells 1 and 2
- Properties improve understanding of fluid flow

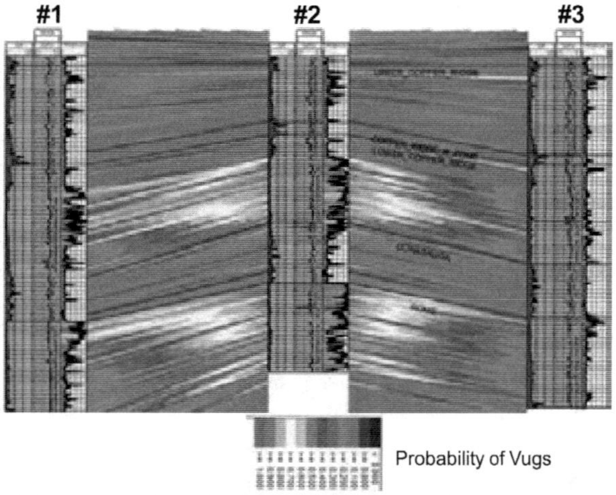

Battelle
The Business of Innovation

Regional Mapping of Potential Storage Targets

Sandy Facies Map for Conasauga Group

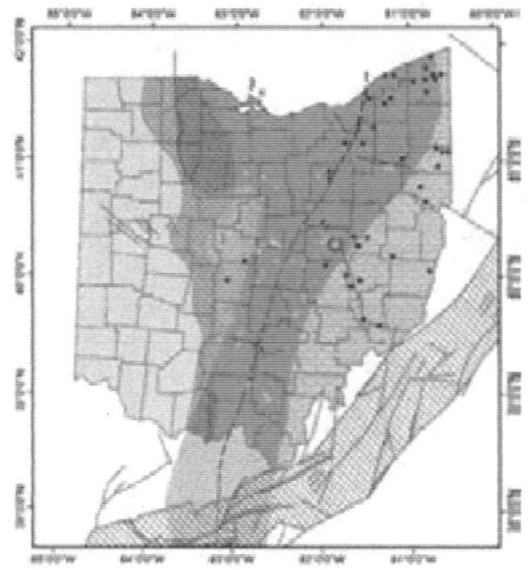

Extent of the sandy facies in the Conasauga (orange) and how it relates to known structure

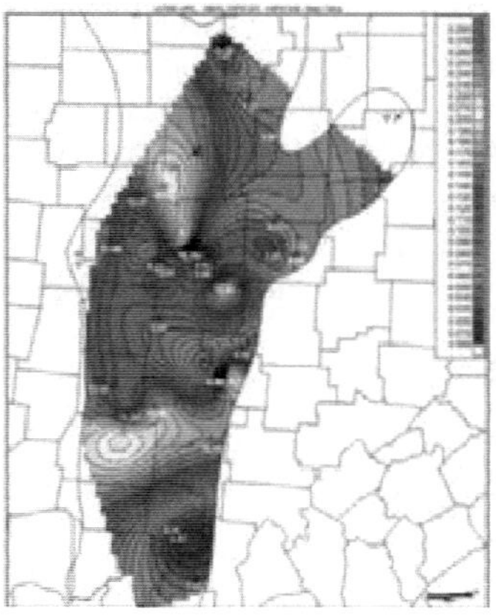

Porosity map of the sandy facies showing high porosity (yellow) in the center

Co-Funded by ODSA/OCDO

Battelle
The Business of Innovation

CO_2 Utilization for EOR and Geologic Storage in Ohio's Depleted Oil Fields

- **Research goals**
  - Develop process understanding and evaluate technical and economic feasibility of CO_2 utilization and storage in Ohio's depleted oil fields
  - Focus on Clinton sandstone and Knox dolomite formations (under-pressured, low permeability reservoirs with poor primary recovery)

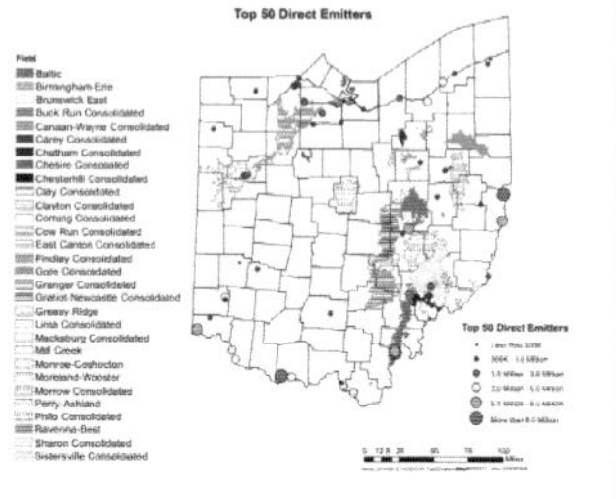

- **Current focus**
  - Source-sink matching
  - Production history assessment
  - Geologic model development
  - Fluid property characterization
  - Reservoir simulation

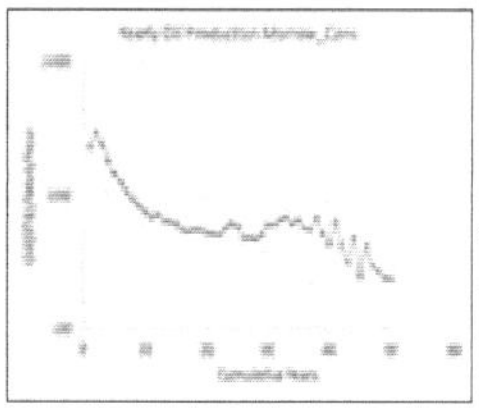

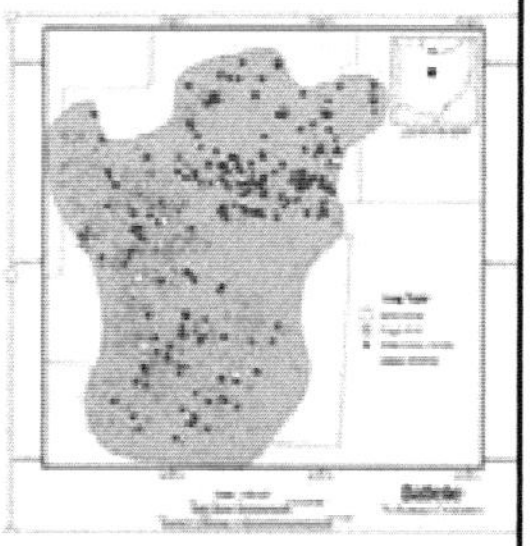

Co-Funded by ODSA/OCDO

Battelle
The Business of Innovation

CO_2-EOR/Storage Assessment

Fluid Property Prediction Tool for oil-gas-water-CO_2 Systems

Inputs:

Shading indicates needed input		
Input Bubble Point Property		
Select Known Bubble Point Property:	Pressure	
Pressure, pB	[illegible]	psia
Additional Reservoir Properties		
Oil gravity, γ_{API}	[illegible]	API
Gas gravity, γ_g	0.75	-
Reservoir Temperature, T	[illegible]	°F
Single Point Pressure of Interest Evaluation at right		
Pressure, p	1250	psia
Pressure Range of Interest See next sheets for tables and plots		
Maximum Pressure	[illegible]	psia
Min Pressure	15	psia
Brine-Gas Calculations		
Calculate Brine-Gas Properties?	Yes	
If Yes, Salinity =	[illegible]	Weight % Solids
CO_2 Calculations		
Calculate Pure CO_2, CO_2-Oil and CO_2-Brine Properties?	Yes	
*Pure CO2 properties only available for [illegible]		
*CO2-Oil and CO_2-Brine Calculations assume dead oil		
If Yes, Avg. Oil Molecular Weight (MW) =	[illegible]	lb/lbmol
If Yes, Salinity =	[illegible]	Weight % Solids

Outputs at Single Point of Pressure:

At:	Pressure, p	1250	psia
	Bubble Point Pressure, p_b	2057	psia
	Solution Gas Oil Ratio at Bubble Pt, R_{sb}	794	SCF/BBL
Oil-Gas	Formation Volume Factor, B_o	1.21	RB/STB
	Solution Gas Oil Ratio, R_{so}	[illegible]	SCF/BBL
	Viscosity, μ_o	0.468	cP
	Compressibility, c_o	[illegible]	1/psi
	Density, ρ_o	46.2	lbm/ft3
Gas	Formation Volume Factor, B_g	[illegible]	ft3/SCF
	deviation factor, Z	0.802	-
	Viscosity, μ_g	0.014	cP
	Compressibility, c_g	[illegible]	1/psi
	Density, ρ_g	5.64	lbm/ft3
Brine-Gas	Formation Volume Factor, B_w	[illegible]	RB/STB
	Solution Gas Water Ratio, R_{sw}	[illegible]	SCF/BBL
	Viscosity, μ_w	[illegible]	cP
	Compressibility, c_w	[illegible]	1/psi
	Density, ρ_w	[illegible]	lbm/ft3
Pure CO_2	Formation Volume Factor, B_{CO2}	0.01	ft3/SCF
	Viscosity, μ_{CO2}	[illegible]	cp
	Density, ρ_{CO2}	21.88	lbm/ft3
	Compressibility Factor, Z_{CO2}	[illegible]	-
CO_2 - Oil	CO_2-Oil Solubility, Sol	[illegible]	mole fraction
	Solution CO2 Oil Ratio, R_{so-CO2}	[illegible]	SCF/STB
	CO_2-Oil Swelling, SF	[illegible]	-
	CO_2-Oil Viscosity Ratio, [illegible]	[illegible]	-
CO_2 -Brine	Solution CO_2-Water Ratio, R_{sw-CO2}	[illegible]	SCF/STB

Battelle
The Business of Innovation

Summary of Recent MRCSP Progress

- Large-scale Test in Michigan
  - Completed baseline monitoring and site preparation for multiple reefs
  - ~244,000 tonnes injected in late stage reef
  - >175,000 tonnes net CO_2 in active EOR reefs
  - Operational and subsurface monitoring underway
  - Reservoir analysis shows closed reservoir conditions
  - Phase changes and compressibility affect pressure
  - Initial geologic and reservoir models prepared
  - Injection in a second new EOR reef likely to start in late 2015
- Local, regional, national, and global outreach
- Regional mapping/characterization across ten states
- Geologic storage and EOR assessment in Ohio

Many CCS Challenges Remain to be Addressed

- Matching global storage capacity with commercial scale injectivity – *balancing large-scale testing with broader geologic exploration*
- Effect of seismicity, pressure constraints, and stakeholder issues on deployment – *will we need to discount total capacity?*
- Validation of monitoring technology across geologic settings – confidence in current technologies; technology evolution
- Regulatory, financial, stakeholder issues for saline storage field tests
- Re-engaging industrial stakeholders for CCS in saline storage and EOR settings – low oil/gas prices, policy uncertainties, capture cost
- Sufficient representation in mitigation portfolio
- Development and retention of human capital

Battelle
The Business of Innovation

Acknowledgements

Battelle's MRCSP team members from 2003 to present

DOE/NETL for their continued trust in Battelle's work under MRCSP and other R&D Projects since 1996

Core Energy, LLC our host site and CO_2 supplier for 10 years of collaboration under MRCSP Phase II and Phase III

The **Ohio Coal Development Office** in the Ohio Development Services Agency for co-funding of the regional characterization efforts

MRCSP's **industrial partners and sponsors**

The ten State **Geology Surveys and Universities** have been essential in expanding the results into regional implementation plans.

31

Battelle
The Business of Innovation

Questions?

Please visit www.mrcsp.org

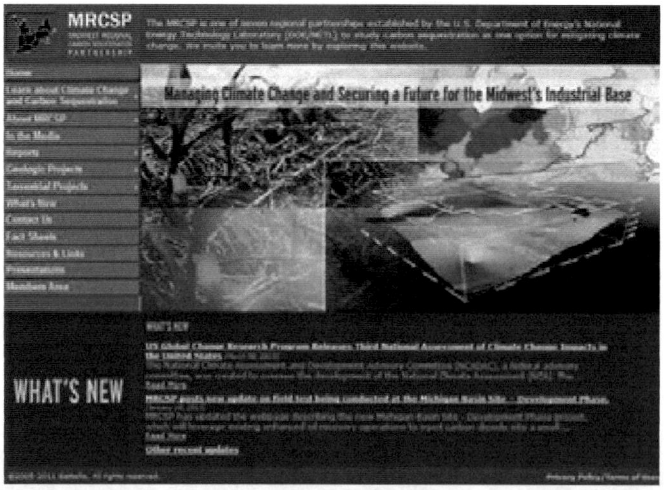

32

Battelle
The Business of Innovation

431539 Post Injection Site Care (PISC) at the Secarb Anthropogenic Test

Abstract:

The Southeast Regional Carbon Sequestration Partnership (SECARB) Anthropogenic Test is a demonstration of the deployment of CO_2 capture, transport, geologic storage and monitoring technologies. This project was the largest fully integrated commercial prototype coal-fired carbon capture and storage project in the USA. The demonstration reached the 100,000 metric ton (tonnes) injection milestone and injection operations were terminated in September 2014 after reaching this goal. This paper provides an assessment of the CO_2 injection program and the strategy for site closure.

Two Underground Injection Control (UIC) Class V Experimental Injection Well permits were issued in November 2011 and the project's injection wells were installed soon thereafter. Surface and subsurface monitoring was initiated prior to start of injection to establish background conditions and trends and continued throughout the injection period. The injection target was the lower Cretaceous Paluxy formation that occurs at a depth of 9,400 feet below ground surface. Post injection monitoring will be conducted into 2018 to track in-situ CO_2 movement.

Per the UIC permit requirements, the Area of Review (AoR) for the injection well was re-evaluated annually since the onset of injection operations via monitoring results and reservoir simulation. The AoR is defined as the region surrounding the geological storage project that may be impacted by the injection activity. The AoR is further defined as the farthest reach of the tangible CO_2 plume and/or the region of 'significant' reservoir pressure increase.

The monitoring program was developed and implemented to track the CO_2 plume and the associated pressure field. Methods used to do so include downhole pressure/temperature measurements, surface CO_2 flux, shallow and deep fluid geochemistry, tracers, cross-well seismic, vertical seismic profiling (VSP) and cased-hole geophysical well logs. The results from these monitoring tools were used to inform the reservoir simulation, which is used to render three dimensional assessments of CO_2 position versus time. The simulation results demonstrate effective containment of CO_2 by the reservoir architecture of the formation, consisting of multiple, stacked, highly porous and permeable sandstones, separated by low permeability flow barriers.

Other observations from the reservoir simulation include the following: 1) the dip of the Paluxy Formation influences the migration of mobile CO_2 to a small degree after injection operations cease, 2) the high transmissivity of the Paluxy results in a maximum CO_2 extent that is greater than the extent of significant pressure build up, and 3) consequently, the AoR determination is based on the extent of the CO_2 movement.

Permit closure is contingent on using the monitoring data and modelling assessments to demonstrate that the CO_2 is safely confined within the injection zone and that underground sources of drinking water (USDWs) are not endangered by the fugitive movement of the in-situ CO_2. This presentation will focus on the strategy to acquire PISC data and the model updates to build the case for permit closure.

422846 Hydrocarbon Mobilization and Potential CO2 Storage Mechanisms in the Middle Bakken, Bakken Shales, and Three Forks

Abstract:

Recent estimates of the original oil in place (OOIP) in the Bakken range from 160 to 900 billion barrels (Nordeng and Helms, 2010; Continental Resources, 2013, Kurtoglu et al., 2013). However, since recoveries are typically low (3 to 6% according to Bohrer et al. 2008 and Nordeng and Helms, 2010), even a small incremental increase in recoveries could yield many billions of barrels of additional produced oil. In addition, there are large organic rich members of these formations which contain substantial oil that is not currently considered to be amenable to production. Efforts to increase oil recovery in these unconventional formations could include carbon dioxide (CO_2) enhanced oil recovery (EOR) with incidental storage of large quantities of CO_2.

The processes and mechanisms which enhance oil production and trap CO_2 in conventional oil reservoirs are expected to be very different from those in tight unconventional reservoirs (Hawthorne et al., 2013). In conventional reservoirs, CO_2 flows through the permeable rock, and oil is mobilized by a combination of oil swelling, reduced viscosity, hydrocarbon stripping, and CO_2 flushing, especially when above the minimum miscibility pressure. In tight unconventional oil reservoirs, CO_2 flow will be dominated by fracture flow, and not significantly through the rock matrix. Fracture-dominated CO_2 flow could essentially eliminate the "flushing" mechanisms responsible for increased recovery in conventional reservoirs. As such, other mechanisms must be optimized in these unconventional oil reservoirs.

Conceptual mechanisms that may occur when CO_2 interacts with these tight formations include: (1) CO_2 flows through the fractures, (2) unfractured rock is exposed to CO_2 at fracture surfaces, (3) CO_2 permeates the rock driven by pressure, carrying some hydrocarbon inward; however, the oil is also swelling, which forces oil out of the pores, (4) oil migrates to the bulk CO_2 in the fractures via swelling and reduced viscosity, (5) as the CO_2 pressure gradient gets smaller, oil production is driven by concentration gradient diffusion from pores into the bulk CO_2 in the fractures, and (6) some fraction of the injected CO_2 is trapped in the irreducible fluids that remain in the reservoir after the production phase.

To investigate these concepts, rock samples from the Bakken Middle Member (low permeability, oil-saturated siltstone), Bakken Upper and Lower Shale Members (very low permeability, oil-saturated shale), Three Forks (low permeability, oil-saturated muddy dolostone), and a conventional reservoir (high permeability, oil-saturated sandstone) were exposed to CO_2 at typical Bakken conditions of 110 C and 5000 psi (230 F, 34.5 MPa) to determine the effects of CO_2 exposure time on hydrocarbon removal. Varying geometries of each rock ranging from small (mm) "chips" to 1 cm-diameter rods were exposed for up to one week, and mobilized hydrocarbons were collected for analysis. Nearly complete (>95%) hydrocarbon recovery occurs in hours with the more permeable matrices, while several days of exposure are required for the upper and lower Bakken shale samples. For example, while oil recoveries from a 1-cm round rod of middle Bakken and Three Forks rock were >95% after 24 hours, the oil recoveries from 1-cm diameter rods of upper and lower Bakken shales were only ca. 55 % after 24 hours (as shown in Figure 1). However, the recoveries from the same shales crushed to pass a 3.5 mm screen approached 95% after 24 hours, demonstrating that there is sufficient pore connectivity in middle Bakken and Three Forks rock, and even in upper and lower Bakken shales, to achieve increased CO_2-enhanced production of oil as well as to achieve CO_2 storage. While these laboratory results are encouraging, a better understanding of the controlling mechanisms is needed to allow exploitation in the Bakken play. We are currently performing additional rock/CO_2 exposure scenarios and developing models to describe the processes that control oil recovery and potential CO_2 storage. The results and implications for CO_2 EOR processes in unconventional reservoirs will be presented.

References

Bohrer, M., Fried, S., Helms, L., Hicks, B., Juenker, B., McCusker, D., Anderson, F., LeFever, J., Murphy, E., and Nordeng, S., 2008, State of North Dakota Bakken Resource Study Project, North Dakota Department of Mineral Resources, 23 p.

Nordeng, S.H., and Helms, L.D., 2010. Bakken Source System – Three Forks Formation Assessment, North Dakota Dept. of Mineral Resources, April, 2010.

Continental Resources, Inc., 2012. Bakken and Three Forks, website http://www.contres.com/operations/bakken-and-three forks, accessed May 30, 2013.

Kurtoglu, B., Sorensen, J., Braunberger, J., Smith, S., and Kazemi, H., 2013. Geologic Characterization of a Bakken Reservoir for Potential CO_2 EOR. Paper URTeC 1619698 presented at 2013 Unconventional Resources Technology Conference, Denver, Colorado, USA, 12-14 August 2013.

Hawthorne, S., Gorecki, C., Sorensen, J., Steadman, E., Harju, J., Melzer, S., FILLIN "What is your paper title?" * CHARFORMAT Hydrocarbon Mobilization Mechanisms from Upper, Middle, and Lower Bakken Reservoir Rocks Exposed to CO_2, Paper presented at the 2013 Unconventional Resources Conference, Calgary, Alberta, Canada, 5-7 November, 2013.

Acknowledgments

Financial support from the U.S. Department of Energy, National Energy Technology Laboratories (NETL) Cooperative Agreement No. DE-FE0024454, and the North Dakota Oil and Gas Research Council are also gratefully acknowledged.

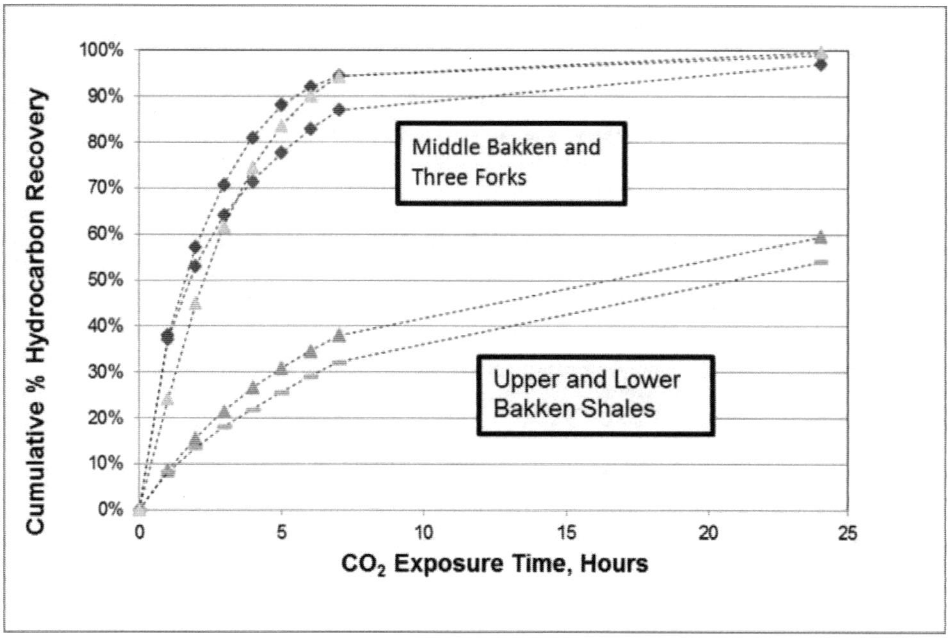

\s

Figure 1: Recovery of crude oil hydrocarbons from 11-mm diameter rock core samples from a McKenzie County (North Dakota) well with CO2 at 110 C and 5000 psi.

431337 How Do You Make Key Risk Management Decisions for Engineered Geologic Carbon Storage Systems in Face of Uncertainties?

Abstract:

Risk assessment and risk management of engineered geologic CO_2 storage systems is an area of active investigation. The potential geologic CO_2 storage systems currently under consideration primarily include deep saline aquifers, depleted oil/gas reservoirs and deep un-mineable coal seams, etc. These geologic systems are inherently heterogeneous and have limited to no characterization data. Effective risk management decisions to ensure safe, long-term CO_2 storage requires assessing and quantifying risks while taking into account the uncertainties in a storage site's characteristics. The key decisions are typically related to definition of area of review, effective monitoring strategy and monitoring duration, potential of leakage and associated impacts, etc. To date, risk assessments performed at sequestration field sites (both pilot and large-scale) have been mostly qualitative, based primarily on FEPs (Features, Events and Processes) analysis, or semi-quantitative. A quantitative methodology for predicting a sequestration site's long-term performance is critical for making key decisions necessary for successful deployment of commercial scale geologic storage projects where projects will require quantitative assessments of potential long-term liabilities. Quantitative assessment typically incorporates site-specific conceptual models for critical scenarios which are defined based on FEPs analysis followed by quantitative predictions of long-term evolution of a storage site in response to CO_2 injection incorporating computational models that can range from process level models to system level models.

The National Risk Assessment Partnership (NRAP) is a US-Department of Energy (US-DOE) effort focused on developing a defensible, science-based methodology for quantifying long-term performance of geologic CO_2 storage sites. The quantitative approach is based on integrated assessment modeling (IAM) paradigm which treats a geologic CO storage site as a system made up of various linked subsystems. The subsystems include storage reservoir, seals, potential leakage pathways (such as wellbores, natural fractures/faults) and receptors (such as shallow groundwater aquifers). CO_2 movement within each of the subsystems and resulting interactions are captured through reduced order models (ROMs). The ROMs capture the complex physical/chemical interactions resulting due to CO_2 movement and interactions but are computationally extremely efficient. The computational efficiency allows for performing Monte Carlo simulations necessary for quantitative probabilistic risk assessment. The approach used to develop ROMs as part of NRAP included performing detailed numerical simulations of CO_2 storage reservoirs, wellbores and aquifers using multiple models and numerical simulators. The simulators included LBNL's TOUGH2 suite, LANL's FEHM, PNNL's STOMP and LLNL's NUFT. Multiple (100s to 1000s) sets of simulation runs with detailed numerical models were performed to capture effects of various uncertain parameters. The numerical simulation results for each of the subsystems were used to develop reduced order models in form of abstracted functions or lookup tables. The resulting ROMs were subsequently incorporated in an IAM using a system modeling platform.

We have used the IAM to predict long-term performance of geologic CO_2 sequestration systems and to answer questions related to probability of leakage of CO_2 through wellbores, impact of CO_2/brine leakage into shallow aquifer, etc. Answers to such questions are critical in making key risk management decisions. A systematic uncertainty quantification approach has been used to understand how uncertain parameters associated with different subsystems (e.g., reservoir permeability, wellbore cement permeability, wellbore density, etc.) impact the overall site performance predictions and ultimately affect key risk management decisions.

424968 CO2 Sequestration Capacity Estimations for Jacksonburg-Stringtown Oil Field, West Virginia, USA

Abstract:

In order to control or reduce the emissions of carbon dioxide (CO_2), the main component of greenhouse gas, geological capture, utilization and storage (CCUS) of CO_2 in depleted oil and gas reservoirs is considered a favorable near-term approach with potential economic benefits. Coupled CCUS with enhanced oil recovery (EOR) is regarded as a cost-effective and environmentally safe method for long-term storage of anthropogenic CO_2. Several critical issues impact evaluation and deployment of an integrated system of CO_2 capture, and coupled CO_2 storage and EOR. These include source-sink matching, storage capacity, injection and breakthrough rates, recycling costs, oil recovery factors, and project duration.

The Upper Devonian fluvial sandstone reservoirs in Jacksonburg-Stringtown oil field, which has produced over 22 million barrels of oil since 1895, is an ideal candidate for CO_2 sequestration coupled with EOR because of supercritical depth (>2500 feet), low minimum miscible pressure (941 psi), favorable API gravity (46.5°) and good water flood response. Moreover, Jacksonburg-Stringtown oil field is adjacent to a large concentration of CO_2 sources located along the Ohio River, which potentially supports enough CO_2 to sequestration and EOR without constructing major long-distance new pipelines.

This work provides a more detailed and comprehensive approach to construct a 3-D geological model to estimate CO_2 sequestration capacity. A detailed geological characterization was performed incorporating abundant legacy data with limited modern data including well log and core data using artificial neural network and support vector regression. Depending on the proposed 3-D geological model, the best regions for coupled CCUS-EOR are located in southern portions of the field, and the estimated CO_2storage capacity for Jacksonburg-Stringtown oil field vary between 24 to 383 million metric tons.

Impact of model complexity for carbon dioxide migration in structured heterogeneous domains

Karl W. Bandilla • Michael A. Celia
Princeton University

Outline

- Carbon storage modeling
- Model setup
- Results
- Conclusions

Carbon storage modeling

Model complexity

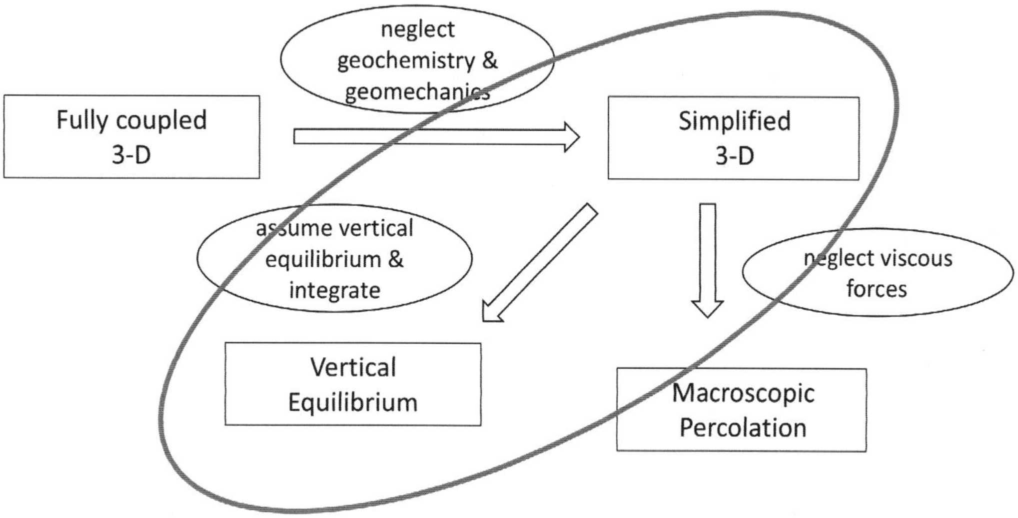

Successful application at Sleipner

Subsurface Hydrology RESEARCH GROUP

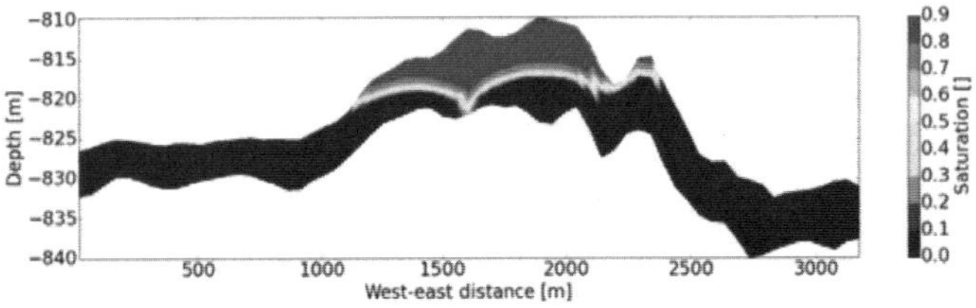

3D

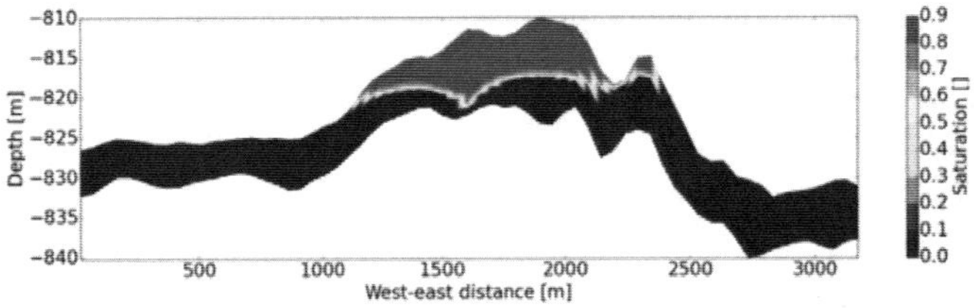

Vertical equilibrium

Fluvial deposits

Subsurface Hydrology RESEARCH GROUP

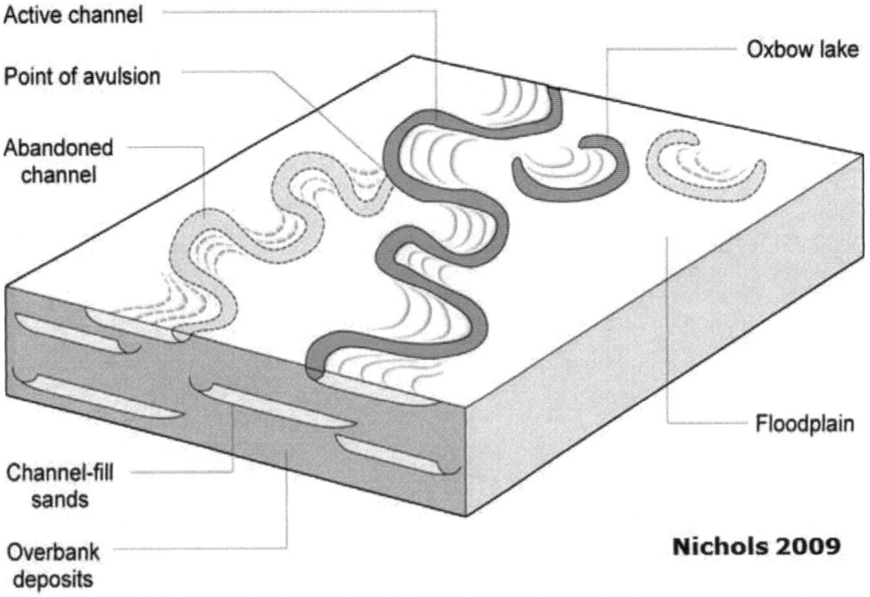

Subsurface Hydrology RESEARCH GROUP

Model setup

15AIChE Annual Meeting, Salt Lake City, UT

Bandilla & Celia, AIChE Annual Meeting, Salt Lake City, UT (11/13/15) 7

Subsurface Hydrology RESEARCH GROUP

Model layout

24 m thick

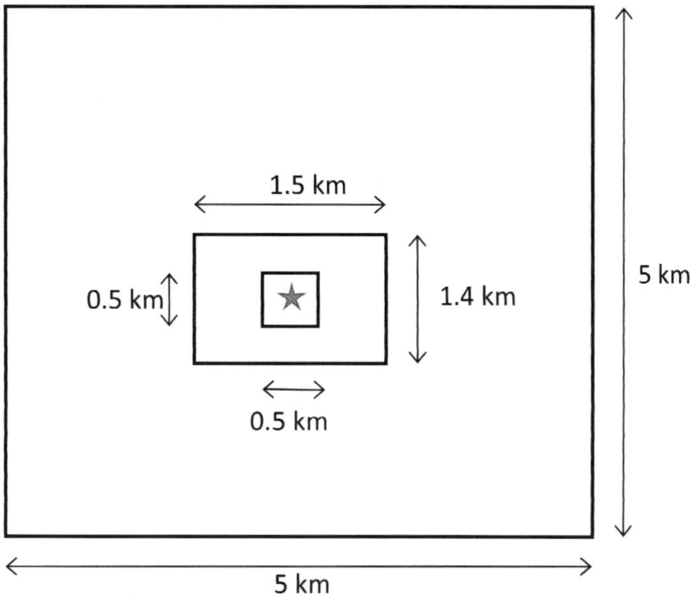

15AIChE Annual Meeting, Salt Lake City, UT

Bandilla & Celia, AIChE Annual Meeting, Salt Lake City, UT (11/13/15) 8

Parameters

- Stream permeability: 72 mD
- Stream porosity: 0.2
- Floodplain permeability: 2 mD
- Floodplain porosity: 0.18
- Injection rate: 0.017 Mt/yr for 4 years

- Lateral discretization: 10m, 50m, 100m
- Vertical discretization: 0.5m

Two cases - streams

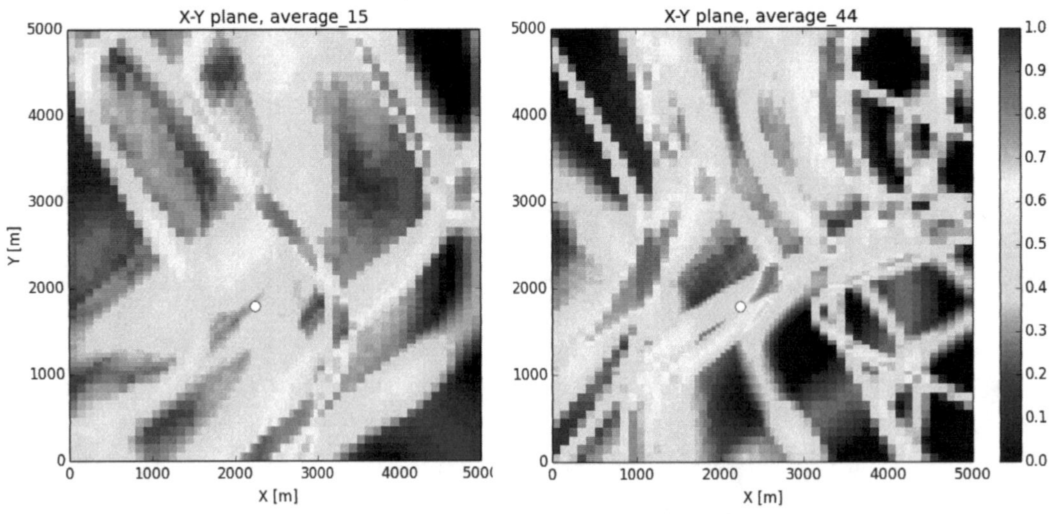

Two cases – cross-section

Large sand body

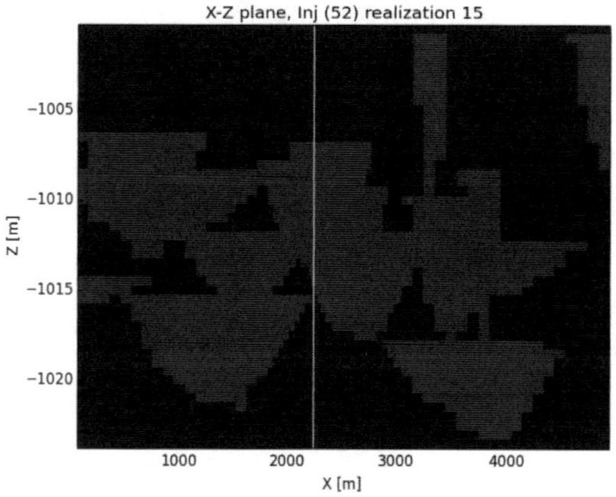

Stacked sand bodies

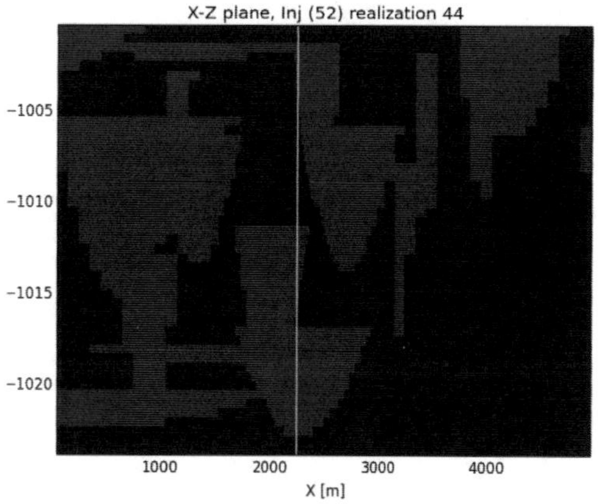

Vertically-integrated parameters

- 3D model: vertical representation
- Vertically-integrated model: permeability and porosity averaged over the thickness

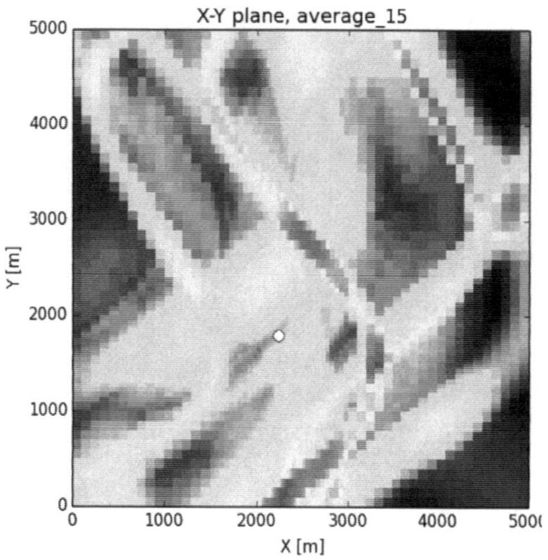

Subsurface Hydrology RESEARCH GROUP

Results

Big sand body

Subsurface Hydrology RESEARCH GROUP

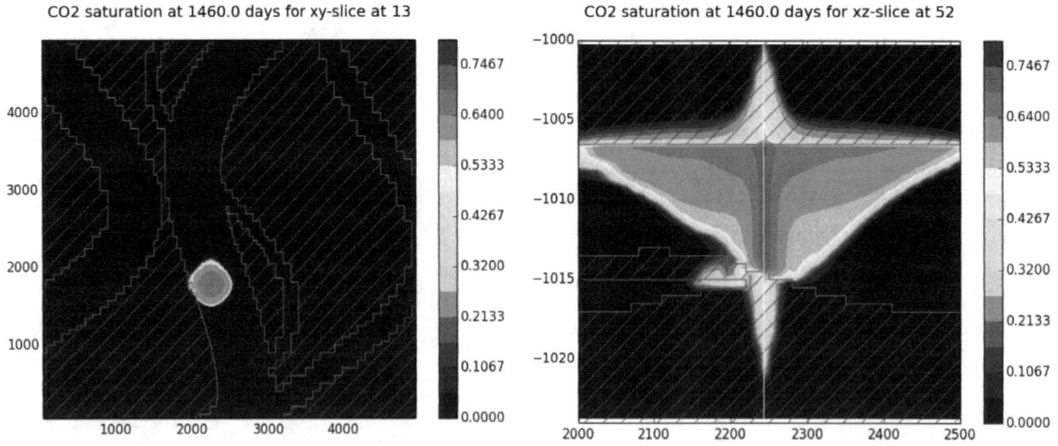

Big sand body – plumes

Subsurface Hydrology RESEARCH GROUP

3D

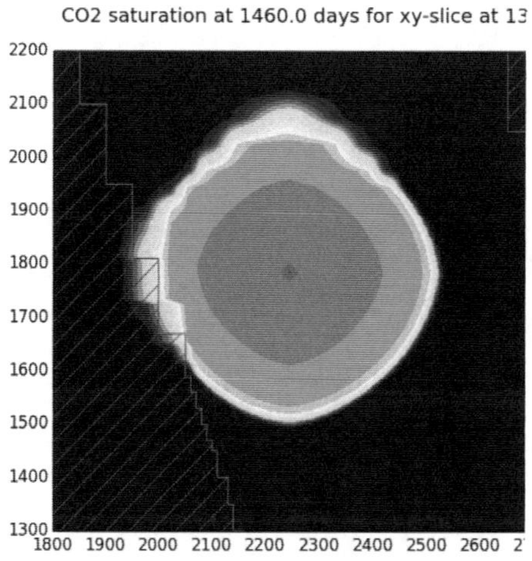

Vertical equilibrium

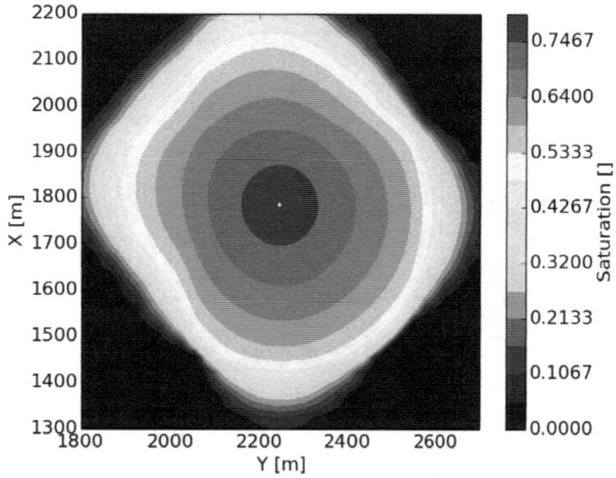

Big sand body – profiles

Subsurface Hydrology RESEARCH GROUP

3D

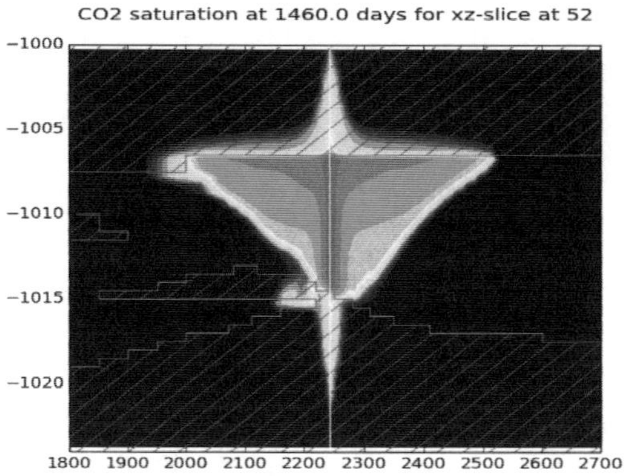

Vertical equilibrium

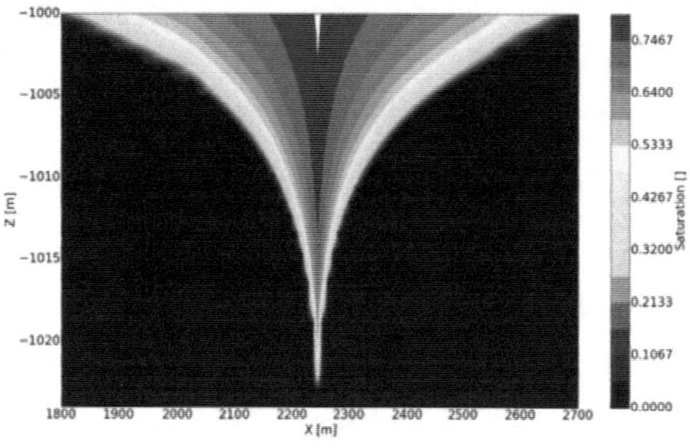

Stacked small bodies

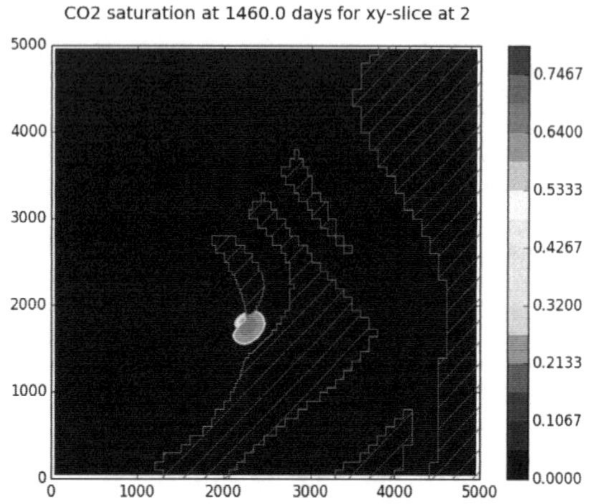

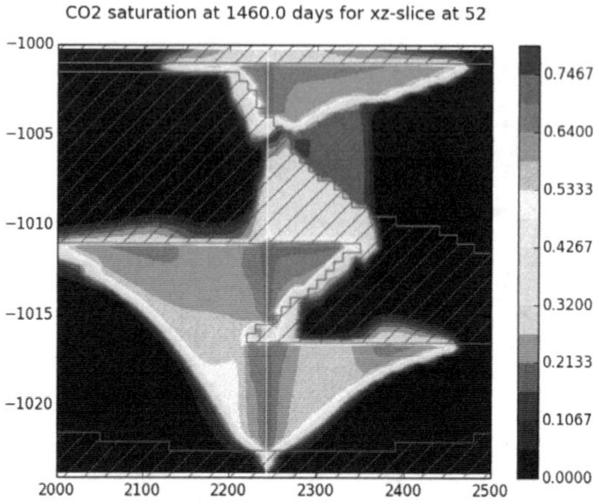

Stacked small – plumes

3D

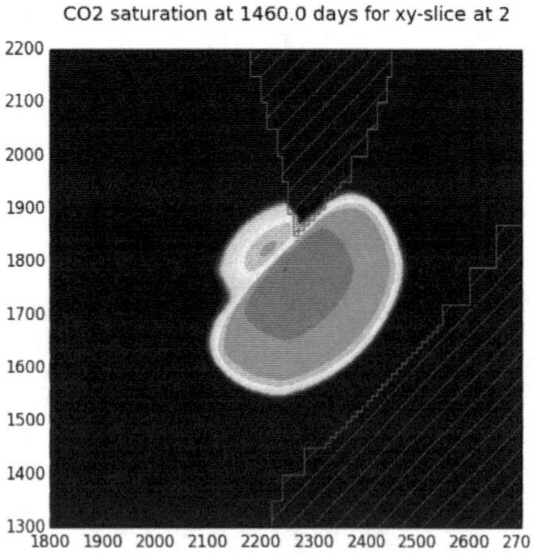

Vertical equilibrium

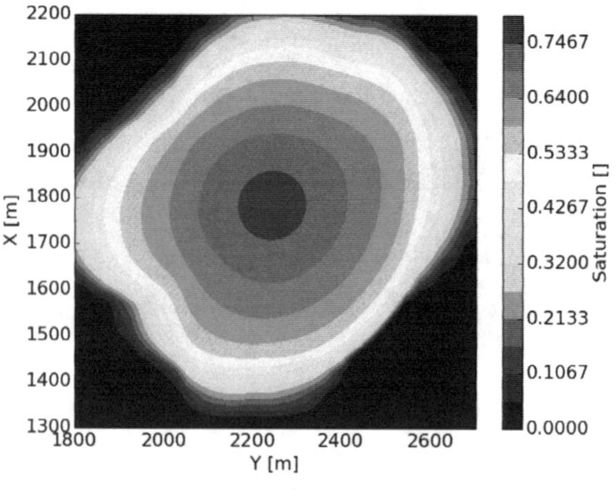

Conclusions

- We are able to create models with random fluvial deposits
- Simple vertically-integrated models are not able to capture CO_2 migration adequately
- Need to investigate more involved approaches such as dynamic reconstruction and multiple layers

Impact of model complexity for carbon dioxide migration in structured heterogeneous domains

Karl W. Bandilla • Michael A. Celia
Princeton University

Thank you

bandilla@prineton.edu

Partially funded by the US Department of Energy (DE-FE009563) and the Carbon Mitigation Initiative of Princeton University

LARGE-SCALE RELEASE AND DISPERSION OF CO_2: EXPERIMENTS AND SIMULATION

Vagesh D. Narasimhamurthy[1], Lorenzo Mauri[1], Sunil Lakshmipathy[1], Trygve Skjold[1], Shaoyun Chen[2], Yong Chun Zhang[2]

[1]Gexcon AS, P.O. Box 6015, Bergen NO-5892, Norway
[2]School of Chemical Engineering, Dalian University of Technology, Dalian 116023, People's Republic of China

Abstract

Carbon capture and sequestration (CCS) has emerged as a viable technique to mitigate climate change, reduce greenhouse gases and ocean acidification, and to enhance oil or gas recovery. An important part of the CCS chain is pipeline transport of dense-phase carbon dioxide (CO_2) from large point sources to suitable storage sites. Since CO_2 is an asphyxiating gas, deliberate or accidental leaks from pipelines represent a serious hazard, and realistic consequence modelling of such scenarios is of paramount importance for safety. This paper presents results from numerical simulations of an industrial-scale pipeline release and the subsequent dispersion of CO_2 in the atmosphere. The experiment and flow model development was part of the CO2PipeHaz project in the European Commission's Seventh Framework Programme for Research and Technological Development (FP7). The length and internal diameter of the pipeline used in the experimental setup was 260 m and 0.233 m, respectively. Pure CO_2 was discharged through a 0.05 m diameter orifice located 0.5 m above ground at the end section of the pipeline. The initial temperature was 20 °C and the initial pressure 53 bar. The measurements included wind speed, wind direction and temperature of the CO_2 plume. Comparison between experimental results and numerical simulations with the computational fluid dynamics (CFD) software FLACS shows good agreement. The numerical study provides useful insight into the dispersion behaviour of the CO_2 plume, and highlights the sensitivity towards the use of various turbulence models, surface roughness and wind conditions. The paper represents a significant contribution to the understanding of hazards associated with pipeline transport of carbon dioxide.

Introduction

Carbon capture and sequestration (CCS) entails long-term storage of large quantities of carbon dioxide (CO_2) collected from industrial sources such as coal- and gas-fired power plants. The main objective is to limit CO_2 emissions to the atmosphere. In addition, injection of CO_2 in oil reservoirs may enhance oil recovery. Pipeline transport of CO_2 is particularly relevant for situations where there are large geographical distances between the source and the storage site. In the entire CCS chain, *pipeline transport* has received less attention than the *capture* and *storage* technologies. This is unfortunate, since impurities present in the CO_2 may have significant effect on the thermodynamic properties and the rate of corrosion of the pipeline material [1]. Accidental pipeline failures can result in massive releases of CO_2. This represents a significant hazard, since CO_2 is an asphyxiating gas – about 4 vol.%

[1] Email: vagesh@gexcon.com, lorenzo@gexcon.com, sunil@gexcon.com, trygve@gexcon.com,
[2] Email: zalidy5518@vip.sina.com, chen_shy@sina.com, chensy@dlut.edu.cn

CO_2 in air is immediately dangerous to life and health [2]. To this end, realistic consequence modelling is an essential part of risk assessments. This paper describes experimental and numerical results from large-scale pipeline release experiments performed as part of the CO2PipeHaz project in the Seventh Framework Programme for Research and Technological Development (FP7) funded by European Commission (EU) [3, 4, 5].

Experimental setup

The experimental setup included a pipeline with length 260 m and internal diameter 0.233 m. The pipeline was initially filled with pure CO_2 at a temperature of 20 ± 2 °C and pressure 53.04 bar (gas/liquid equilibrium). The 0.05 m diameter orifice at the end of the pipeline was located 0.5 m above the ground. Figure 1 (left) summarizes the wind speed and wind direction measured 10 m above ground during the release. The release direction was from North to South, i.e. more or less perpendicular to the wind direction. The ambient temperature was 10 °C. Thermocouples mounted at vertical masts measured the temperature of the CO_2 plume 0.5 m and 4.0 m above ground downstream of the orifice. The variation in temperatures between various sensor positions prior to the release was about 5 °C. The vertical temperature profiles were not consistent with a possible ambient stratification, and the differences in temperature are assumed to be due to calibration errors. Hence, as part of the post-processing of the thermocouple readings, all measurements were adjusted to the ambient temperature (10 °C) 30 seconds prior to the start of the release.

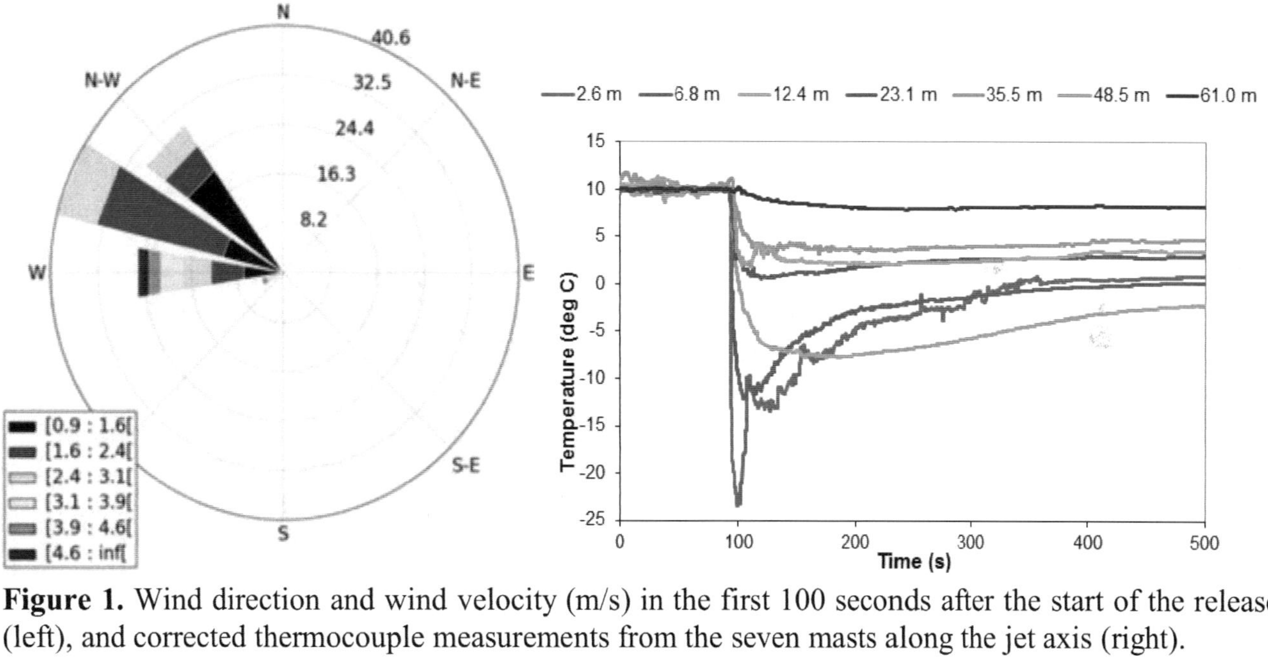

Figure 1. Wind direction and wind velocity (m/s) in the first 100 seconds after the start of the release (left), and corrected thermocouple measurements from the seven masts along the jet axis (right).

Figure 1 (right) shows sample temperature measurements at seven positions along the jet axis (i.e. 0.5 m above ground). The release started at time $t = 93$ seconds. Figure 2 (left and center) shows the experimental location and a snapshot of the release. The ground downstream of the release is not smooth, but has large scattered rocks. The jet is pure vapour phase without any CO_2 particles. The phase diagram for CO_2 shows that solid-vapor equilibrium at atmospheric pressure is attained for a temperature of -78.45 °C. The fact that no solid deposition was observed in the experiment suggests immediate sublimation of the solid phase after the release.

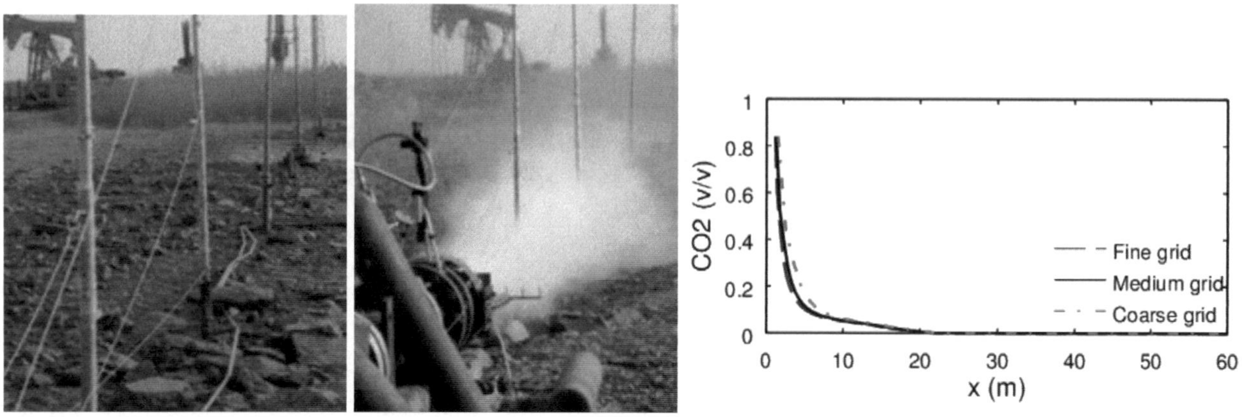

Figure 2. Snapshot of the experimental site (left), release (center), and results from grid convergence study illustrating CO_2 volume fraction along the jet axis (right).

Numerical method

The simulations were performed with the computational fluid dynamics (CFD) tool FLACS [6, 7], which solves the three-dimensional (3D) time-dependent Favre-averaged Navier-Stokes equations on a staggered Cartesian grid using the finite volume method. The code uses a second-order central differencing scheme for diffusive fluxes and a second-order "kappa" scheme, i.e. a hybrid scheme with weighting between upwind and central difference (with delimiters for some equations), for convective fluxes. Time-marching is carried out using a first-order implicit scheme (backward Euler) and the discretized equations are solved using the BICGStab iterative method with the SIMPLE pressure correction algorithm. The code is parallelized with OpenMP. In addition, FLACS uses the porosity/ distributed resistance (PDR) concept to embed solid geometries within the Cartesian grids (using special *subgrid* turbulence models). In this study, four turbulence models were used: the standard k-ε model [8], the RNG k-ε model [9], a realizable k-ε model [10], and the k-$\sqrt{kl}$ (kskL) model [11].

The experimental conditions seven seconds after the start of the release are modelled, which corresponds to 100th second in Figure 1 (right). Jet inlet conditions (CO_2 mass fraction, velocity, temperature, k and ε) corresponding to this time instant were obtained from a near-field source model resolving the shock region [12]. This source data stems from a location 20 orifice diameters downstream of the orifice, where CO_2 is in the gaseous phase. The mass flow rate obtained by integrating this source data results is 6.5 kg/s. Extrapolating the data obtained from the source model to the FLACS setup resulted in a mass flow rate of 6.72 kg/s. It is however not known whether this is an accurate value, since the mass flow rate was not measured in the experiment. The corresponding jet axial velocity and temperature is 258 m/s and -94.15 °C, respectively.

Based on the wind rose data in Figure 1, wind direction and speed in the simulations are set 288° and 3.1 m/s, respectively. Inflow wind condition comprised of Pasquill stability class D (neutral) and ground roughness value 0.03. The ambient temperature was set to 10 °C, as in the experiments. At the outlets, a non-reflecting boundary condition (*Nozzle*) was imposed. For the wall-boundary, the default wall-function in FLACS was used. Table 1 shows the numerical domain size and grid details. The time-step chosen for the current simulation was such that the CFL condition of approximately one was satisfied, and the unsteady simulations converged to steady-state after 160 seconds of CPU time. Figure 2 (right) shows the grid convergence results, where it is clear that the medium and fine grids yield nearly identical results. The medium grid is therefore chosen for all other sensitivity studies.

Apart from using different turbulence models, two other sensitivity studies are performed. In the first case, wind direction is hypothetically set to the release direction (0°), i.e. a co-flow instead of a cross-wind. In the second case, the effect of large scattered rocks in the experimental test site is modelled by mounting 'roughness' blocks of size [x, y, z] = [1 m × 0.8076 m × 0.325 m] on the surface in a staggered pattern (with pitch 6 m in x and y directions).

Table 1. Number of grid points N_i and minimum grid dimension Δi_{min} used in various cases. Numerical domain size in all cases is 205 m × 170 m × 10 m in x, y, and z directions, respectively.

Case	N_X	N_Y	N_Z	ΔX_{min} (m)	ΔY_{min} (m)	ΔZ_{min} (m)
Fine grid	181	166	62	0.25	0.05	0.05
Medium grid	118	102	53	0.50	0.05	0.05
Coarse grid	85	74	53	1.00	0.05	0.05

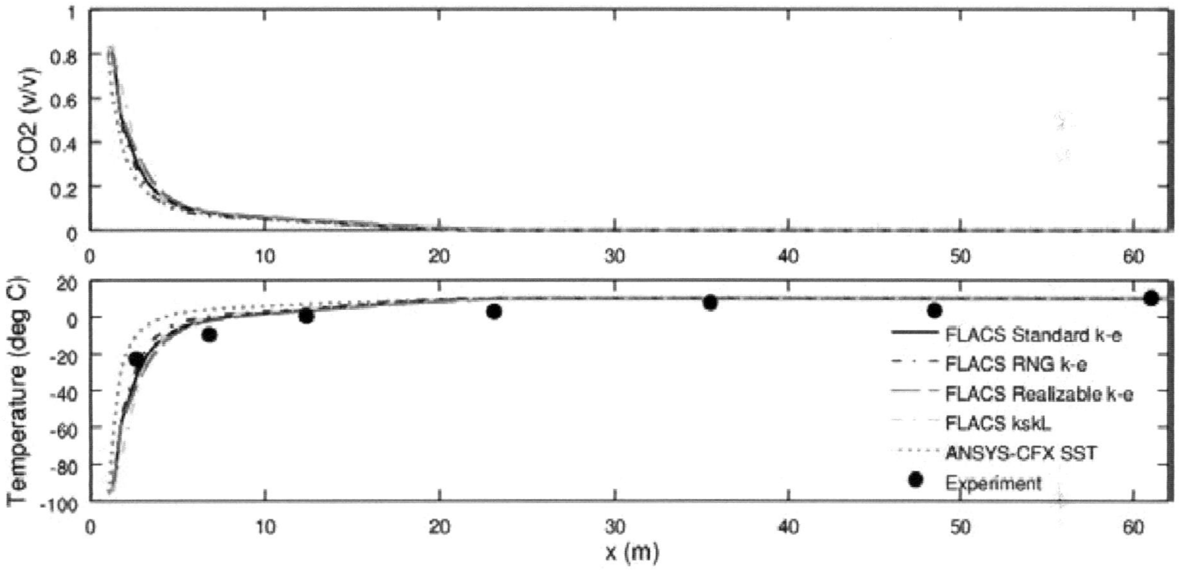

Figure 3. Comparison between different turbulence models from FLACS and ANSYS-CFX [12] with experimental data along the jet axis (x-direction).

Results

Turbulence models

The first part of the analysis addresses the sensitivity of the model system with respect to choice of turbulence model. Figure 3 shows the CO_2 volume fraction (v/v) and temperature along the jet axis for the various turbulence models. All the four turbulence models in FLACS are predicting nearly identical results, probably because this is a free-jet case without any flow-separation, impingement, unsteadiness or rotation. The ANSYS-CFX SST model prediction [12] is also shown in Figure 3 for cross-comparison. The results obtained with FLACS are in closer agreement with the experiment than the results obtained with ANSYS-CFX. Further downstream (x > 20 m), all models predict the jet to be deflected sideways by the cross-wind and miss the measuring masts, whereas the experiments show a small temperature reduction (about 5° C) at these locations. This is clearer in Figure 4, where temperatures are compared at individual masts. The results from FLACS are from the nearest grid point, and not necessarily the exact mast location (typically about 0.05 - 0.5 m offset). Further, the experimental data downstream is not exhibiting a monotonic trend, i.e. the thermocouples have recorded a raise in temperature between mast 4 and mast 5, but then drops to a lower value at mast 6 before reaching the ambient temperature at mast 7. This may indicate some measurement uncertainty.

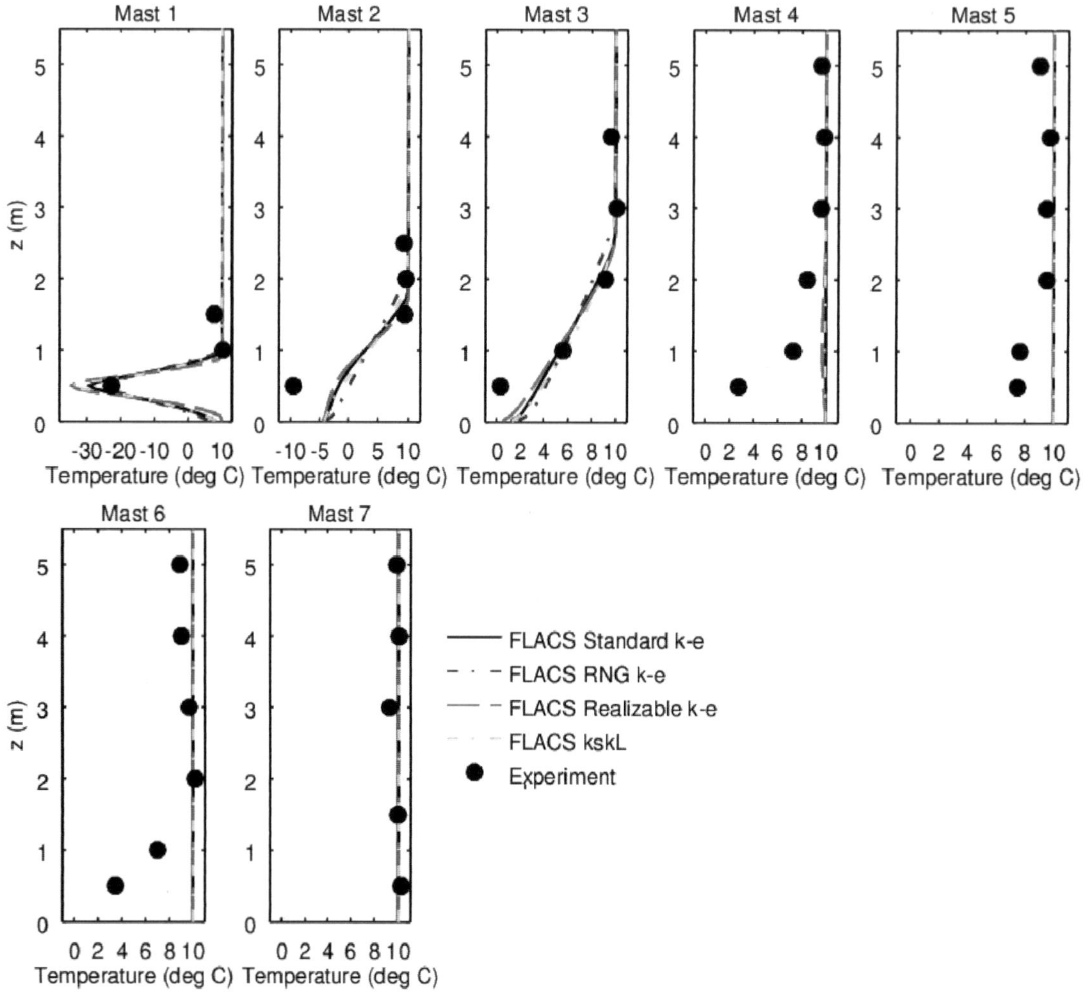

Figure 4. Comparison between different turbulence models from FLACS and experimental data in the wall-normal (z) direction. See Figure 1 for the streamwise (x) position of all the seven masts.

Surface 'roughness' and wind

Figure 5 shows the effect of modelling large scattered rocks as 'roughness' blocks. Note that here the 'roughness' blocks act as bluff-bodies rather than as roughness due to its larger size (relative to the boundary-layer). The obstructions produce flow separation and increased turbulence resulting in widening of the jet in comparison to the smooth ground case. Figure 5 shows a dramatic change in the dispersion behaviour. Figure 6 compares the experimental values to the smooth ground case and three 'roughness' cases. In the first case, 'roughness' blocks are considered only near the masts. In the second case, 'roughness' blocks are staggered throughout the computational domain. Finally, the third case represents a combination of the first and second. In addition, a hypothetical scenario is considered, where the wind direction in the smooth ground case is set to 0°, i.e. along the release (x) direction instead of the default cross-wind. All cases are simulated using standard k-ε model. Figure 6 shows that 'roughness' blocks have an effect on the temperature profiles in the near-field (x < 25 m), but show no variation further downstream. On the contrary, the hypothetical wind-direction case shows negligible differences with the default wind case in the near-field, but exhibits significant deviations farther downstream. Recall that this is a high momentum release case (258 m/s) and a wind speed of 3.1 m/s (regardless of the direction) will have a little impact on the near-field. Downstream, however, wind effects appear to dominate the dispersion. Figure 7 demonstrates this observation more clearly at each mast location.

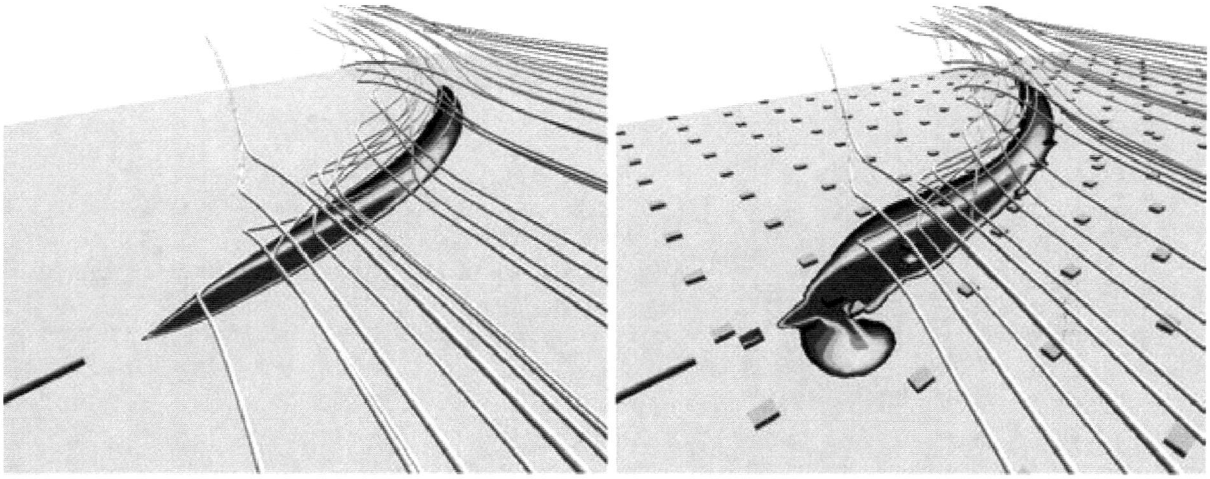

Figure 5. Steady state CO_2 concentration (above 1% v/v) near the ground: (left) smooth wall case; (right) 'roughness' blocks near masts and in staggered pattern elsewhere. Corresponding 3D wind field is also appended. All simulations use standard k-ε model.

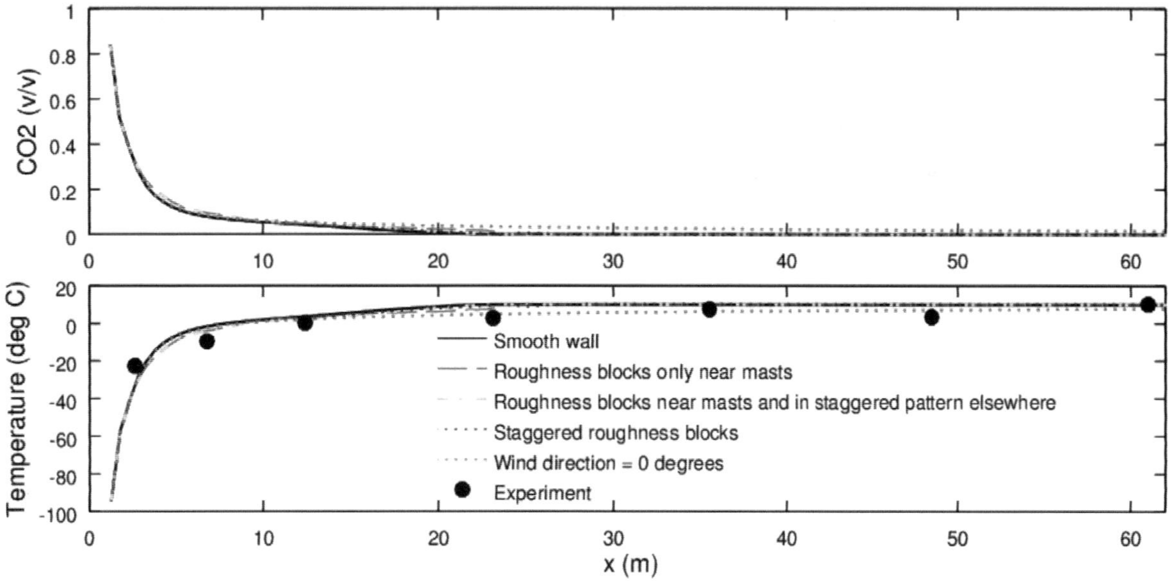

Figure 6. Comparison of 'roughness' blocks and wind effects along the jet axis (x-direction). All simulations use standard k-ε model.

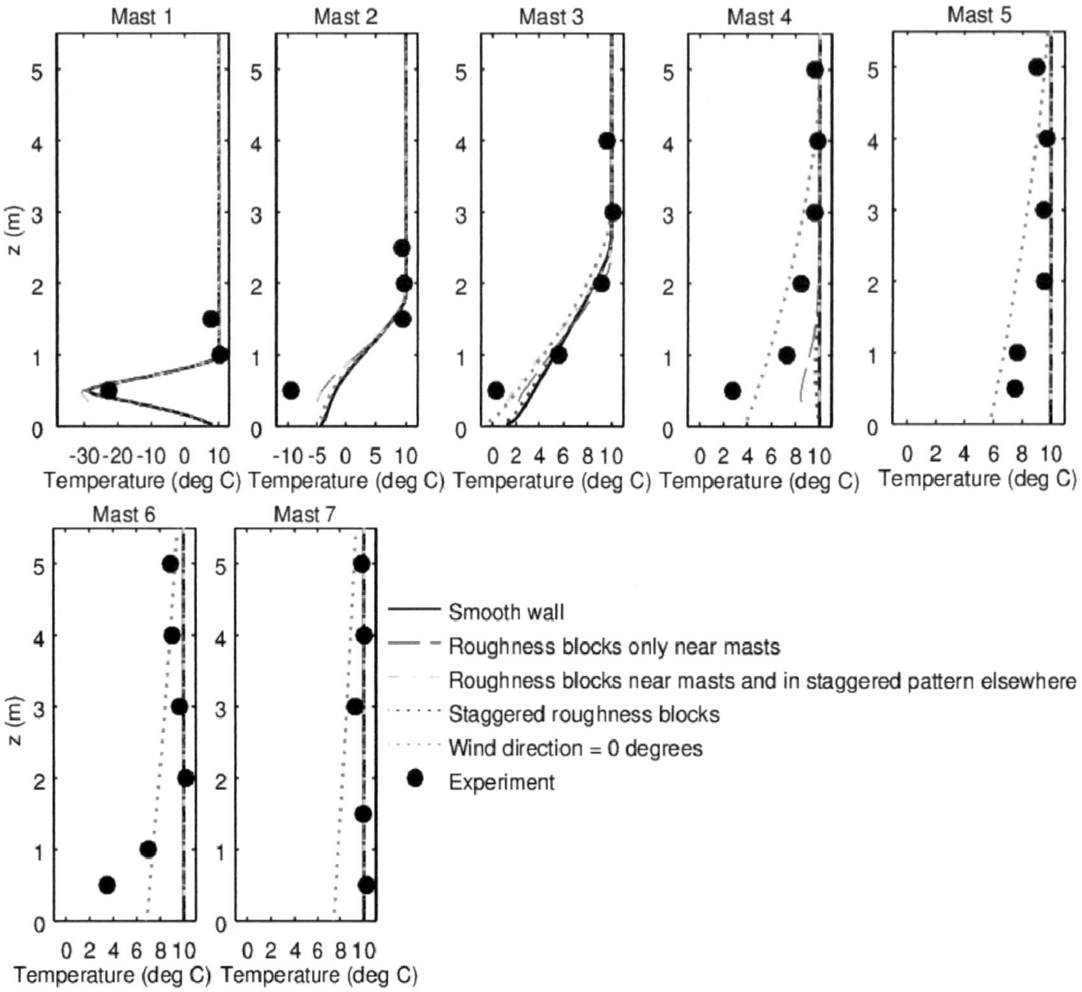

Figure 7. Comparison of 'roughness' blocks and wind effects in the wall-normal (z) direction. See Figure 1 for the streamwise (x) position of all the seven masts. All simulations use standard k-ε model.

Conclusion

This paper reports numerical and experimental results for a large-scale release of CO_2 from a pipeline. The CFD simulations with FLACS include four different turbulence models, and all showed good agreement with the experimental data. Further, model sensitivity towards various surface 'roughness' conditions and wind boundary conditions demonstrated the importance of both parameters in the near- and far-field dispersion behaviour. Modest deviations observed between simulated and measured data in the far-field region can be attributed to the unknown nature of experimental mass-flux and possible uncertainty in the thermocouple measurements.

Acknowledgements

The authors gratefully acknowledge the financial contribution to the CO2PipeHaz project from the European Commission through the Seventh Framework Programme. Prof. Haroun Mahgerefteh and Dr. Sergey Martynov from University College London, and Dr. Christophe Proust from INERIS, contributed significantly to the design of the experiments.

References

1. IEAGHG (2011), "Effects of Impurities on Geological Storage of CO_2," *IEA Greenhouse Gas R & D Programme*, http://www.globalccsinstitute.com/publications/effects-impurities-geological-storage-co2, accessed 29.09.15.
2. NIOSH (1995), "Documentation for Immediately Dangerous to Life or Health Concentrations (IDLH)," *National Institute for Occupational Safety and Health (NIOSH) chemical listing and documentation of revised IDLH values* (as of 3/1/95), NTIS publication no. PB-94-195047.
3. CO2PipeHaz (2009), "Quantitative Failure Consequence Hazard Assessment for Next Generation CO2Pipelines: The Missing Link," http://www.co2pipehaz.eu/, accessed 29.09.15.
4. Woolley, R. M., Fairweather, M., Wareing, C. J., Proust, C., Hebrard, J., Jamois, D., Narasimhamurthy, V. D., Storvik, I. E., Skjold, T., Falle, S. A. E. G., Brown, S., Mahgerefteh, H., Martynov, S., Gant, S. E., Tsangaris, D. M., Economou, I. G., Boulougouris, G. C. and Diamantonis, N. (2014), "An Integrated, Multi-Scale Modelling Approach for the Simulation of Multiphase Dispersion from Accidental CO2 Pipeline Releases in Realistic Terrain," *International Journal of Greenhouse Gas Control*, Vol. 27, pp. 221-238.
5. Gant, S. E., Narasimhamurthy, V. D., Skjold, T., Jamois, D. and Proust, C. (2014), "Evaluation of Multi-Phase Atmospheric Dispersion Models for Application to Carbon Capture and Storage," *Journal of Loss Prevention in the Process Industries*, Vol. 32, pp. 286-298.
6. GexCon AS (2015), "FLACS v10.4 User's Manual," http://www.flacs.com, accessed 29.09.15.
7. Narasimhamurthy, V. D., Hisken, H., Atanga, G., Skjold, T. (2015), "Porosity/distributed resistance modeling for industrial CFD applications," *Proc. 8th National Conference on Computational Mechanics (MekIT'15)*, B. Skallerud and H. I. Andersson, ed. Trondheim, Norway, CIMNE.
8. Launder, B. E. and Spalding, D. B. (1974), "The numerical computation of turbulent flows," *Comput. Meth. Appl. Mech. Eng.*, Vol. 3, pp. 269-289.
9. Yakhot, V., Orszag, S. A., Thangam, S., Gatski, T. B. and Speziale, C. G. (1992), "Development of turbulence models for shear flows by a double expansion technique," *Phys. Fluids A*, Vol. 4, pp. 1510-1520.
10. Shih, T.-H., Liou, W. W., Shabbir, A., Yang, Z. and Zhu, J. (1995), "A new k-ε eddy viscosity model for high Reynolds number turbulent flows," *Computers & Fluids*, Vol. 24, pp. 227-238.
11. Menter, F., Egorov, Y. and Rusch, D. (2005), "Steady and unsteady flow modelling using the k-$\sqrt{k}$l model," *Turbulence, Heat and Mass Transfer*, Vol. 5, pp. 403-406.
12. Narasimhamurthy, V. D., Skjold, T. and Gant, S. E. (2013), "On the Validation Studies of Two-Phase Models in FLACS and CFX," *Technical Report*, CO2PipeHaz Project Deliverable 1.5.4: Basic theory, code and the validation of the FLACS and CFX models.

Abstract ID# 403711

PLUGGED WELL LEAKAGE RISK ASSESSMENT IN A CARBON SEQUESTRATION PROJECT

Ben Li, University of Louisiana at Lafayette
Boyun Guo, University of Louisiana at Lafayette
Hui Li, University of Louisiana at Lafayette

Abstract

Carbon sequestration has been proposed as a promising way to mitigate the accumulation of greenhouse gases. Depleted oil and gas reservoirs can be considered as desirable vessel for carbon sequestration. It is crucial to maintain the integrity of formations that contain high concentration of CO_2. Plugged wells can be one of the risky leakage pathways which defect the purpose of carbon sequestration. The leak risk of plugged wells should be monitored and evaluated through the whole lifetimes of carbon sequestration activities.

Even though numerous researches have focused on the well integrity assessment for carbon sequestration projects, a systematic manner addressed the plugged wells' CO_2 leakage in a practical way has not been presented. This study will fill this gap. Lean analysis is performed by using Cause and Effect Analysis (CEA) method and Failure Mode and Effects Analysis (FMEA) method. CEA is used to identify all the possible causes that could result to the CO_2 leakage of plugged wells, while FEMA was implemented to build the plugged well failure modes and evaluate the possibility of CO_2 leakage for plugged wells. WH field, a CO_2 EOR oil field located in the Gulf of Mexico, is studied as an example to explain how CEA and FMEA could be functioned as effective methodologies to evaluate the CO_2 leak potentials of the plugged wells. The plugged wells with high risk of leakage in this field are predicted. Lean analysis is proven to be an effective way to evaluate the risk of CO_2 leakage for plugged wells in carbon sequestration activities. The proposed method can also be used in evaluating the integrity of CO_2 injection wells and oil production wells in an oil field where CO_2 is or will be injected in the future.

Introduction

Carbon sequestration is achieved by injecting CO_2 into the depleted oil filed for storage purpose. This process would reduce greenhouses effect on the global environment. However, depleted oil field has its own features that will defect the goal of carbon sequestration. This mainly lies in the existing wells in the carbon sequestration area. CO_2 may migrate along the wellbore and finally come to the underground drinking water formation or atmosphere. Plugged well in the carbon sequestration field is one of the risky CO_2 leakage pathways. This is because that plugged wells should have some construction defects or integrity drawback before plugged. Plugged wells are previously drilled for exploration and production of oil or gas and abandoned. Older plugged wells were constructed with lower annular cement tops, allowing many zones to be in communication behind casing (Watson and Bachu, 2009).[1] The cement class used in these wells was not initially designed for CO_2 corrosion resistant. Carbon sequestration site must be leak free to the atmosphere and to other geological formations such as underground drinking water layer (Watson and Bachu, 2009).[1] Quantifying the well integrity during carbon sequestration has gained considerable interest. CO_2 leakage rate along a wellbore has been studied by Huerta et al. (2009) and Tao et al. (2010). [2] [3] D'Alesio et al. (2011) investigated well integrity of three CO_2 storage fields in Italy.[4] SCP is an indicator of gas leakage that could occur through the failure of the cement sheath between the casing and the formation. Rocha-Valadez et al. (2014) proposed a model to assess wellbore integrity in sustained casing pressure annulus.[5]

However, Gas migration outside the casing is relative difficult to detect. If gas migrates to underground drinking water formation, the drinking water would be contaminated not only by the injected CO_2 but also by other hydrocarbon gas migrated from the same routine. Early detection of this leakage is crucial to protect the drinking water formation as well as the human health. This paper will focus on the leakage potential of plugged wells in carbon sequestration projects. CEA and FEMA methods are used to investigate the leakage risk by various factors based on the assembled database. Numerous of parameters are involved in the study, including well age, well history, surface casing depth, underground drinking water formation depth, well depth, well density, regulatory changes. The level of the impact of factors are also identified and verified. The leakage risk of plugged wells is evaluated in a qualitative manner.

Methodology

Cause Effective Diagram (CEA)

CEA is known as a causal diagram which was firstly proposed by Kaoru Ishikawa, a pioneer of quality management (Ishikawa, 1976).[6] The objective of CEA is to break down root causes that potentially contribute to a particular effect. The CEA can reveal key relationships among various variables, and the possible causes provide additional insight into process behavior. The common uses of the CEA are product design and quality defect prevention, to identify potential factors causing an overall effect (Ishikawa, 1976).[6] It was originally developed as a quality control tool, however CEA is currently used in numerous ways such as discover a root cause of a problem, detect the bottlenecks in a process, and identify the malfunctioned part in a process. CEA includes four major steps: (1) identify the problem; (2) work out the major factors involved; (3) identify possible causes; (4) analyze the proposed diagram.

CO_2 leakage prediction of plugged wells is a key issue to evaluate the successfulness of a carbon sequestration project. CO_2 could leak into the underground drinkable water formations or atmosphere, which will cause catastrophic effects both on human and other creatures of the world. Even though numerous researchers focused on the well leakage issues, fewer articles were published related to the plugged well CO_2 leakage in carbon sequestration projects. Root causes resulted to CO_2 leakage of plugged wells has not been described in a systematic manner. This is due to the lack of effective tools in petroleum engineering scope which can cover all the possible causes of the CO_2 leakage problem. This paper will fill this gap by the implement of CEA to identify the likely causes of CO_2 leakage issues thoughtfully. CEA provides a visual way to address the cause and effect in a more structured approach than some other tools. The proposed method will be used to identify root causes and effects in a clear and specific manner.

Failure Mode and Effect Analysis(FMEA)

FMEA is one of the first systematic techniques for failure analysis, which was developed by reliability engineers in the 1940s to study problems that might arise from malfunctions of military systems (Fadlovich, 2007).[7] FEMA helps to identify potential failure modes based on the common physics of failure logic. Although initially developed by the military, FMEA methodology is now extensively used in a variety of industries including semiconductor processing, food service, plastics, software, and healthcare. Risk Priority Number (RPN) is used to evaluate the model failure possibility, which is shown in Equation (1).

$$RPN=S*P*D \qquad (1)$$

where S is severity, meaning the consequence of a failure mode. Severity describes the worst potential consequence of a failure, determined by the degree of injury, system damage, and time lost to repair the failure; P means probability, probability ranking considering the likelihood of occurrence of the failure; D stands for detectability, which addresses how the failure mode can be discovered. Table 1 shows the rating and meaning of S, P, and D, where "N/A" means not applicable.

Table 1 Rating and Meanings of S, P, and D

Rating	Meaning		
	S	P	D
1	No Relevant	Extremely Unlikely	Certain
2	Very Minor	Unlikely	Almost Certain
3	Minor	Occasional	High
4	Moderate	Reasonably Possible	Moderate
5	Critical	Frequent	Low
6	Catastrophic	N/A	Undetected

When causes and effects of plugged wells CO_2 leak are identified by CEA methodology, the impact factor levels of failure on plugged wells can be determined by FMEA. The CO_2 leakage risk is affected by a list of failure modes ranked according to the seriousness, the degree of detectable, and the likelihood of occurrence. Each leakage scenario will be represented by RPN, and the CO_2 leakage possibility caused by the combination of failure modes of a single plugged well can be described by Equation (2).

$$LP = \sum_{i=1}^{n}(\mathrm{RPN}_i) \tag{2}$$

Where LP is the leakage possibility, n is the total number of causes of CO_2 leak. In order to rank the reliabilities of plugged well in carbon sequestration projects, normalized leakage possibility (NLP) of plugged wells are calculated by Equation (3).

$$NLP_j = \frac{\mathrm{LP}_j - \mathrm{LP}_{min}}{\mathrm{LP}_{max} - \mathrm{LP}_{min}} \tag{3}$$

Where NLP_j is the normalized leakage possibility of j well, and LP_{max} is the maximum LP value in the studied plugged wells, LP_{min} is the minimum LP value in the studied plugged wells. The range of NLP is between 0 and 1. The higher the NLP value, the higher risk of the plugged well will leak. This investigation method should benefit to the decision-making, leak prediction and remediation activities.

Leakage Scenarios

Numerous reasons can result to the leakage of CO_2 from the existing wellbore. Well construction defects (WCD), and well age (WA), well plugged age (WPA), number of plugs (NOP), number of CO_2 injection wells in one-quarter mile radius (NCIW) of a plugged well, mechanical integrity of cement sheath (MICS), regulation changes on plugging wells (RCOPW) are addressed as the leakage scenarios based on the availability of current database.

Well Construction Deficiency

The U.S. Environmental Protection Agency (EPA) is mandated under the Underground Injection Control (UIC) Program of the Safe Drinking Water Act (SDWA) to protect Underground Source of Drinking Water (USDW) and the health of people from underground injection. USDW means an aquifer or its part that currently supplies any public water system or contains a sufficient quantity of ground water to supply a public water system; and currently supplies drinking water for human consumption or contains fewer than 10,000 mg/L total dissolved solids. Based on the Texas Administrative Code (TAC) Rule 5.203, Geologic Storage and Associated Injection of Anthropogenic Carbon Dioxide, the operator of a geologic storage facility must ensure that all anthropogenic CO_2 injection wells are constructed and completed to prevent the movement of injection CO_2 into any areas where it could endanger USDW.

Carbon dioxide can leak along the interfaces between casing and cement sheath, between cement sheath and formation, and fractures inside the cement sheath. The leakage can lead to the impacts to USDW or atmosphere. If the surface casing depth is below the USDW, CO_2 can escape to the atmosphere through the annuli between the production casing and formation, and between the production casing and surface casing. This leakage should be due to the de-bonding or poor bonding of the cement sheath as well as cement failure. The CO_2 could migrate to the USDW due to the defect of cement sheath set in the annulus between surface casing and formation, and between the production casing and formation. The CO_2 could also migrate to the surface rather than the USDW due to the cement sheath fail in the casing and formation annulus along the whole wellbore. The surface casing depth of some plugged wells may be above the USDW due to the fact that the regulations at that time did not consider USDW or the USDW was not clear when these wells were drilled. If the cement sheath of production casing fails to cover the whole column, CO_2 leak through different pathway at the same time, which means CO_2 could leak to USDW and atmosphere potentially. The risk of CO_2 leakage is more likely to happen on this sort of wells because only the failure of the production casing cement sheath will lead to the leakage of CO_2. There is no barrier to postpone the CO_2 escaping to the surface or USDW.

WCD can be another risk for CO_2 leakage. When insufficient surface casing is set to protect all usable quality water strata, usable quality water strata are exposed to the wellbore. A cement plug shall be a minimum of 100 feet in length and shall extend at least 50 feet above and 50 feet below the base of the deepest usable quality water stratum. The risk of CO_2 leakage into these strata is higher due to the lack of or poor barriers between usable quality strata and CO_2 injection zones. The production casing shall be perforated at the required depths to place cement outside of the casing by squeeze cementing through casing perforations as shown in Figure 4. The P of WCD is determined by the method proposed in the author's former paper as shown in Appendix B. Most of construction details of plugged wells are available, so the D of WCD is set to 2. If a well scheme does not fulfill the requirement of carbon sequestration, the leak risk can be higher. The S is set to 4 due to this consideration.

Well Age and Well Plugged Age

WA is the age from the well completion year to the reference year. WPA is the age from the well plugged year to the reference year. In this study, 2015 is set as the reference year for WA and WPA. Since almost all the plugged wells were oil well before being plugged, the difference between WA and WPA is the well production age (PA). Both WA and WPA are normalized. Equation (4) and (5) are the methods for calculating P for WA and WPA, respectively.

$$P_{WAN} = \left(\frac{WA - WA_{min}}{WA_{max} - WA_{min}}\right) * 5 \tag{4}$$

$$P_{WPAN} = \left(\frac{WPA - WPA_{min}}{WPA_{max} - WPA_{min}}\right) * 5 \tag{5}$$

In these equations P_{WAN} is the P of WA. WA_{max} and WA_{min} are the maximum and minimum WA of the studied plugged wells, respectively. P_{WPAN} is the P of WPA, WPA_{max} and WPA_{min} are the maximum and minimum WPA of studied plugged wells, respectively. P_{WAN} and P_{WPAN} are rescaled from 0 to 5 and rounded up to the nearest integer. P_{WAN} and P_{WPAN} are the Ps of WA and WPA, respectively. S and D of WA and WPA failure modes are set to 2 and 2, respectively. This is because WA and WPA of most plugged wells are well documented and the relationship between well leakage and WA/WPA is not clear.

NCIW

NCIW describes how many CO_2 injection wells could affect a plugged well in one-quarter mile radius. The Railroad Commission of Texas Rule 46 requires that CO_2 injection projects must attach a map with surveys marked showing the location and depth of all wells of public record within one-quarter (1/4) mile radius of the proposed injection well(s) . The concentration of CO_2 will increase if NCIW is higher. Both cement sheath and casing would be more seriously corroded when exposed to a high CO_2 concentration environment. Appendix A provides more detail information on NCIW of plugged wells. Equation (6) is used to determine P for NCIW.

$$P_{NCIWN} = (\frac{NCIW - NCIW_{min}}{NCIW_{max} - NCIW_{min}}) * 5 \tag{6}$$

In this equation P_{NCIW} is the P of NCIW. $NCIW_{max}$ and $NCIW_{min}$ are the maximum and minimum NCIW of studied plugged wells, respectively. P_{NCIW} is rescaled from 0 to 5 and rounded up to the nearest integer. The distances between a plugged well and CO_2 injection wells are well documented, so D of NCIW is set to 1. Since the influence of NCIW on the leak risk of plugged wells is moderate, the S of NICW is set to 4.

RCOPW

RCOPW is a parameter considering the regulation changes during the well plugged year. The RRC firstly promulgated well plugging rules in 1919 when S.B. 350 was passed. In 1934, the RRC issued specific plugging instructions, which required that plugging operations must be started within 20 days on all dry and abandoned wells or when production operations ceased. The protection of fresh water sands was not a requirement until 1957. In 1966, the RRC promulgated a special order which contained many of the specific well plugging requirements of the current rule, including requirements regarding the setting of cement plugs to protect fresh water sands. In 1992, the RRC amended its plugging rules which required that older wells should be tested for the eligibility of plugging extensions. In addition, beginning in 1997, all wells that were both more than 25 years old and inactive for more than 10 years were required to be tested for mechanical integrity every five years. Table 2 is the P of RCOPW. D is set to 1 considering the well plugged date is a reliable record. The S of RCOPW is set to 3 due to the improvement of government regulations.

Table 2 Failure Probability Assignment of RCOPW

Plugged Year	Failure Probability
1934-1957	5
1957-1966	4
1966-1992	3
1992-1997	2
1997-2015	1

Cement Type

Most cement used in oil industry is Portland cement, a burned blend of limestone and clay. American Petroleum Institute (API) classify the cement into A, C, G etc., based on the composition and properties of Portland cement. Table 3 gives the API designated classes for cements. "HTHP" denotes high temperature and high pressure in Table 3.

Cement type of the plugged wells should also be considered in carbon sequestration field. Improperly cement type will cause cement sheath failure due to the high pressure and dynamically temperature change. The cement sheath with suitable cement class will be identified as reliable cement design and the cement type indicator is set to 1. If the cement type is set deeper than the usable depth as shown in Table 3, the indicator is set to 0, which means improperly cement design. And the cement type indicator is the sum of surface casing cement type indicator and production casing cement type indicator. The P of cement type will be rescaled from 0 to 5 using the same method as NICW.

Table 3 API Cement Classes

API Class	Usable Depth feet	Feature
A	0-6000	
B	0-6000	Sulfate Resistance
C	0-6000	High Early Strength
D	6000-10000	HTHP Resistance
E	10000-14000	HTHP Resistance
F	10000-16000	Extremely HTHP Resistance
G	0-8000	
H	0-8000	
I	12000-16000	Extremely HTHP Resistance

NOP

NOP is a parameter described the number of plugs in a plugged well. Figure 1 shows the number and position of plugs in a properly plugged well. Proper position and sufficient number of plugs can reduce CO2 leak possibility from wellbore to the surface. All the plugged wells of WH field are plugged in this manner. The minimum requirement of NOP is 4, if surface casing is above the base of usable water. Equation (7) is used to determine P for NOP. In this equation P_{NOP} is the P of NOP. NOP_{max} and NOP_{min} are the maximum and minimum NOP of studied plugged wells, respectively. P_{NOP} is rescaled from 0 to 5 and rounded up to the nearest integer. D is assigned to 3 because not all the NOP data are well documented. If a well was plugged with insufficient plugs or the position of the plugs was not set at the designed positions, the leak risk of the well is higher. For this reason, S is set to 5, which means moderate severity.

$$P_{NOP} = \left(\frac{NOP - NOP_{min}}{NOP_{max} - NOP_{min}}\right) * 5 \quad (7)$$

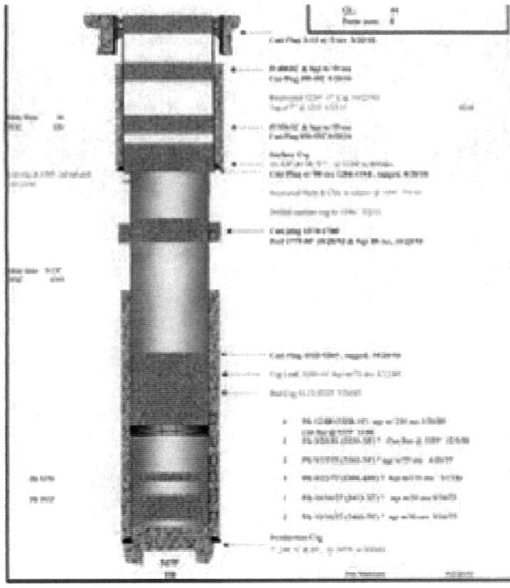

Figure 1 Plugging Details of a Plugged Well

Case Study

WH field started CO_2-EOR process in December 2010. CO_2 is injected into Frio sandstone formation at 5,700 ft with average pay zone thickness 100 ft. The average porosity is 30%, and average horizontal permeability is 900md. The gravity of the oil is 31°. Forty-three plugged wells of WH field were used in this study. These wells are plugged and exposed to CO_2 via injection with various well designs and operation histories. Figure 2 is the CO_2 injection area of WH field. Plugged wells are displayed with their locations. Most of the plugged wells in WH field are depleted oil wells.

As shown in Figure 2, some plugged wells have the same well number. The reason is that these plugged wells were held by different stakeholders, and the stakeholders just used the simple number such as 1, 2 to identify their wells. However, it is inconvenient to identify so many plugged wells with the same well number. In this case, a more specific and effective well number is assigned. This study renames the plugged wells in WH field.

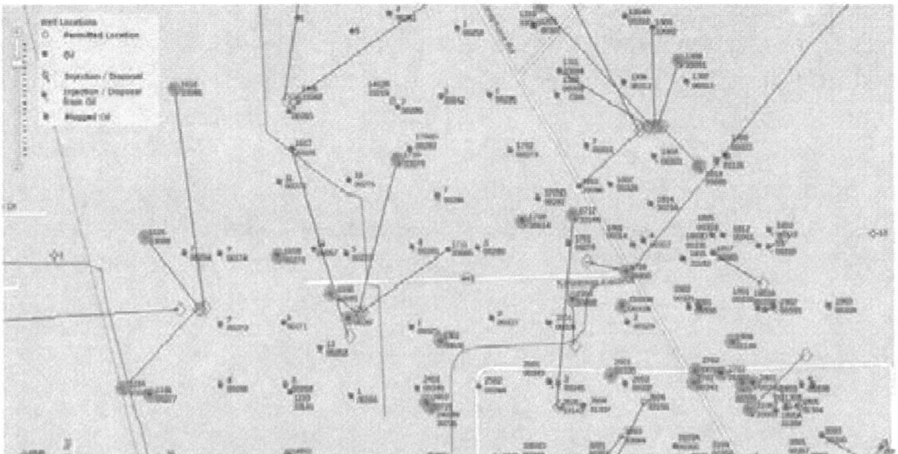

Figure 2 CO_2 Injection Area of WH Field

CEA Diagram

The first step is to use CEA to analyze the causes of CO_2 leakage in plugged wells. Figure 3 shows typical causes of CO_2 leakage in plugged wells. The main reasons of CO_2 leakage include[8-11],

(1) impropriety of well construction design and well plugged methods;
(2) in-situ stress variation due to oil production or water/gas injection activities;
(3) mechanical properties failed to maintain the integrity of wellbore;
(4) specific rules and regulations changed for CO_2 exposed wells;
(5) well history influences, such as change of well type and temperature/pressure change along the wellbore;
(6) cement sheath or casing corrosion due to high concentration of CO_2.

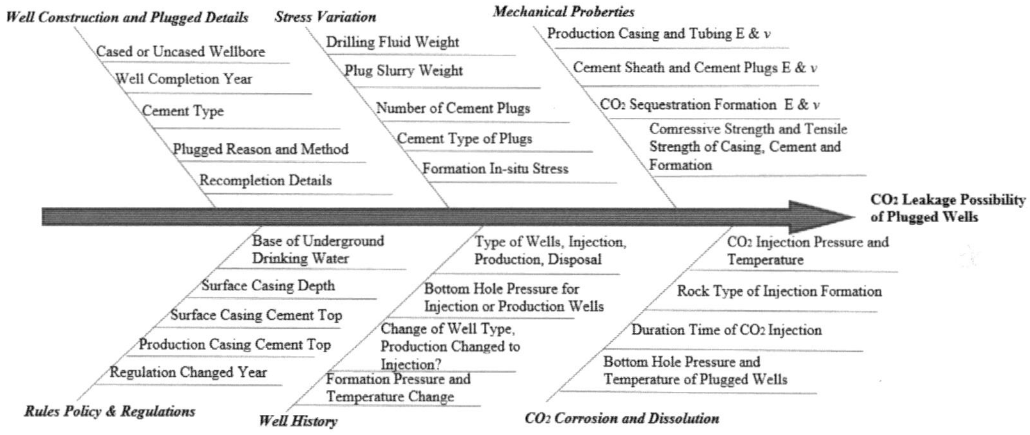

Figure 3 CEA of Plugged Wells Leakage

Figure 3 is the CEA diagram for plugged well leakage. It identifies the root causes of CO_2 leakage in a qualitative manner. This analysis helps to build failure modes for plugged wells in WH field. S, P, D can be determined by the results of CEA.

FMEA

Figure 4 shows the well plugged year and well completion year of WH field. Thirty-four wells were drilled and completed between 1930 and 1960, and thirty-seven wells were plugged between 1960 and 1980. Figure 5 is the NLP results of FMEA method. The greater the NLP value, the higher possibility the plugged well will leak. In this paper, if the NLP of a plugged well is greater than 0.8, this well is identified as high risk of leakage well. NLP of eleven plugged wells are between 0.8 and 1. These wells are considered as high potential leakage wells which need well integrity test immediately. Plugged wells with NLP value 0.4 to 0.6 and 0.6 to 0.8 are fair wells. The leakage risk of these wells is moderate. If the NLP of the plugged wells are below 0.4, these wells are relative safe wells. Even wells drilled and plugged at almost the same year, the leakage risk are different. This is because other parameters such as NCIW and MICS can contribute to the leakage risk. FMEA method considers all the possible causes of CO_2 leakage based on the data availability.

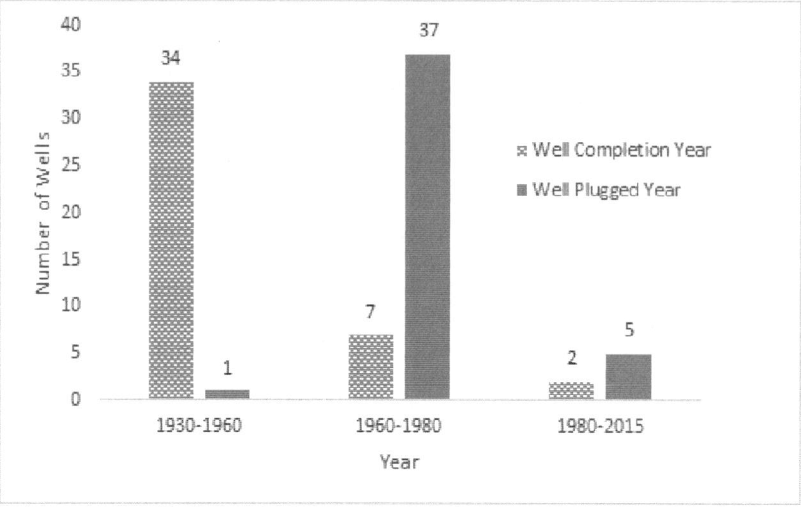

Figure 4 Well Completion Years and Plugged Year of WH Field

FMEA method can predict the leakage risk of plugged wells. Plugged wells with higher NLP (0.8 to 1) may not really leak. The leak risk of these wells is higher than other wells, and the priority of inspection is the highest. Table 4 is the plugged wells with highly leak risk in WH field.

Table 4 High Leak Risk Plugged Wells of WH Field

Well ID	Completion Year	Plugged Year	NLP
WH1454	1938	1971	0.87
WH1485	1937	1975	0.8
WH2407	1937	1975	1
WH34210	1937	1976	0.99
WH3462	1969	1974	0.94
WH4493	1937	1955	0.95
WH6426	1938	1975	0.94
WH7427	1938	1972	0.85
WH14426	1968	1975	0.94
WH21675	1937	1975	0.8

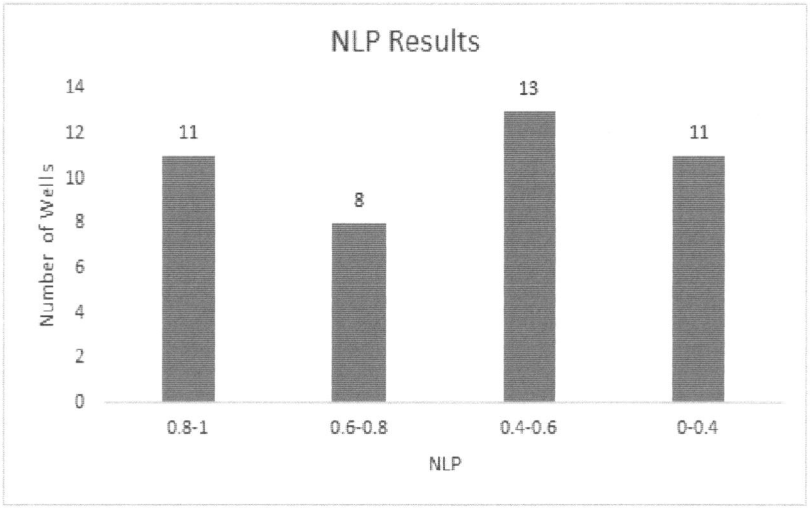

Figure 5 FMEA Results of Plugged Wells in WH Field

Conclusions

This paper is the first attempt to apply lean concept to evaluate the plugged well's leakage risk in carbon sequestration projects. Lean analysis is performed by CEA method and FMEA method. CEA is used to identify all the possible causes that could result to the CO_2 leakage of plugged wells, while FMEA is implemented to build the plugged well failure modes and to evaluate the possibility of CO_2 leakage for plugged wells. WH field, a CO_2 EOR oil field located in the Gulf of Mexico, is taken as an example to explain how CEA and FMEA could be used to evaluate the CO_2 leak potentials of the plugged wells.

The leakage scenarios of plugged wells exposed to CO_2 have been developed by the implement of CEA method. The proposed leakage scenarios are not only useful in WH field, but also can also be used to evaluate leakage potentials in other CO_2 EOR fields. FMEA can be functioned as an effective way to predict the CO_2 leakage risk based on the data availability of WH field. The plugged wells with high risk of leakage are identified.

CEA and FMEA are effective ways to evaluate the risk of CO_2 leakage for plugged wells in carbon sequestration projects. The proposed method can also be used in evaluating the integrity of CO_2 injection wells and oil production wells in oil fields where CO_2 is or will be injected in the future.

Acknowledgment

The work present in this paper was supported by the United States Department of Energy's National Energy Technology Laboratory (NETL), Federal Grant ID Number: DE FE0009284. Any opinions, findings and conclusions or recommendations expressed in this paper are those of the authors and do not necessarily reflect the views of the NETL.

Nomenclature

CEA = cause and effect analysis
FMEA = failure mode and effects analysis
RPN = risk priority number
S = severity, the consequence of a failure mod
P = failure occurrence probability
D = detectability of specific failure
LP = leakage possibility of a plugged well

n = total number of causes of CO_2 leak
i = specific cause of CO_2 leakage
NLP = normalized leakage possibilities
NLP_j = normalized leakage possibility of well j
LP_j = the leakage possibility of well j
LP_{max} = the maximum LP value in the studied plugged wells
LP_{min} = the minimum LP value in the studied plugged wells
WA = well age
WPA = well plugged age
NOP = number of plugs
NCIW = number of CO_2 injection wells in one-quarter mile radius of a plugged well
MICS = mechanical integrity of cement sheath
WCD = well construction defects
RCOPW = regulation changes on plugging wells
PA = well production age
P_{WAN} = P of WA
WA_{max} = the maximum WA of the studied plugged wells
WA_{min} = the minimum WA of the studied plugged wells
P_{WPAN} = P of WPA
WPA_{max} = the maximum WPA of the studied plugged wells
WPA_{min} = the minimum WPA of the studied plugged wells
P_{NOP} = P of NOP
NOP_{max} = the maximum NOP of the studied plugged wells
NOP_{min} = the minimum NOP of the studied plugged wells
P_{NCIWN} = P of $NCIW$
$NCIW_{max}$ = the maximum NCIW of the studied plugged wells
$NCIW_{min}$ = the minimum NCIW of the studied plugged wells
RRC = Railroad Commission of Texas

References

1. Watson, T.L., and Bachu, S. 2009. Evaluation of the Potential for Gas and CO_2 Leakage Along Wellbores. SPE Drilling & COmpletion, 24(1): 115-126
2. Huerta, N.J., Checkai, D.A., amd Bryant, S.L. 2009. Utilizing Sustained Casing Pressure Analog to Provide Parameters to Study CO_2 Leakage Rates Alnong a Wellbore. Paper SPE 126700 Presented at the SPE International Conference on CO_2 Capture, Storage, and Utilization, San Diego, California.
3. Tao, Q., Checkai, D., Huerta, N., 2010, Model to Predict CO_2 Leakage Rates Along a Wellbore. paper SPE 135483 presented at the SPE Annual Technical Conference and Exhibition, Florence, Italy.MS
4. D'Alesio, P., Poloni, R., Valente, P. 2011.Well-Integrity Assessment and Assurance: The Operational Approach for Three CO_2-Storage Fields in Itlay. SPE Production & Operation, 26(2): 140-148.
5. Rocha-Valadez, T., Hasan, A.R., Mannan, S. and Kabir, C.S., 2014, Assessing Wellbore Integrity in Sustained-Casing-Pressure Annulus, SPE 169814, SPE Drilling & Completion, pp131-138.
6. K. Ishikawa, 1976, Guide to Quality Control, Asian Productivity Organization, ISBN 92-833-1036-5, pp34-37
7. Fadlovich, E., 2007, Performing Failure Mode and Effect Analysis, Embedded Technology, pp74-79.
8. Loizzo, M., Akemu, O.A.P., Jammes, L. 2011. Quantifying the Risk of CO_2 Leakage Through Wellbores. SPE Drilling & Completion, 26(3): 324-331.
9. Ladva, H.K.J., Craster, B., Jones, T.G.J., Goldsmith, G.and Scott, D., 2011, the Cement-to-formation Interface in Zonal Isolation, SPE Drilling & Completion, pp 186-196.
10. Le Guen, Y., Le Gouevec, J., Chammas, R., Gerard, B., Poupard, O., Beken, A.V.D. and Jammes, L., 2009, CO_2 Storage: Managing the Risk Associated with Well Leakage Over Long Time Scales, SPE Projects, Facilities & Construction, pp 87-96.

11. D'Alesio, P., Poloni, R., Valente, P.and Magarini, P.A., 2011, Well-integrity Assessment and Assurance: the Operational Approach for Three CO_2-Storage Fields in Italy, SPE Production & Operations, pp 140-148.

418594 Effects of Reservoir Temperature and Percent Levels of Methane and Ethane on CO2/Oil MMP Values As Determined Using Vanishing Interfacial Tension/Capillary Rise

Abstract:

Geological CO_2 storage combined with enhanced oil recovery (EOR) can be an economically-viable approach to storing CO_2 while gaining the benefit of increasing oil recoveries from a reservoir which has undergone primary (and secondary) production. CO_2 is widely understood to mobilize otherwise immobilized oil via processes involving several mechanisms including hydrocarbon solvation into the CO_2 phase, physical displacement, oil swelling, lowered oil viscosity, and the formation of a separate "miscible" CO_2/hydrocarbon mixed phase as the CO_2 contacts lighter hydrocarbons. This mixed phase is characterized to form at the minimum miscibility pressure (MMP). MMP is a major input variable for models used to optimize EOR efficiencies. Long-established methods to determine MMP of crude oil (e.g., slim tube, rising bubble) can be costly, slow, and subject to operational variations. In contrast, newer vanishing interfacial tension (VIT) methods rely on a more fundamental definition of miscibility, i.e., the conditions at which there is no interfacial tension (IFT) between the two fluids. IFT is measured at different pressures by observing the height of oil in a capillary tube (with accurately known internal diameter) and the density of both fluids at each experimental pressure, a requirement that greatly increases the cost and complexity of the instrumentation and operation [1].

Fortunately, it is not necessary to measure IFT at several pressures to determine when the IFT goes to zero since a plot of the oil height in the capillary tube versus pressure will intercept the pressure axis at the same value as the IFT data [2]. That is, both zero IFT and zero capillary rise height occur at the same pressure. Using the simple view cell apparatus (and a suitable pump and temperature bath) shown in Figure 1, this approach yields linear plots, and no density measurements are needed, which greatly reduces instrument cost and experimental complexity. With this simplified approach, MMP values can be rapidly and cost-effectively determined allowing the effects of multiple parameters on MMP to be investigated including (but not limited to) the effects of various reservoir conditions such as temperature, changing gas composition (e.g., the effect of methane in recycle CO_2), and changes in crude oil composition on MMP. For example, methane mixed with CO_2 increases MMP, while the presence of ethane in CO_2 lowers the MMP.

Similarly, the MMP of a crude oil increases greatly with increasing reservoir temperature.

These tests show that dramatic oil/CO_2 interactions occur both below and above MMP, and led us to focus on such processes by visually observing the behavior of crude oil/CO_2 mixtures at reservoir conditions as the CO_2 pressure is raised from below to above MMP, followed by a controlled depressurization to observe the oil behavior as CO_2 pressure is reduced (e.g., as could occur between the injection and production wells).

High-pressure view cell videos of CO_2/oil interactions show that achieving MMP is not a "magic" condition required for oil mobilization, and that significant mobilization is occurring at pressures below the crude's MMP. In addition, when the pressure is slowly lowered to mimic the drop in pressure from an injection to production well, as much as 1/2 of the mobilized oil is lost from the CO_2-mobilized phase, *even at pressures well above MMP.* For example, when crude from a reservoir that is currently undergoing a CO_2 flood is exposed to 5000 psi CO_2, ca. 2/3 of the hydrocarbon is suspended in the overlying (upper) CO_2 "miscible" phase. However, lowering the pressure to 4000 psi causes much of the mobilized oil to precipitate into the bulk crude, *even though the pressure is still higher than the 2800 psi MMP of that oil.* Methods have also been developed to sample and analyze the hydrocarbon composition in the upper "miscible" CO_2/oil phase, and have shown dramatic changes both in the mass of mobilized oil, as well as the carbon number distribution of the oil in the "miscible" phase. For example, the mass of oil hydrocarbons from a conventional crude oil at 1500 psi (MMP is 1450 psi) was only ca. 1/2 that mobilized at the expected CO_2 injection pressure of 2300 psi, and the lower pressure showed a much higher fraction of lower-carbon-number hydrocarbons. Similarly, when the pressure was dropped from the 2300 psi injection pressure to just above MMP, substantial losses of oil occurred from the "miscible" phase with the higher molecular weight hydrocarbons showing the most deposition.

Experiments are continuing to qualitatively and quantitatively describe CO_2/oil mobilization processes that occur under different reservoir scenarios, including the effects of changing gas composition on MMP, as well as video investigations on including the CO_2 mobilization of residual oil following a water flood. The MMP method and associated effects of changing reservoir conditions will be presented, and videos of CO_2/oil interactions at reservoir conditions and their interpretations will be discussed

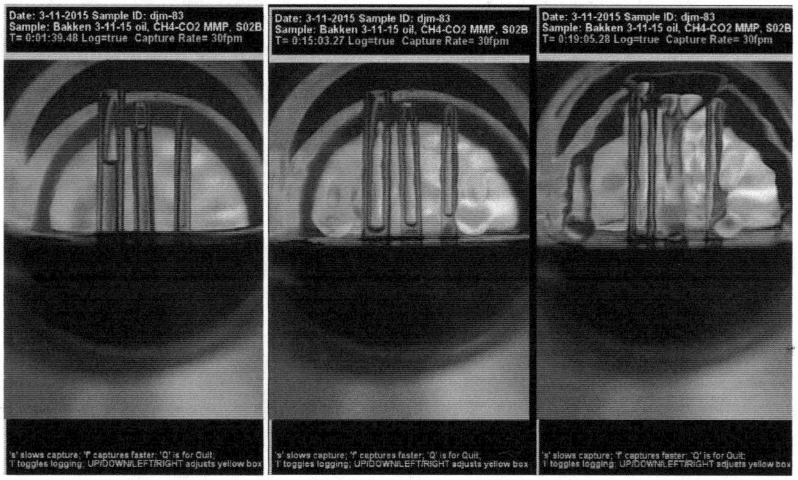

Figure 1: High pressure view cell with the capillary tubes suspended in crude oil before pressurization with mixed CO_2/methane (left), as the injected CO_2/methane pressure approaches MMP (middle), and at pressures above the MMP (right). The capillary tubes are ca. 15 mm long.

REFERENCES

[1] S.C. Ayirala and D.N. Rao, *J. of Canadian Petroleum Technology, 50,* **2011**, pp. 71-81 (SPE 99606-PA).

[2] EERC patent pending.

ACKNOWLEDGEMENTS

This material is based on work supported by the U.S. Department of Energy National Energy Technology Laboratory under Award No. DE-FC26-05NT42592.

2015 AIChE Annual Meeting

The Effects of Mineral composition of sandstone on the Formation of Residual Water

Jingxia Wang, Yi Li, Qingchun Yu

Speaker: Jingxia Wang

School of Water Resources and Environment
China University of Geosciences (Beijing)

OUTLINE

- Introduction
- Experimental Materials
- Rock Cores
- Experimental Methods
- Experimental Results
- Analysis and Discussion
- Conclusions

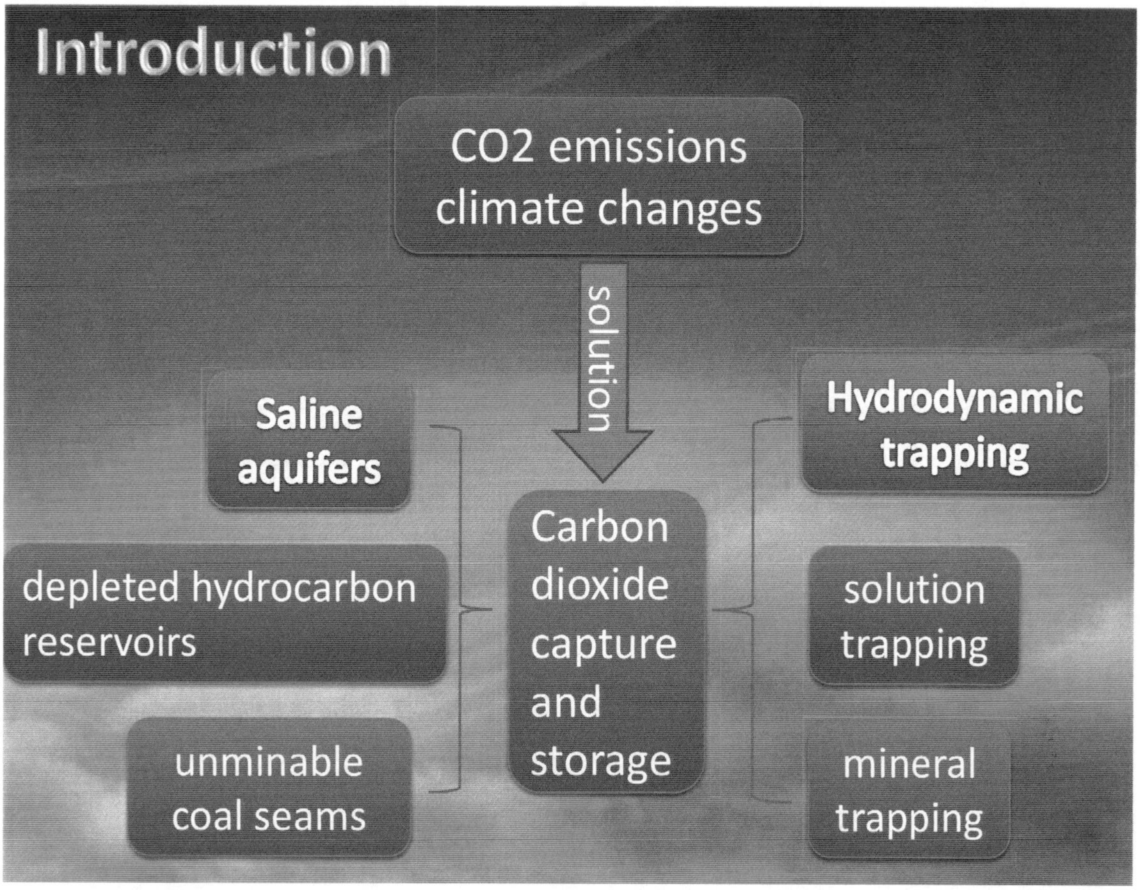
Introduction
CO2 emissions
climate changes
solution
Saline
aquifers
depleted hydrocarbon
reservoirs
unminable
coal seams
Carbon
dioxide
capture
and
storage
Hydrodynamic
trapping
solution
trapping
mineral
trapping

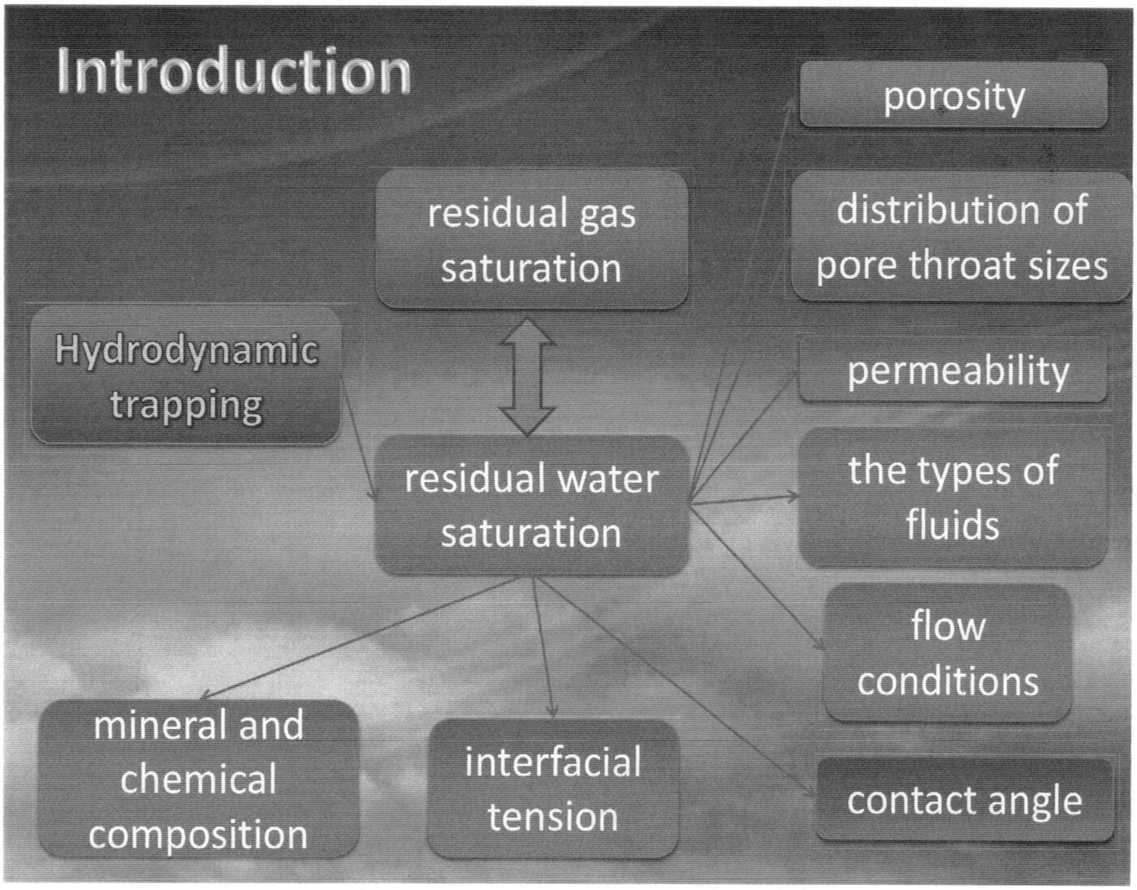
Introduction
residual gas
saturation
Hydrodynamic
trapping
residual water
saturation
porosity
distribution of
pore throat sizes
permeability
the types of
fluids
flow
conditions
contact angle
mineral and
chemical
composition
interfacial
tension

Experimental Apparatus

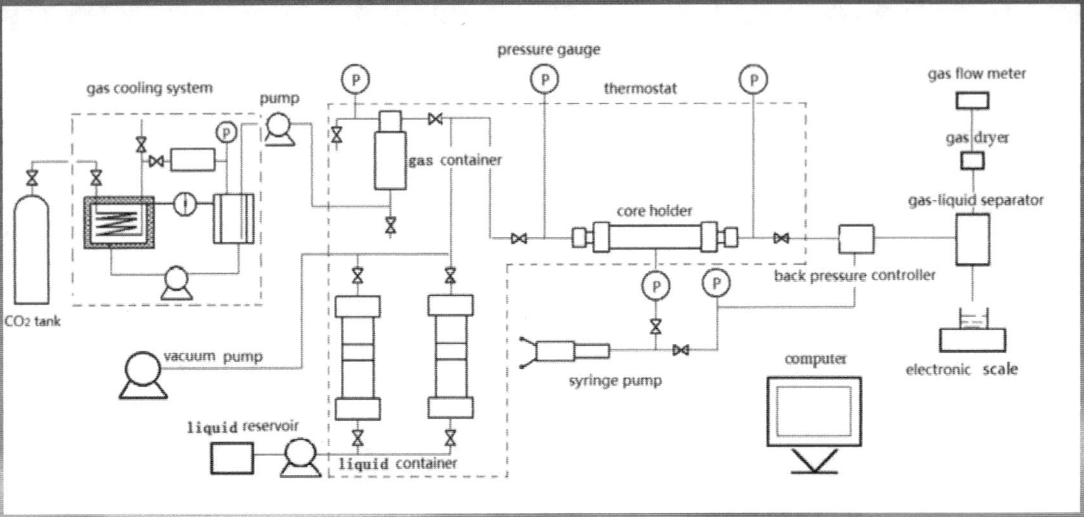

Schematic diagram of experimental apparatus

Rock Cores

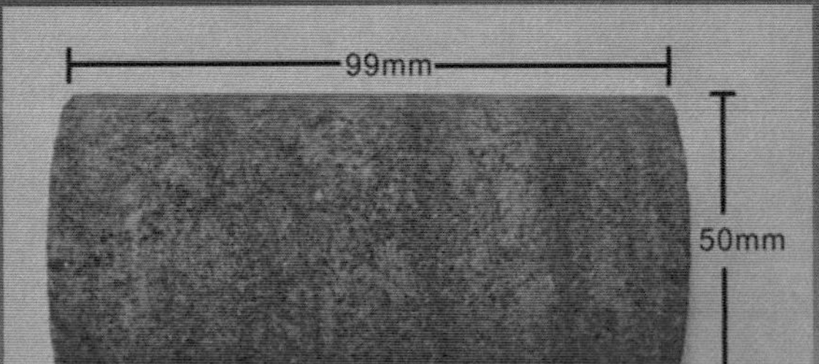

Sample #1 extracted from the Triassic Heshanggou Group

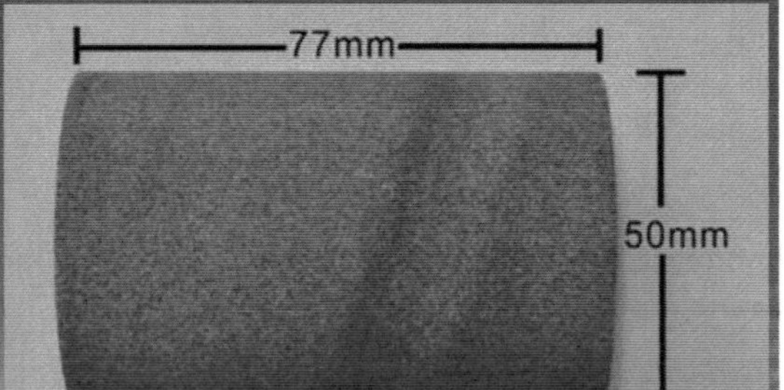

Sample #1 extracted from the Triassic Liujiagou Group

Sample #3 the artificial rock core

Parameters for the three rock core samples

Sample	#1	#2	#3
Porosity	12.6	13.1	13.0
Permeability	0.3	0.2	0.2
Depth	1540.3	1690.5	-----------

Experimental Methods

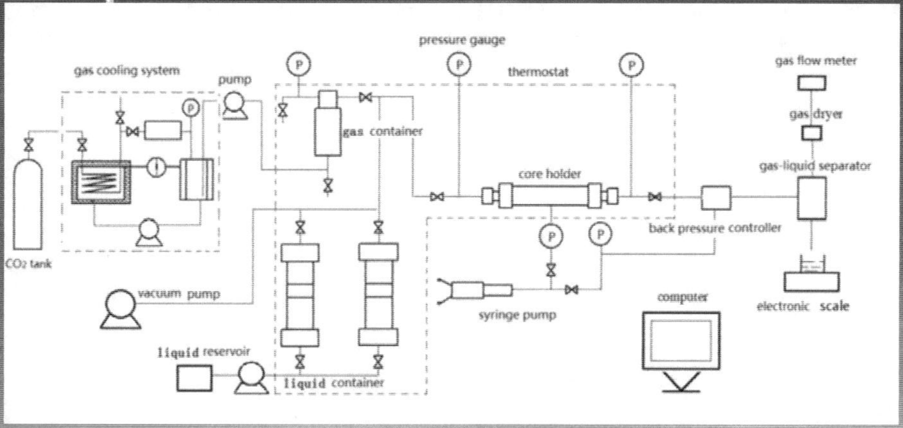

Schematic diagram of experimental apparatus

$$S_{wrr} = \frac{m_2 - m_0}{m_1 - m_0}$$

Experimental conditions:
Supercritical CO_2
Steady flow
Temperature: 40℃
Pressure: 8MP
35g/L NaCl brine

S_{wrr}: the saturation of the residual water
m_0: the dry weight of the rock core
m_1: the weight of the saturated rock core
m_2 : the residual weight of the rock core after CO_2 drainage

Experimental Results

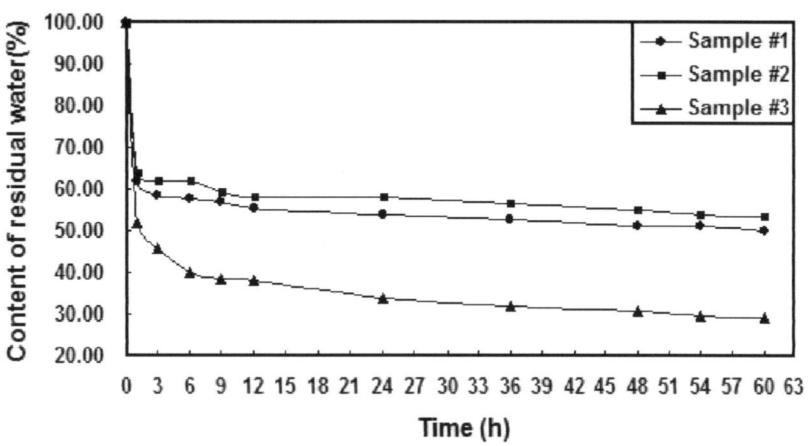

Residual water content of the three rock core samples during the experimental runs

The residual water saturations for the three rock core sample

Sample	#1	#2	#3
Irreducible water saturations	50.12%	53.28%	29.05%

Analysis and Discussion

pore characteristics of sample #1

pore characteristics of sample #2

The analysis table of the experimental results

Sample	Porosity	Permeability (mD)	Mineral composition				
			Quartz	Albite	Microcline	Laumontite	Kaolinite
#1	12.6%	0.3	58%	3%	32%	7%	----------
#2	13.1%	0.2	31%	52%	10%	2%	2%
#3	13.0%	0.2	98%	-------	2%	-----------	----------

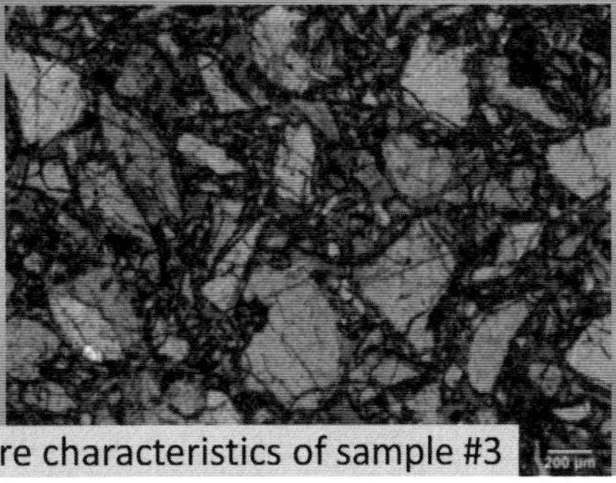

pore characteristics of sample #3

Distribution of pores for three rock core

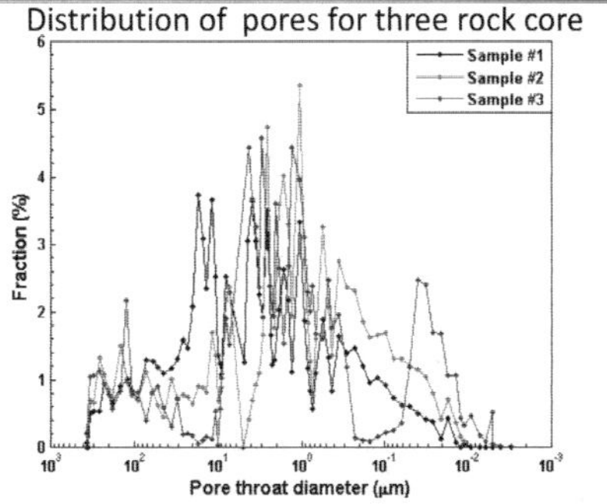

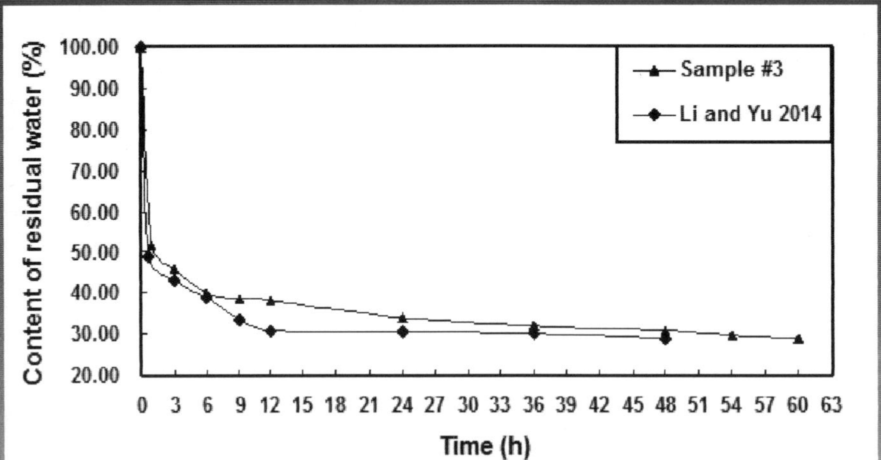

Comparison of residual water content between sample #3 and Li and Yu (2014) during the experimental runs

The table of analysis results for sample #3 and Li and Yu (2014)

Sample	#3	Li and Yu 2014	relative error
Permeability (mD)	0.2	2.2	
Irreducible water saturations	29.05	28.85	0.69%

quartz
Clay mineral
Albite
Polarized optical microphotograph
of sample #1 rock core
Clay mineral
quartz
albite
of sample #1 rock core
Clay mineral
Albite
quartz
Polarized optical microphotograph
of sample #2 rock core
Albite
quartz

quartz
Albite
microphotograph
rock core

Scanning electron microscopic image of three rock sample magnified 1000 times

Scanning electron microscopic image of three rock sample magnified 10000 times

$$\cos\theta = \frac{\gamma_{sv} + \gamma_{sl}}{\gamma_{lv}}$$

- θ is the contact angle
- γ_{sv} is the interfacial tension of the rock and CO_2
- γ_{sl} is the interfacial tension of the rock and the brine
- γ_{lv} is the interfacial tension of the brine and CO_2

$$P_c = \frac{2\gamma_{lv}\cos\theta}{R}$$

P_c is the CO_2 capillary breakthrough pressure
R is the radius of the largest pore throat in the porous medium

conclusions

- The mineral composition in a rock core has a measurable effect on the formation of residual water.
- The clay mineral content impacts on the formation of residual water.
- The types of clay mineral influence the formation of residual water.

Thank you !

中国地质大学
CHINA UNIVERSITY OF GEOSCIENCES

Impacts of Relative Permeability on Subsurface CO_2 Mineralization and Storage

Brian McPherson, Wei Jia, Nathan Moodie,
Feng Pan, Vivek Patil

Department of Civil and Environmental Engineering
University of Utah

2015 AIChE Annual Meeting
Salt Lake City, Utah USA
November 13, 2015

Acknowledgements

- U.S. Department of Energy
- National Energy Technology Laboratory
- Utah Science, Technology and Research Initiative (USTAR)

Wei Jia Vivek Patil Nate Moodie

OUTLINE

- Premise: CO_2-EOR and Deep Saline Storage
- Sources and Extent of Uncertainty
- Greatest Source of Uncertainty: Heterogeneity
- Often Overlooked Source of Heterogeneity: Relative Permeability
- Uncertainty of Storage Forecasts Due to Heterogeneity (CO_2-EOR)
- Uncertainty in Storage Forecasts Due to Relative Permeability (CO_2-EOR)
- Variability in Storage and Trapping Due to Relative Permeability (Deep Brine Storage)
- Uncertainty in Mineralization Due to Relative Permeability
- Summary of Most Important Results

OUTLINE

- Premise: CO_2-EOR and Deep Saline Storage
- Sources and Extent of Uncertainty
- Greatest Source of Uncertainty: Heterogeneity
- Often Overlooked Source of Heterogeneity: Relative Permeability
- Uncertainty of Storage Forecasts Due to Heterogeneity (CO_2-EOR)
- Uncertainty in Storage Forecasts Due to Relative Permeability (CO_2-EOR)
- Variability in Storage and Trapping Due to Relative Permeability (Deep Brine Storage)
- Uncertainty in Mineralization Due to Relative Permeability
- Summary of Most Important Results

Premise: CCS via CO_2-EOR and Deep Storage

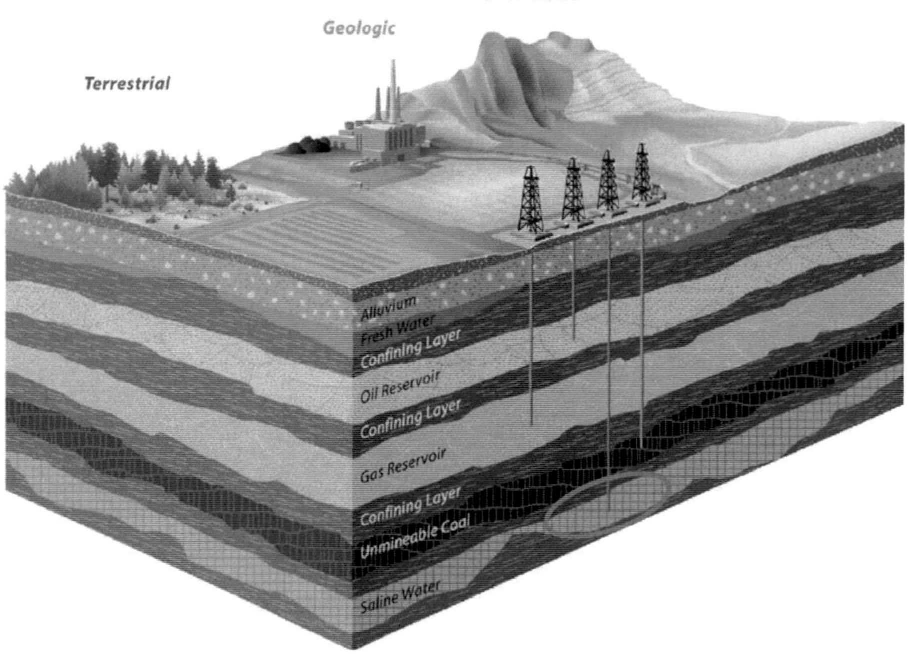

Graphic: Genevieve Young

OUTLINE

- Premise: CO_2-EOR and Deep Saline Storage
- Sources and Extent of Uncertainty
- Greatest Source of Uncertainty: Heterogeneity
- Often Overlooked Source of Heterogeneity: Relative Permeability
- Uncertainty of Storage Forecasts Due to Heterogeneity
  (CO_2-EOR)
- Uncertainty in Storage Forecasts Due to Relative Permeability
  (CO_2-EOR)
- Variability in Storage and Trapping Due to Relative Permeability
  (Deep Brine Storage)
- Uncertainty in Mineralization Due to Relative Permeability
- Summary of Most Important Results

Sources and Extent of Uncertainty

1. Many sources of uncertainty in CCS forecasts, including ***parameter uncertainty*** and ***model uncertainty***.
2. Greatest Source of Uncertainty: Heterogeneity
   – ***This is a type of Parameter Uncertainty***
3. Often Overlooked Source of Heterogeneity: Relative Permeability
   – ***This is a type of Model Uncertainty***
4. We examined:
   – Uncertainty in Storage Forecasts Due to Relative Permeability in CO_2-EOR
   – Variability in Storage Forecasts Due to Relative Permeability in Brine Formations
   – Uncertainty in Mineralization Trapping Due to Relative Permeability
5. Extent of uncertainty? Must quantify for each site.

OUTLINE

- Premise: CO_2-EOR and Deep Saline Storage
- Sources and Extent of Uncertainty
- Greatest Source of Uncertainty: Heterogeneity
- Often Overlooked Source of Heterogeneity: Relative Permeability
- Uncertainty of Storage Forecasts Due to Heterogeneity (CO_2-EOR)
- Uncertainty in Storage Forecasts Due to Relative Permeability (CO_2-EOR)
- Variability in Storage and Trapping Due to Relative Permeability (Deep Brine Storage)
- Uncertainty in Mineralization Due to Relative Permeability
- Summary of Most Important Results

Greatest Source of Uncertainty: Heterogeneity

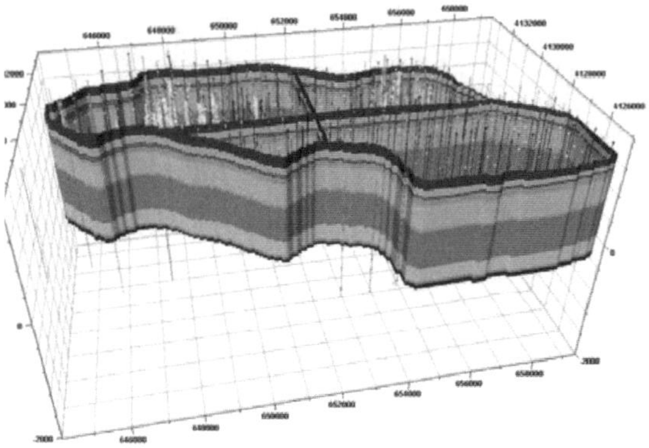

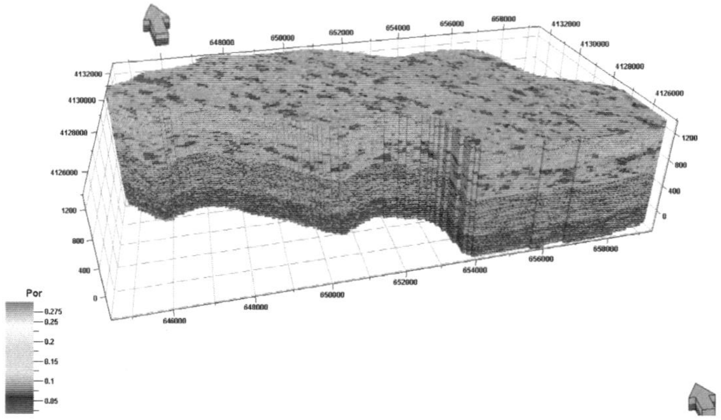

Greatest Source of Uncertainty: Heterogeneity

Greatest Source of Uncertainty: Heterogeneity

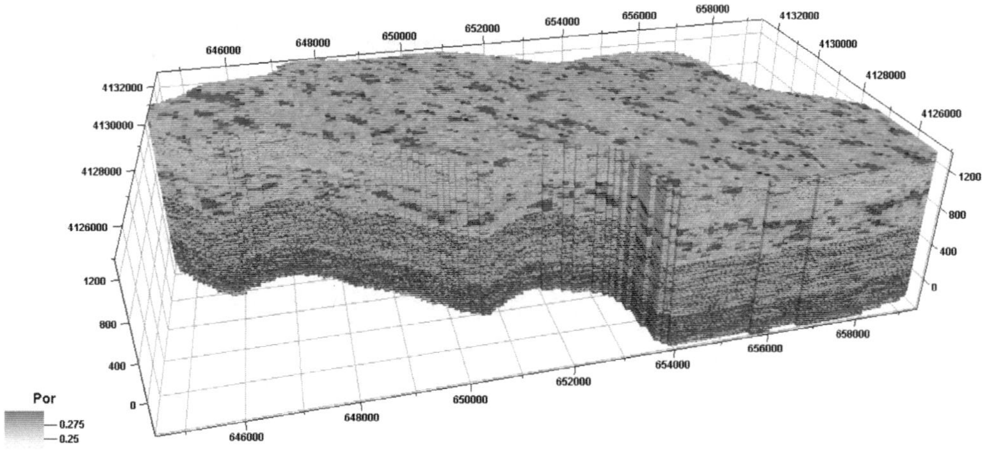

*This is a form of **Parameter Uncertainty**.*

OUTLINE

- Premise: CO_2-EOR and Deep Saline Storage
- Sources and Extent of Uncertainty
- Greatest Source of Uncertainty: Heterogeneity
- **Often Overlooked Source of Heterogeneity: Relative Permeability**
- Uncertainty of Storage Forecasts Due to Heterogeneity (CO_2-EOR)
- Uncertainty in Storage Forecasts Due to Relative Permeability (CO_2-EOR)
- Variability in Storage and Trapping Due to Relative Permeability (Deep Brine Storage)
- Uncertainty in Mineralization Due to Relative Permeability
- Summary of Most Important Results

Often Overlooked Source of Heterogeneity: Relative Permeability

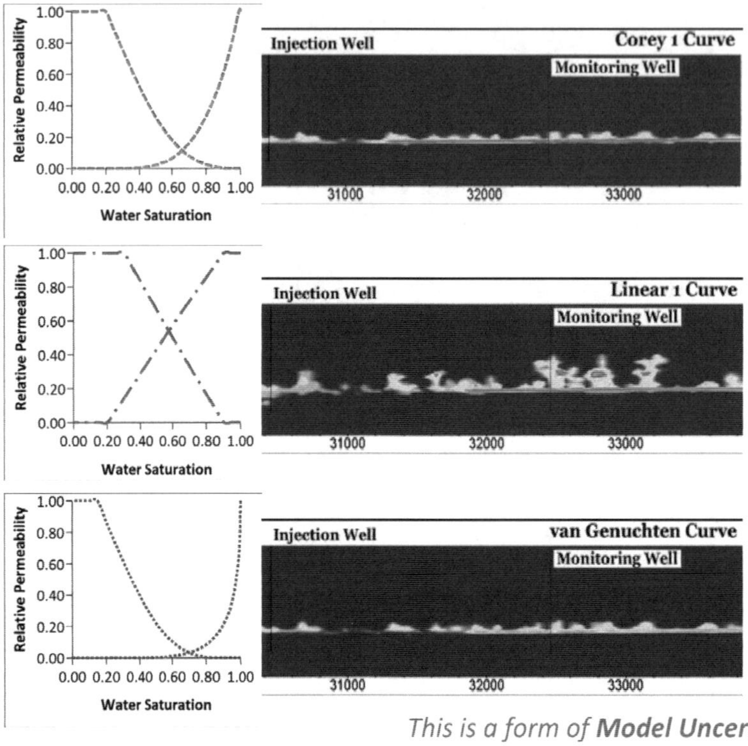

This is a form of ***Model Uncertainty.***

OUTLINE

- Premise: CO_2-EOR and Deep Saline Storage
- Sources and Extent of Uncertainty
- Greatest Source of Uncertainty: Heterogeneity
- Often Overlooked Source of Heterogeneity: Relative Permeability
- **Uncertainty of Storage Forecasts Due to Heterogeneity (CO_2-EOR)**
- Uncertainty in Storage Forecasts Due to Relative Permeability (CO_2-EOR)
- Variability in Storage and Trapping Due to Relative Permeability (Deep Brine Storage)
- Uncertainty in Mineralization Due to Relative Permeability
- Summary of Most Important Results

Uncertainty of Storage Forecasts in CO_2-EOR: Parameter Uncertainty (Heterogeneity)

- SACROC Unit, ongoing CO_2–EOR project of Phase II of the Southwest Regional Partnership on Carbon Sequestration (SWP)
- Model size: 8283 m * 3908 m * 239 m
- Model grid: 13600 cells (34 * 16 * 25)
- Injection scheme:
  - 22 injection wells, 23 production wells
  - 30 years of CO_2-EOR with CO_2-WAG cycles, 20 years of post-EOR CO_2 injection (no prod.), 50 years of monitoring (no inj./prod.)
- Source of uncertainty:
  - Porosity
  - Permeability
- 1000 realizations generated from sequential Gaussian simulation

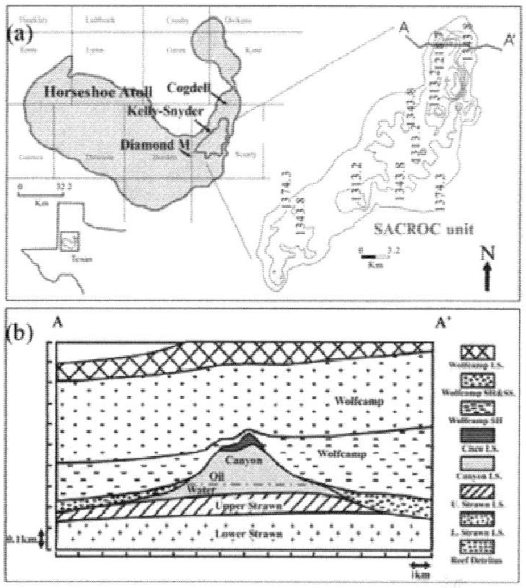

(modified from Han, 2007)

Reduced Order Models

<u>Monte Carlo simulation with ROMs</u>

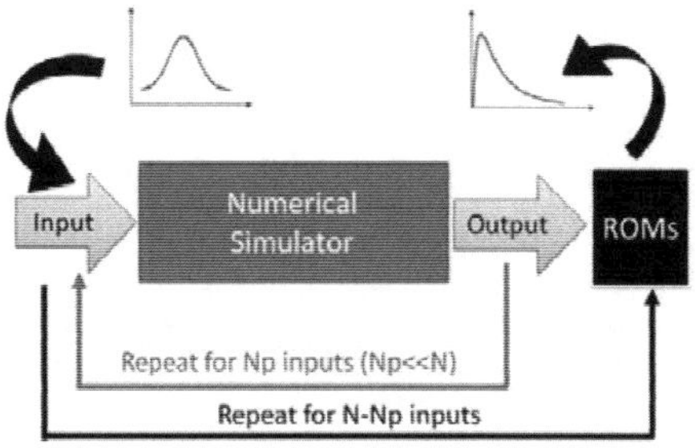

- ❖ Model Inputs:
  - ➢ Porosity, Permeability at 2720 cells
- ❖ Model Outputs:
  - ➢ Mass of CO_2 trapped in oil phase
  - ➢ Mass of CO_2 in supercritical phase
  - ➢ Mass of CO_2 in aqueous phase
  - ➢ Oil saturation

<u>Polynomial Chaos Expansion (PCE)</u>

$$y_i = \mathrm{M}(\mathrm{x}) = \alpha_0 B_0 + \sum_{j=1}^{M} \alpha_j B_1(x_j) + \sum_{j=1}^{M}\sum_{k=1}^{j} \alpha_{jk} B_2(x_j, x_k) + \sum_{j=1}^{M}\sum_{k=1}^{j}\sum_{h=1}^{k} \alpha_{jkh} B_3(x_j, x_k, x_h) \cdots$$

<u>ROMs Regression Fit</u>

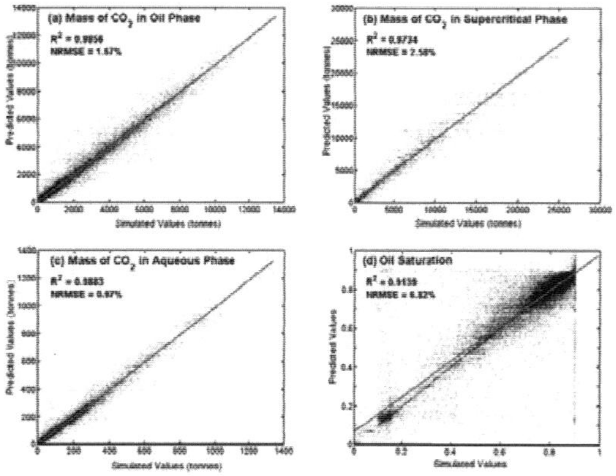

Uncertainty of CO_2 Trapping Mechanisms

- Time 1: end of 30-year CO_2-EOR
- Time 2: end of 20-year post-EOR CO_2 injection
- Time 3: end of 50-year monitoring

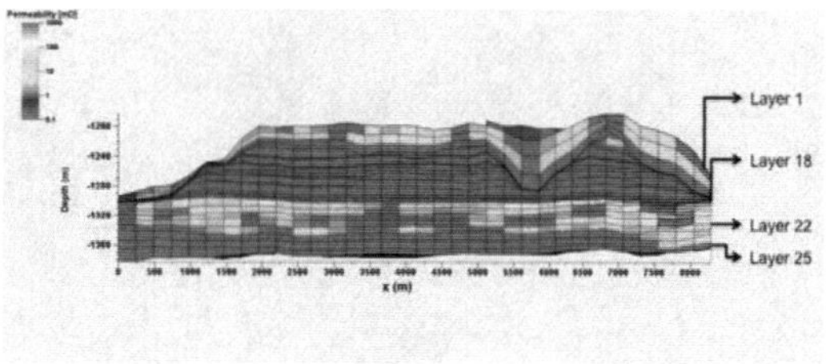

<u>CO_2 Oil Solubility Trapping</u>

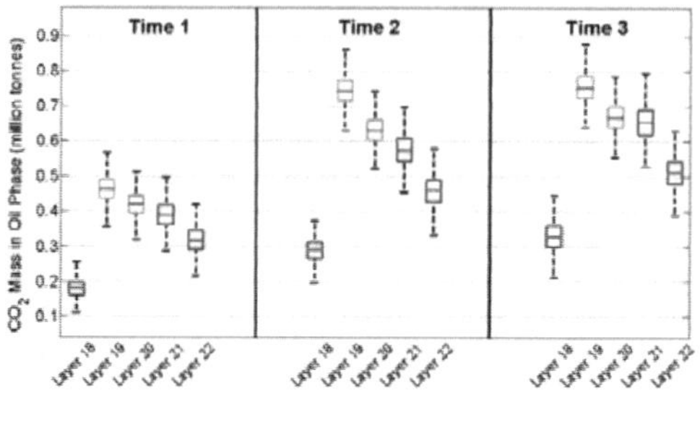

<u>CO_2 Supercritical Trapping (Free Gas + Residual)</u>

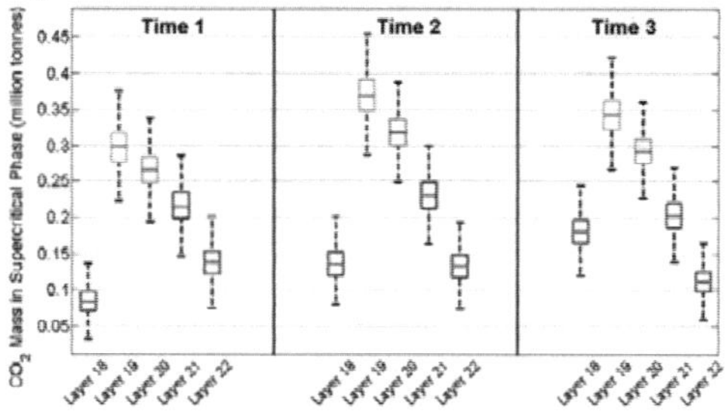

<u>CO_2 Aqueous Solubility Trapping</u>

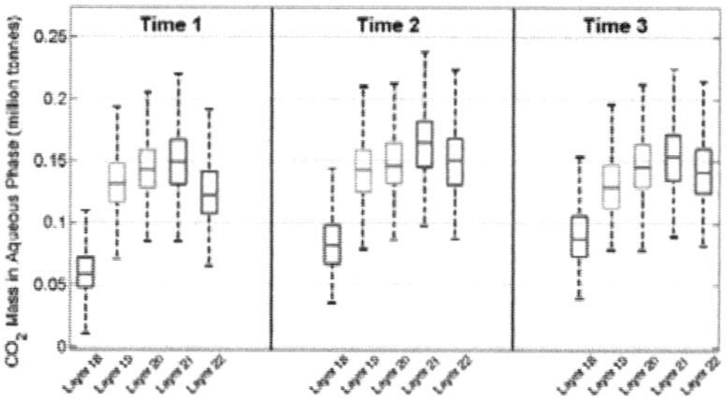

Probabilistic Assessment of CO_2 Trapping Mechanisms

<u>Cumulative Distribution Function of Trapped CO_2</u>

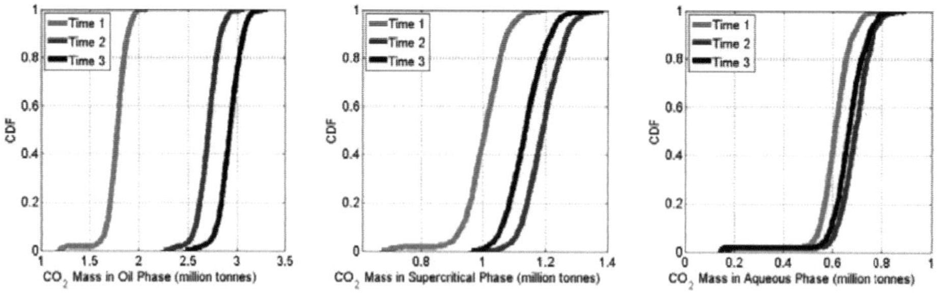

<u>CO_2 Storage Percentage of Trapping Mechanisms</u>

	Time 1		Time 2		Time 3	
	P5	P95	P5	P95	P5	P95
Oil	44.44%	52.54%	50.01%	56.89%	54.23%	61.22%
Supercritical	24.76%	30.08%	21.74%	25.41%	20.56%	24.41%
Aqueous	14.33%	19.32%	11.79%	15.55%	11.28%	15.18%
5 Layers Total:	86.65%	98.76%	85.95%	95.01%	88.71%	97.99%

OUTLINE

- Premise: CO_2-EOR and Deep Saline Storage
- Sources and Extent of Uncertainty
- Greatest Source of Uncertainty: Heterogeneity
- Often Overlooked Source of Heterogeneity: Relative Permeability
- Uncertainty of Storage Forecasts Due to Heterogeneity
  (CO_2-EOR)
- Uncertainty in Storage Forecasts Due to Relative Permeability
  (CO_2-EOR)
- Variability in Storage and Trapping Due to Relative Permeability
  (Deep Brine Storage)
- Uncertainty in Mineralization Due to Relative Permeability
- Summary of Most Important Results

Uncertainty of Trapping Mechanism in CO_2-EOR: Model Uncertainty

- Farnsworth Unit, ongoing CO_2–EOR project of Phase III of the Southwest Regional Partnership on Carbon Sequestration (SWP)
- Model size: 14000 m * 11000 m * 230 m
- Model grid: 12320 cells (28 * 22 * 20)
- Injection scheme:
  - 1 injector-centered five-spot pattern
  - 20 years of CO_2 injection and oil production, followed by 20 years of monitoring (no injection/production)
- Source of uncertainty:
  - 3-phase relative permeability models (M1, M2, M3)
- 20 realizations generated from sequential Gaussian simulation

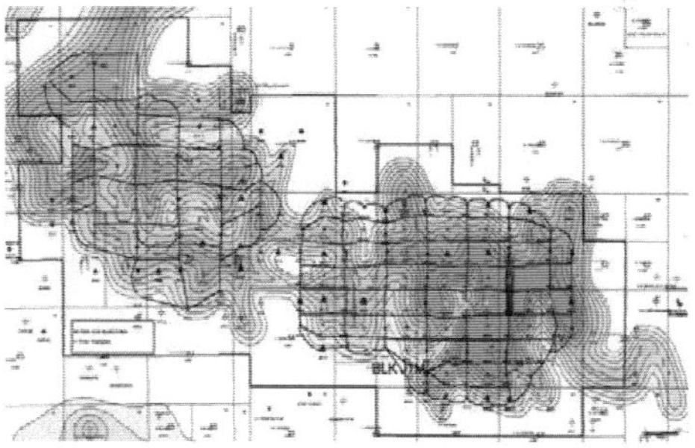

Source: Grigg and McPherson (2013)

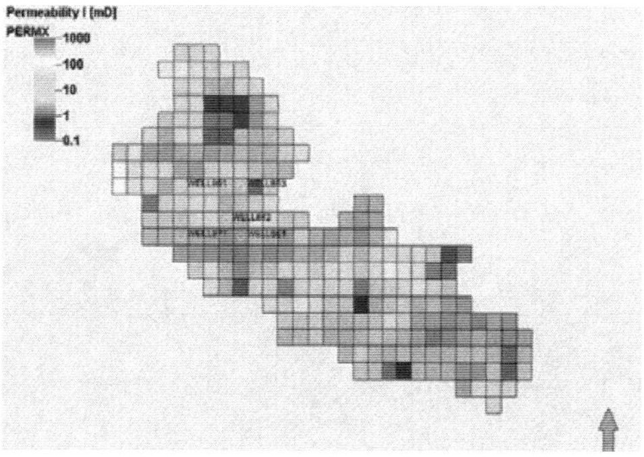

Three-Phase Relative Permeability Models

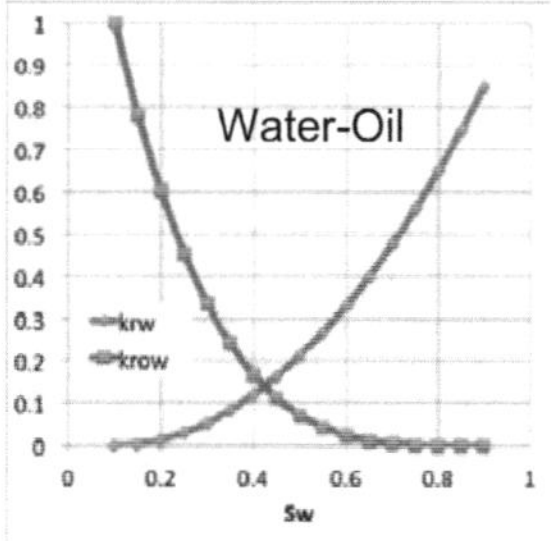

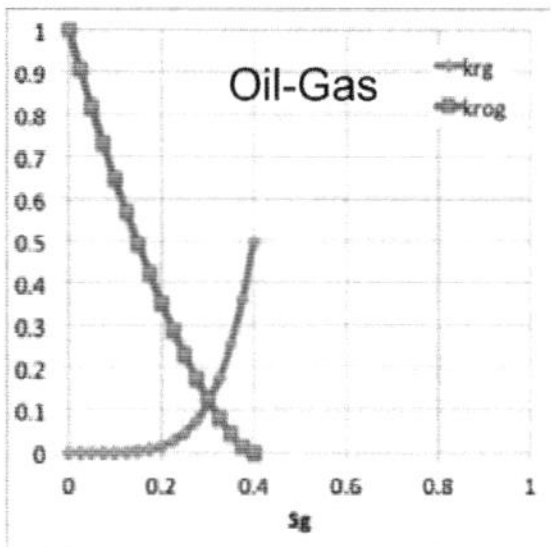

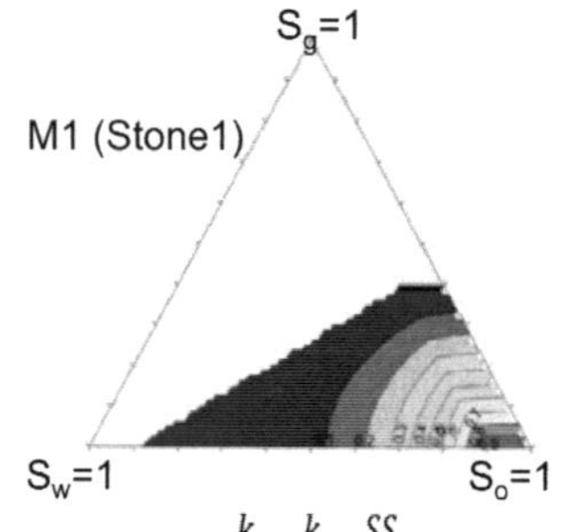

$$k_{ro} = \frac{k_{row}k_{rog}SS_o}{k_{rocw}(1 - SS_w)(1 - SS_g)}$$

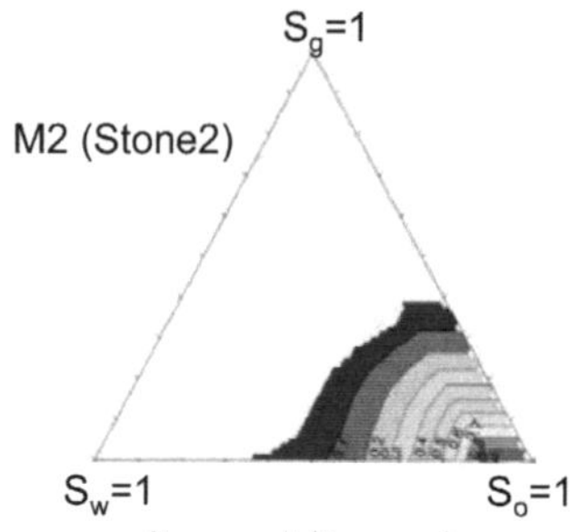

$$k_{ro} = k_{rocw}\left[\left(\frac{k_{row}}{k_{rocw}} + k_{rw}\right)\left(\frac{k_{rog}}{k_{rocw}} + k_{rg}\right) - k_{rw} - k_{rg}\right]$$

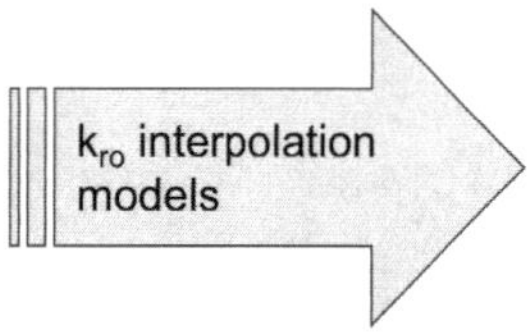

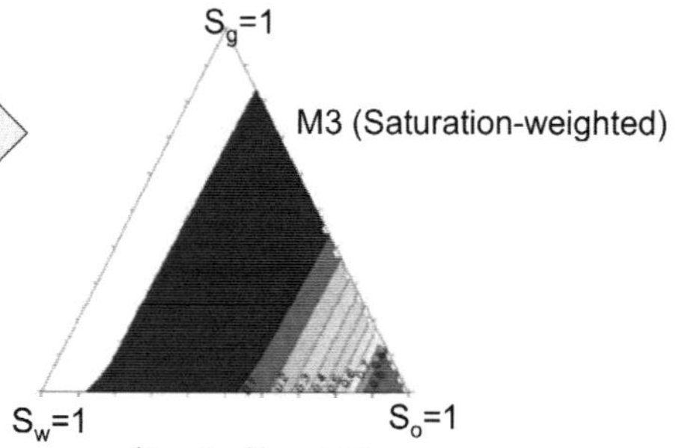

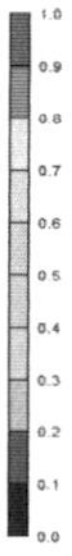

$$k_{ro} = \frac{(S_w - S_{wco})k_{row} + S_g k_{rog}}{S_w + S_g - S_{wco}}$$

Uncertainty of CO_2 Trapping Mechanisms

Free Gas Trapping (Hydrodynamic Trapping)

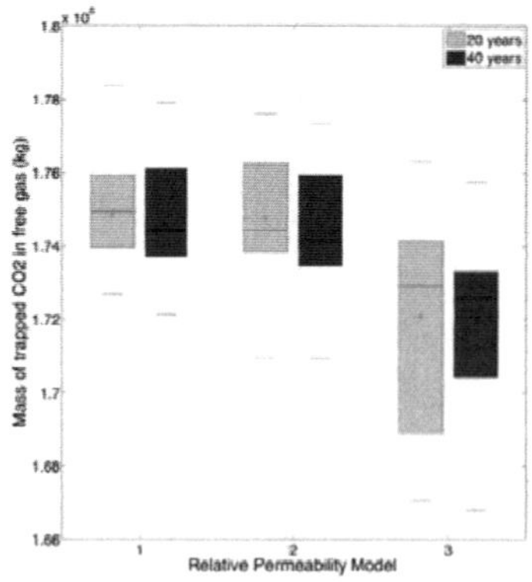

Residual Gas Trapping

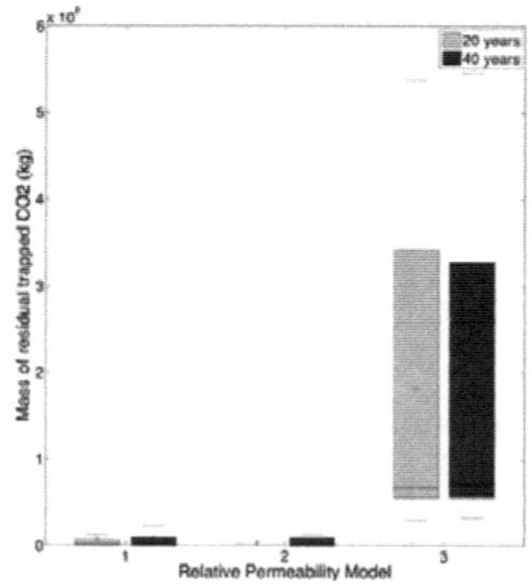

Oil Solubility Trapping

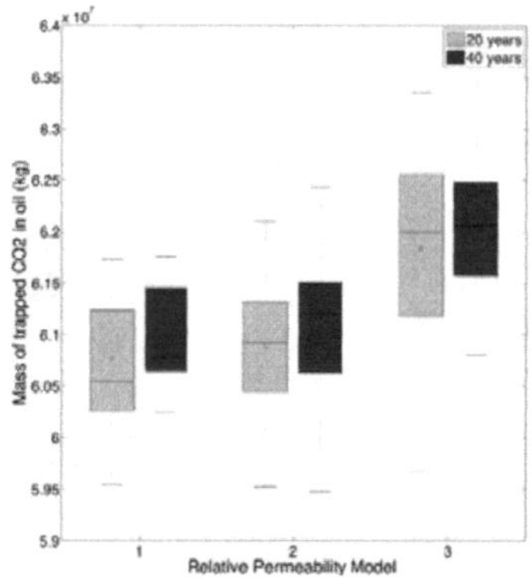

Aqueous Solubility Trapping

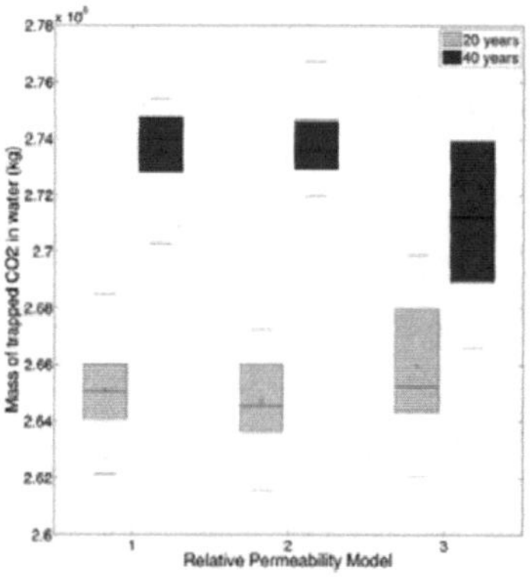

Whiskers: 1.5*(Q3-Q1)
Upper box edge: Q3 (p75)
Lower box edge: Q1 (p25)
Center line: Median (p50)
Circle: Mean

Impacts of Three-Phase Rel. Perm. Models

- M1(Stone 1) similar to M2 (Stone 2) on predictions of CO_2 storage
- **M3** (Saturation-Weighted) **underestimates** CO_2 trapping in **free gas phase**, **overestimates** CO_2 amount in **residual gas trapping, oil solubility trapping, and aqueous solubility trapping**
- M2 yields greater uncertainty in CO_2 oil solubility trapping than M1
- M3 yields **the greatest uncertainty**

CO_2 Storage at the end of simulation:

	M1	M2	M3
Free Gas Trapping (tonnes)	170,190~178,120	170,930~177,330	**167,000~176,460**
Residual Gas Trapping (tonnes)	4~598	12~254	**330~5,458**
Oil Solubility Trapping (tonnes)	60,242~61,755	59,478~62,427	**60,700~63,607**
Aqueous Solubility Trapping (tonnes)	2,702~2,754	2,720~2,767	2,666~2,773

OUTLINE

- Premise: CO_2-EOR and Deep Saline Storage
- Sources and Extent of Uncertainty
- Greatest Source of Uncertainty: Heterogeneity
- Often Overlooked Source of Heterogeneity: Relative Permeability
- Uncertainty of Storage Forecasts Due to Heterogeneity (CO_2-EOR)
- Uncertainty in Storage Forecasts Due to Relative Permeability (CO_2-EOR)
- **Variability in Storage and Trapping Due to Relative Permeability (Deep Brine Storage)**
- Uncertainty in Mineralization Due to Relative Permeability
- Summary of Most Important Results

Relative Permeability Curves

- Corey's function (1954)
  - Corey 1 curve (red) arbitrary
  - Corey 2 curve (green) used in the TOUGH2 manual[1]
- van Genuchten function (1980)
  - Used in the TOUGH2 manual[1] (Pruess et al. 1999)
- Linear function
  - Linear 1 curve (orange) is 'default' curve in Petrasim (grid generation program for TOUGH2)
  - Linear 2 curve(purple) arbitrary

Relative Permeability Parameters	Corey 1	Corey 2	Linear 1	Linear 2	van Genuchten
Water i-sat. (S_{lr})	0.2	0.3	0.2	0.2	0.15
Water sat. (S_{ls})	n/a	n/a	0.9	1	1
Gas i-sat. (S_{gr})	0.01	0.05	0.1	0.05	0.1
Gas sat. (S_{gs})	n/a	n/a	0.7	1	n/a
Lambda - λ	n/a	n/a	n/a	n/a	0.457

van Genuchten function

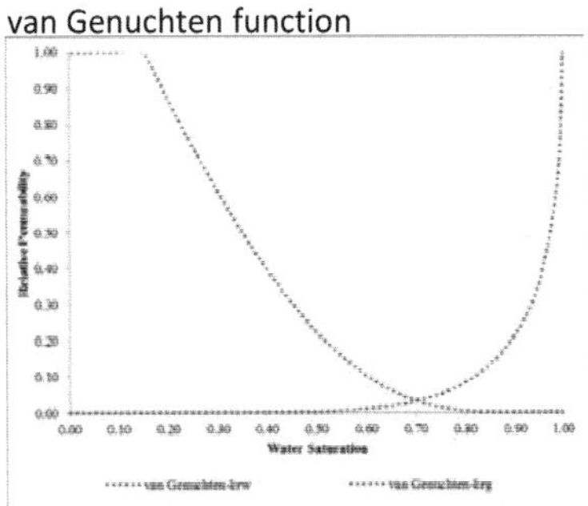

Corey's Function

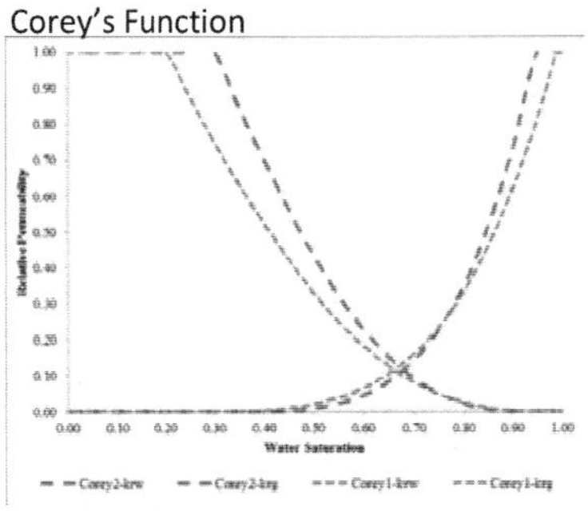

Linear Function

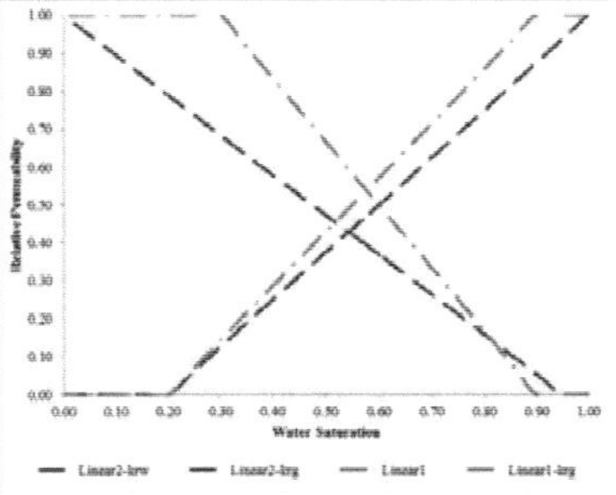

1 - Pruess et al. 1999

Results

Supercritical CO_2 saturation profile

End of 50 year EPA Post Injection Site Care (PISC) monitoring period

80 years simulation time

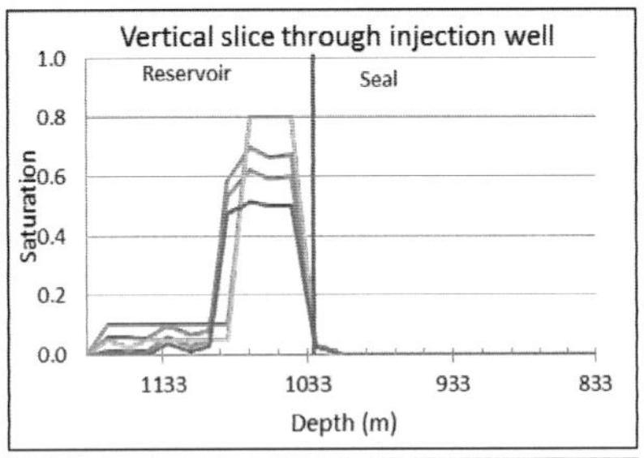

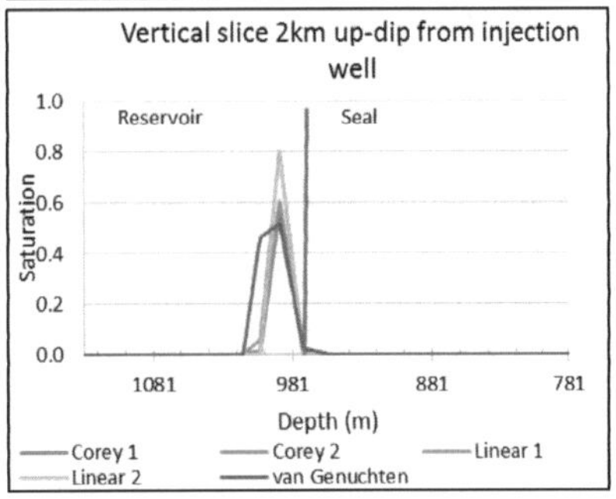

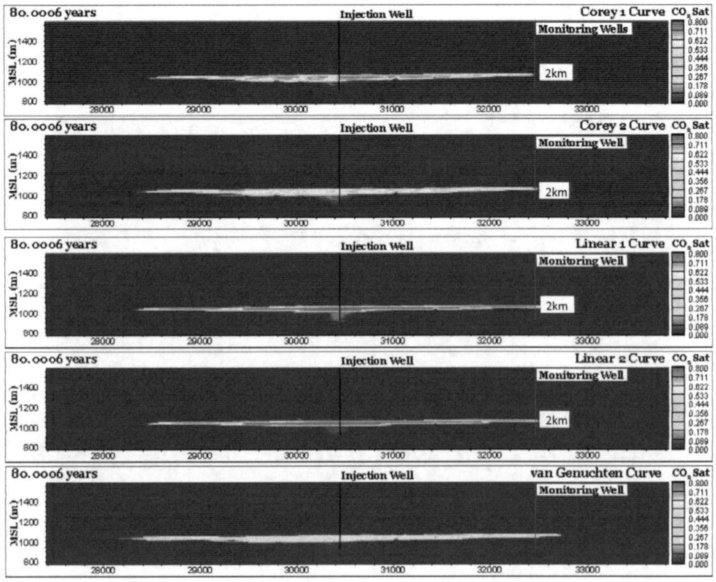

26

Results
Trapping

Relative permeability relationship influences trapping

- Same parameters used in different function can have similar predictions
  - Irreducible gas saturation
- Curve selection is important
  - Parameters + function

CO_2 Mass Distribution by Relative Permeability Curve					
	Monitoring Periods				
	30 years	80 years	140 years	500 years	1000 years
Residual CO_2	Increases with S_{gr}				
Corey1 (S_{lr} = 0.2, S_{gr}=0.01)	18,765	16,943	16,627	14,823	13,888
Corey2 (S_{lr} = 0.3, S_{gr}=0.05)	95,518	92,059	87,689	83,173	74,254
Linear1 (S_{lr} = 0.2, S_{gr}=0.1)	145,869	144,659	142,651	147,454	141,539
Linear2 (S_{lr} = 0.2, S_{gr}=0.05)	76,384	73,253	74,024	72,007	70,287
van Genuchten (S_{lr} = 0.15, S_{gr}=0.1)	203,101	170,573	159,943	138,950	130,594
Dissolved CO_2	Controlled by plume movement				
Corey1 (S_{lr} = 0.2, S_{gr}=0.01)	176,281	245,717	295,626	419,164	537,367
Corey2 (S_{lr} = 0.3, S_{gr}=0.05)	182,905	261,228	309,120	449,044	582,070
Linear1 (S_{lr} = 0.2, S_{gr}=0.1)	162,000	233,267	284,190	438,986	575,358
Linear2 (S_{lr} = 0.2, S_{gr}=0.05)	161,626	232,440	270,253	366,164	444,340
van Genuchten (S_{lr} = 0.15, S_{gr}=0.1)	201,487	282,536	345,437	504,140	631,674
Free CO_2	Inverse relationship with Sgr				
Corey1 (S_{lr} = 0.2, S_{gr}=0.01)	794,344	750,517	715,627	608,516	475,392
Corey2 (S_{lr} = 0.3, S_{gr}=0.05)	710,716	659,055	629,971	505,243	360,476
Linear1 (S_{lr} = 0.2, S_{gr}=0.1)	634,715	634,715	600,770	452,410	298,166
Linear2 (S_{lr} = 0.2, S_{gr}=0.05)	748,755	705,273	677,469	603,909	522,663
van Genuchten (S_{lr} = 0.15, S_{gr}=0.1)	582,464	556,973	517,358	383,919	242,036

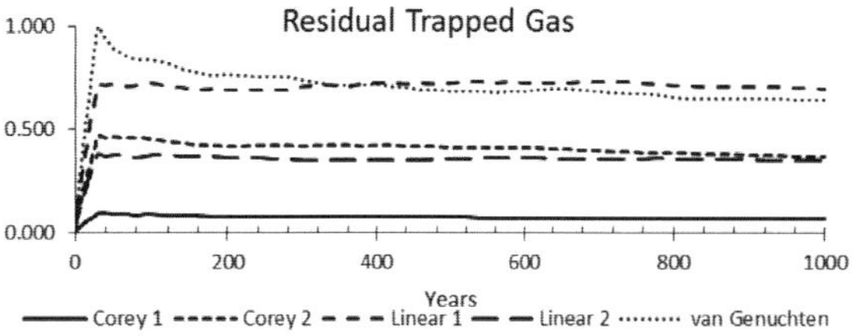

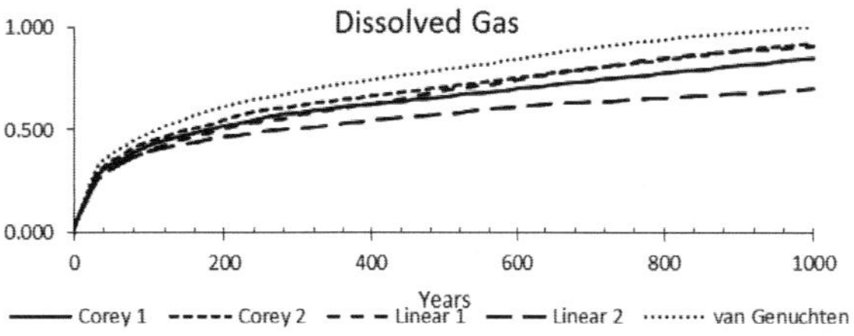

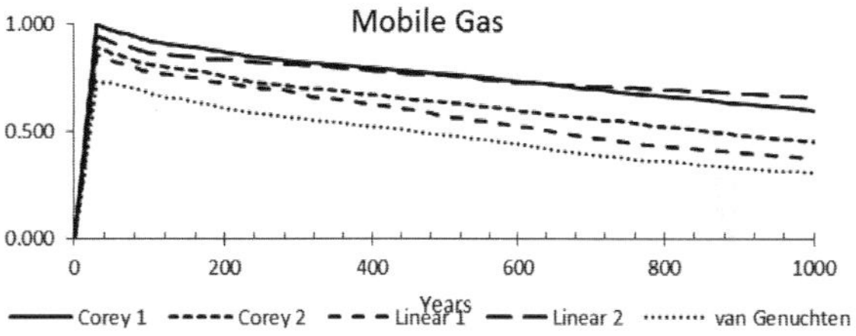

27

Conclusion

1. Assumed relative permeability relationship has a large impact on simulation results and must be given consideration in model development
   1. Estimates of trapping efficiency
   2. Estimates of seal integrity
   3. Plume behavior
2. Assuming a 'default' relationship could lead to estimate of poor reservoir performance.
3. Unless formation specific relationships are available it should be considered a large source of model uncertainty.

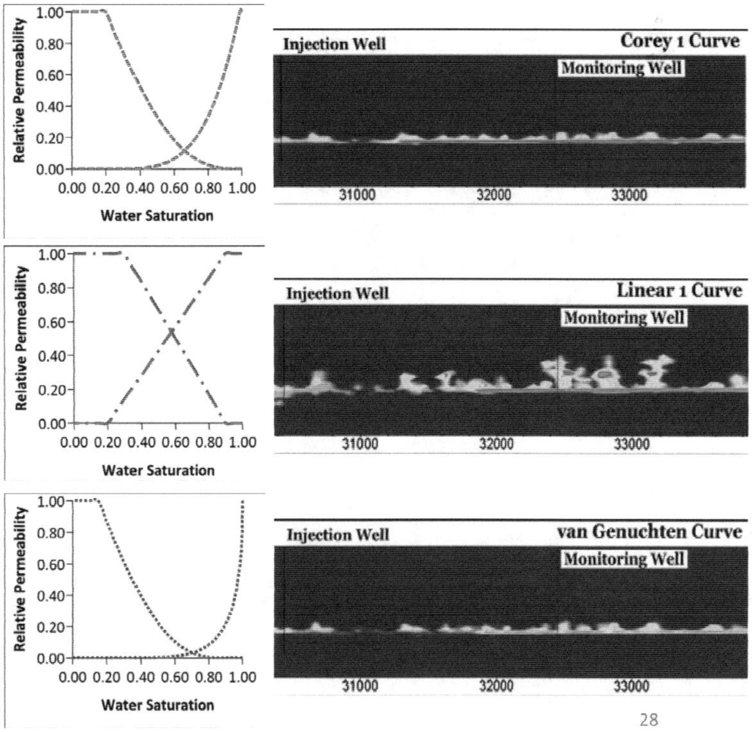

28

OUTLINE

- Premise: CO_2-EOR and Deep Saline Storage
- Sources and Extent of Uncertainty
- Greatest Source of Uncertainty: Heterogeneity
- Often Overlooked Source of Heterogeneity: Relative Permeability
- Uncertainty of Storage Forecasts Due to Heterogeneity
  (CO_2-EOR)
- Uncertainty in Storage Forecasts Due to Relative Permeability
  (CO_2-EOR)
- Variability in Storage and Trapping Due to Relative Permeability
  (Deep Brine Storage)
- Uncertainty in Mineralization Due to Relative Permeability
- Summary of Most Important Results

Little Grand Wash fault-zone (LGWF) in Utah is a natural example of CO_2-leakage

The LGWF system: geographic description

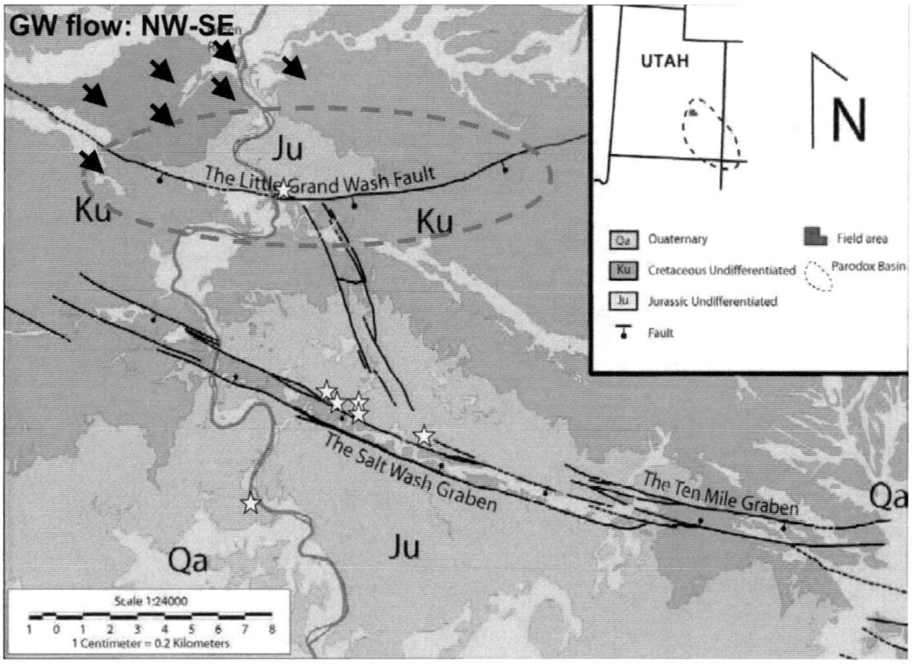

Map Source: Evans et al. 2005

The LGWF system: geographic description

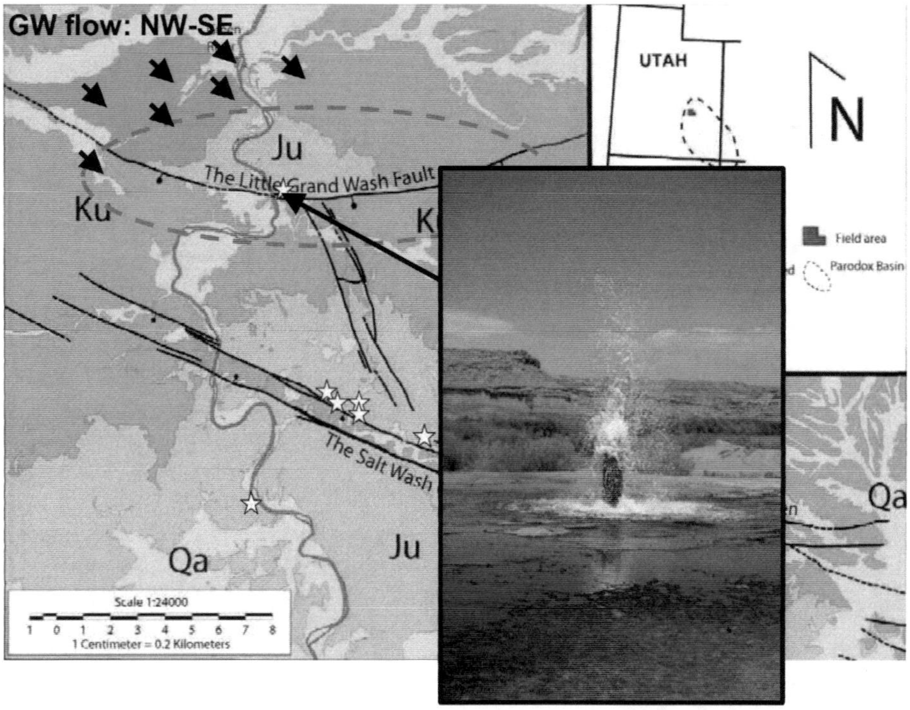

Map Source: Evans et al. 2005

Evidence of CO_2-leakage: Calcite precipitation

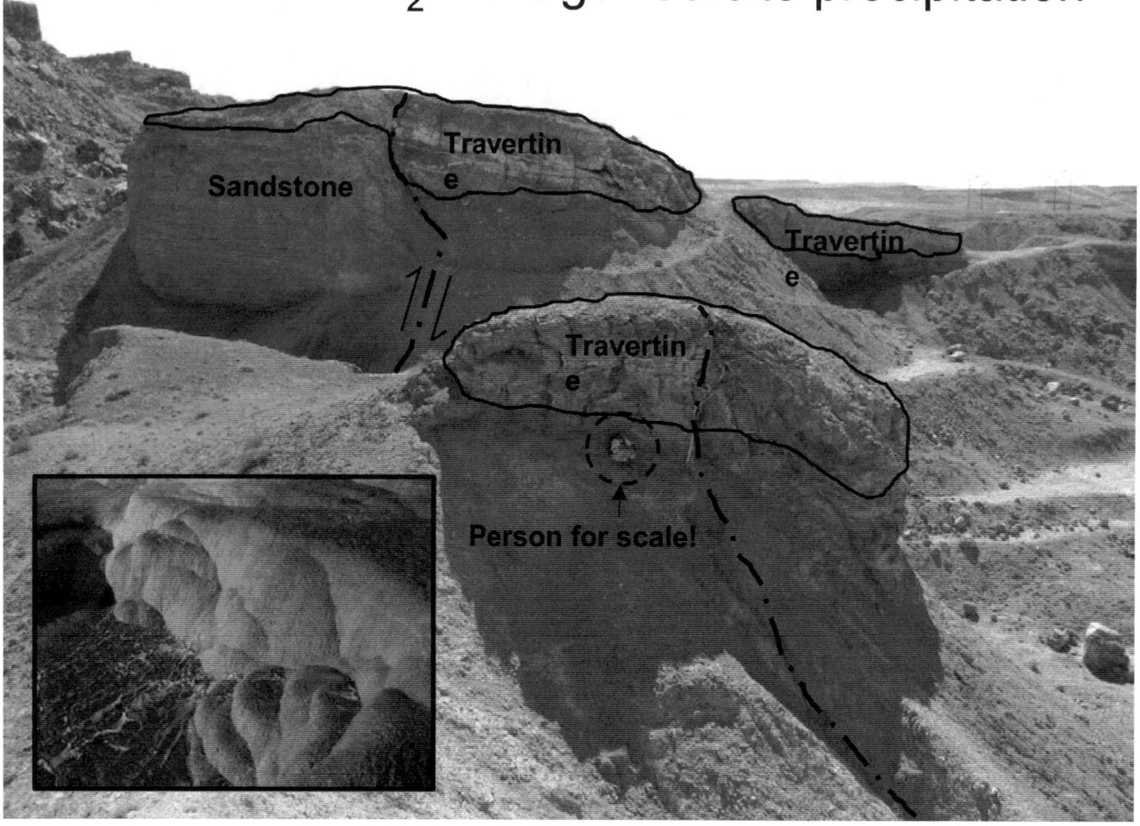

Conceptual flow model of LGWF system

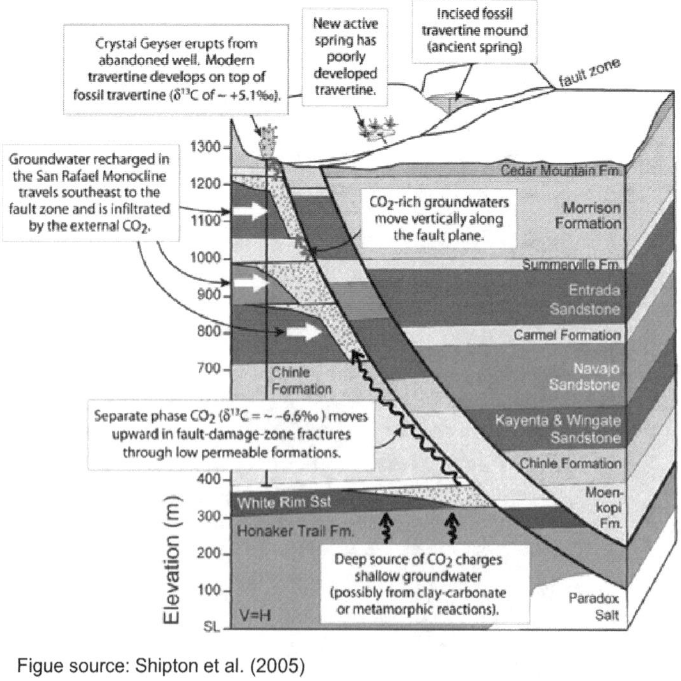

Figue source: Shipton et al. (2005)

- **Hydrologic analysis**: Hood and Patterson (1984)
- **Hydrogeochemical characterization** and **conceptual flow model**: Shipton (2004, 2005), Heath (2004, 2009)
- Estimation of key **geochemical constraints**: Kampman (2009), Wrigley (2012)
- **Carbonate age estimation**: Burnside (2013)
- **Scientific drilling test**: Kampman (2014)
- **Reactive transport modeling**: ***This study***

Simulated Mineralization Rates

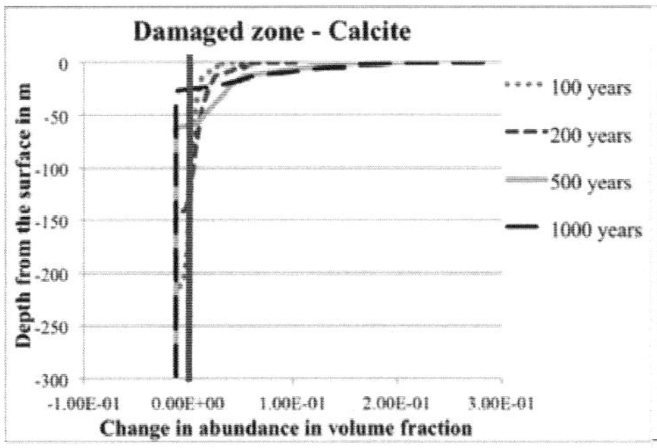

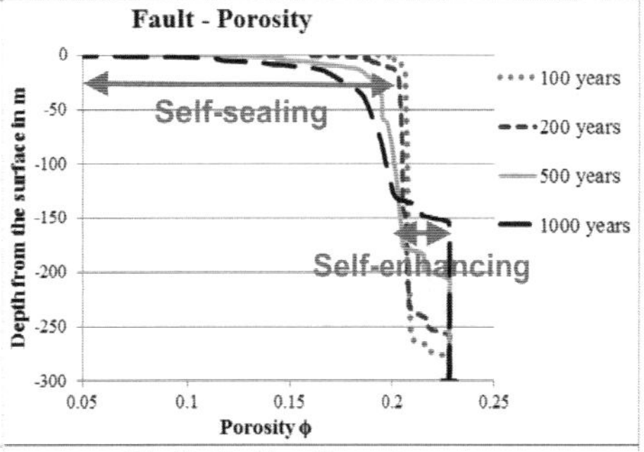

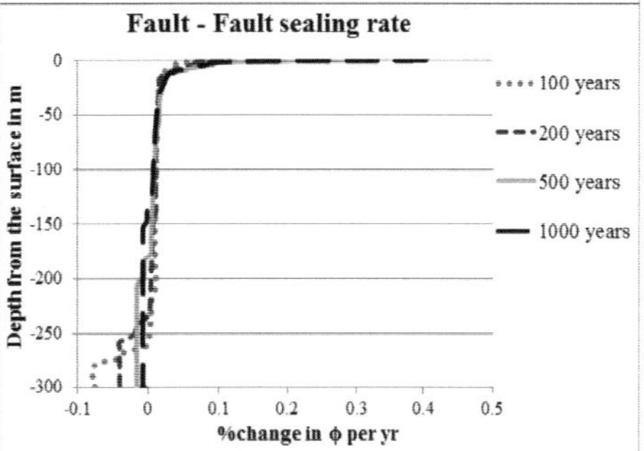

$$Fs = \frac{\left\{\frac{(\emptyset_0 - \emptyset)}{\emptyset}\right\} * 100}{t}$$

Negligible changes in fault core. Overall **fault porosity** evolution in **damaged zone**.

Impact of Different Relative Permeability Choice

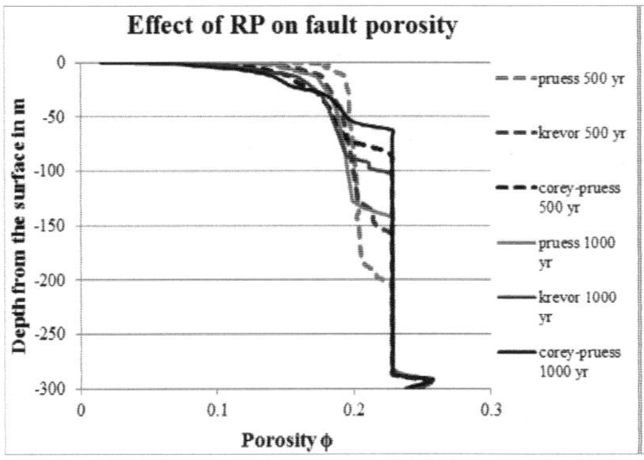

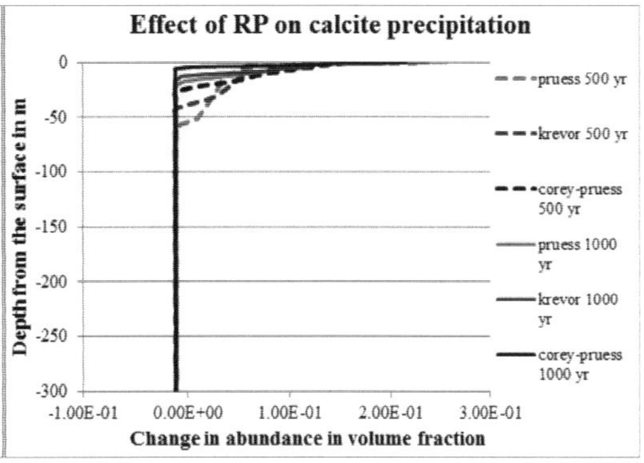

Corey RP curves **predict increased reaction** rates compared to van Genuchten curves

OUTLINE

- Premise: CO_2-EOR and Deep Saline Storage
- Sources and Extent of Uncertainty
- Greatest Source of Uncertainty: Heterogeneity
- Often Overlooked Source of Heterogeneity: Relative Permeability
- Uncertainty of Storage Forecasts Due to Heterogeneity (CO_2-EOR)
- Uncertainty in Storage Forecasts Due to Relative Permeability (CO_2-EOR)
- Variability in Storage and Trapping Due to Relative Permeability (Deep Brine Storage)
- Uncertainty in Mineralization Due to Relative Permeability
- **Summary of Most Important Results**

Most important results:

- choice of relative permeability function is critical

- calibration of relative permeability function is critical

The End

Slides for Q/A

SACROC Cross Sectional View

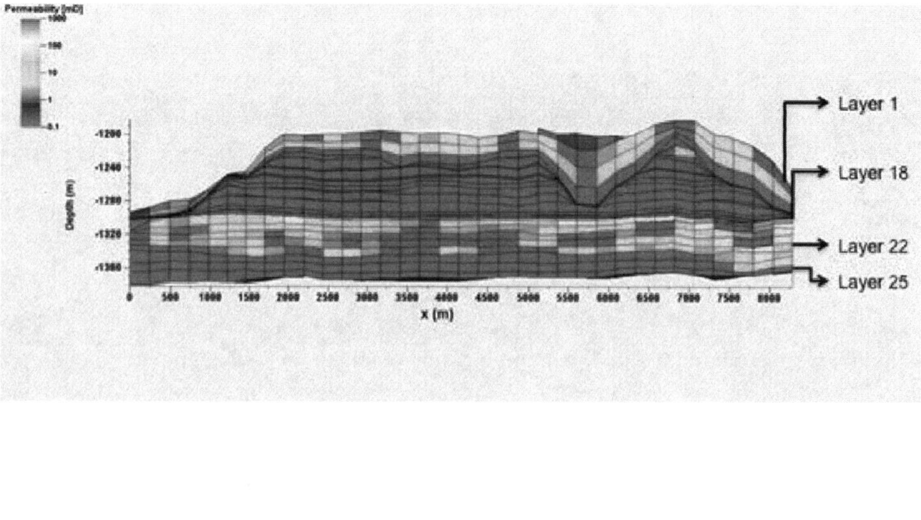

SACROC ROMs Validation

Model outputs of interest		R^2	NRMSE
CO_2 mass in oil phase	TIME 1	0.9891	5.49%
	TIME 2	0.9819	5.73%
	TIME 3	0.9666	5.98%
CO_2 mass in supercritical phase	TIME 1	0.9751	5.75%
	TIME 2	0.8398	6.93%
	TIME 3	0.8932	6.17%
CO_2 mass in aqueous phase	TIME 1	0.9891	4.54%
	TIME 2	0.9906	4.14%
	TIME 3	0.9801	3.93%
Oil Saturation	TIME 1	0.8561	8.42%
	TIME 2	0.8285	8.90%
	TIME 3	0.8596	8.56%

Water chemistry of Crystal Geyser

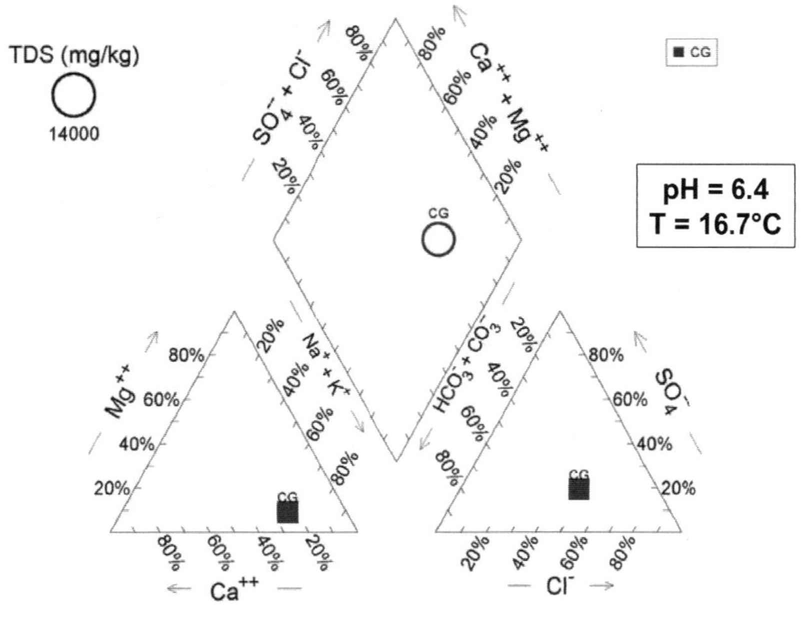

Super saturated with **carbonates, hematite** and **clays**.

$CaCO_3$ precipitation mechanism: pCO_2 loss

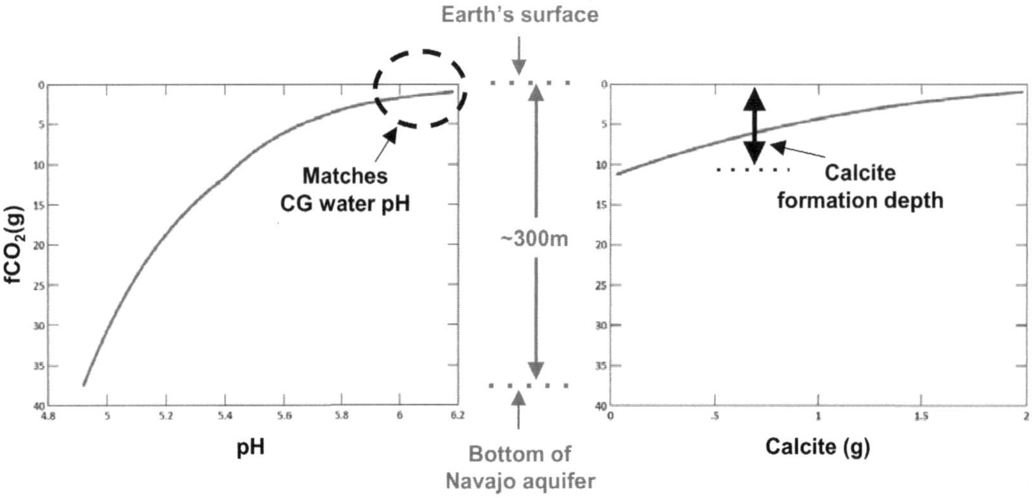

$$Ca^{++} + HCO_3^{-} = CaCO_{3(s)} + H_2O + CO_{2(g)}$$

Fault mineralogy

XRD results: **Clays** (smectites, illite, kaolinite), **carbonates** (calcite, aragonite) and secondary **quartz**

Fault architecture and permeability structure

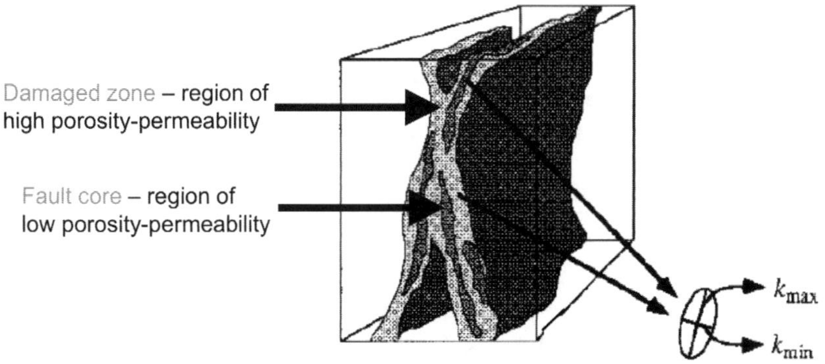

Total fault width = Damaged zone width + Fault core width

$$Fa = \frac{Damaged\ zone\ width}{Fault\ width}$$

Modeling approach: dual permeability domain

Figure adapted from Caine et al. (1996)

Sensitivity analysis: RP-CP curves

Relative permeability sensitivity parameters						
Parameter	**van Genuchten-basecase**	**van Genuchten-pruess**	**van Genuchten-krevor**	**Corey**	**Linear**	**Corey-pruess**
λ	0.457	0.457	0.67	n/a	n/a	n/a
Slr	0.3	0.15	0.2	0.2	0.2	0.2
Sgs	n/a	n/a	n/a	n/a	1	n/a
Sls	1	1	1	n/a	1	n/a
Sgr	0.05	0.1	0.05	0.05	0.05	0.05
Capillary pressure sensitivity parameters						
CP(1)	0.457	0.457	0.67	Set to linear CP curve	1,000,000	0.457
CP(2)	0	0	0.15		0.2	0
CP(3)	5.10E-05	5.219e-6*	5.219e-6*		0.9	5.219e-6*
CP(4)	1.00E+07	1.00E+07	1.00E+07		n/a	1.00E+07
CP(5)	0.999	1	1		n/a	1
*5.29e-6 for damaged zone; 1.613e-7 for fault core						

1D reactive flow model: Grid

Parameters	Value	Units
Fault depth	**300**	m
Fault thickness	5	m
Fa	**0.5**	
Damaged zone thickness	2.5	m
Fault core thickness	2.5	m
No. of grid cells	30	
No. of boundary cells	**2**	
Cell size in z-direction	1	m
Boundary cell size	1.00E-03	m
No. of continua	2	

1D model: Boundary and initial conditions

Parameters	Value	Units
Type	**Dirichlet**	
Top porosity	1	
Top pressure	1.013	bar
Top temperature	10	deg C
Bottom porosity	1	
Bottom pressure	35	bar
Bottom temperature	16.3	deg C
Bottom Sg	0.1	

Overpressure of 5 bars

Left and right boundaries: **No flow**

Hydrostatic initial flow conditions

Overpressure drives fluids up along the fault.

1D model: Fluid flow properties

Parameters	Value	Units
Damaged zone porosity	0.4	
Fault core porosity	0.02	
Effective porosity	0.21	
Damaged zone permeability	1.00E-12	m2
Fault core permeability	1.00E-16	m2
Effective permeability	5.00E-13	m2
Relative permeability	VanGen	
Capillary pressure	VanGen	

Highly permeable

Less permeable

1D model: Geochemistry

Average Mineralogy

Mineral	Wt%
Quartz	77
K-feldspar	6
Plagioclase	2
Calcite	2
Dolomite	2
Hematite	2
Kaolinite	4
Illite	3
Gypsum	2

- **Initial** water chemistry set in **equilibrium with rock composition.**
- Chemistry of **entering water** (bottom boundary) set to **Crystal Geyser**.
- Secondary precipitation: carbonates (**Ankerite, dawsonite, siderite**), clays (**smectites, chlorite**) and miscellaneous (**gibbsite, albite-low**).
- **CO_2** enters with bottom fluid.

AUTHOR INDEX

AUTHOR INDEX

AUTHOR INDEX

AUTHOR INDEX